This book provides a self-contained introduction to magnetohydrodynamics (MHD), with emphasis on nonlinear processes. Chapters 2 to 4 outline the conventional aspects of MHD theory, magnetostatic equilibrium and linear stability theory, which form a natural basis for the topics in the subsequent chapters. The main part, chapters 5 to 7, presents nonlinear theory, starting with the evolution and saturation of individual ideal and resistive instabilities, continuing with a detailed analysis of magnetic reconnection, and concluding with the most complex nonlinear behavior, that of MHD turbulence. The last chapters describe three important applications of the theory: disruptive processes in tokamaks, MHD effects in reversed-field pinches, and solar flares. In the presentation the focus is more on physical mechanisms than on special formalisms. The book is essential reading for researchers and graduate students interested in MHD processes both in laboratory and in astrophysical plasmas.

Cambridge Monographs on Plasma Physics 1

General Editors: W. Grossman, D. Papadopoulos, R. Sagdeev,
K. Schindler

Nonlinear magnetohydrodynamics

Cambridge Monographs on Plasma Physics

1. D. Biskamp: Nonlinear magnetohydrodynamics

NONLINEAR MAGNETOHYDRODYNAMICS

DIETER BISKAMP

Max Planck Institute for Plasma Physics, Garching

CAMBRIDGE
UNIVERSITY PRESS

PUBLISHED BY THE PRESS SYNDICATE OF THE UNIVERSITY OF CAMBRIDGE
The Pitt Building, Trumpington Street, Cambridge CB2 1RP, United Kingdom

CAMBRIDGE UNIVERSITY PRESS
The Edinburgh Building, Cambridge CB2 2RU, United Kingdom
40 West 20th Street, New York, NY 10011-4211, USA
10 Stamford Road, Oakleigh, Melbourne 3166, Australia

First published 1993
First paperback edition, with corrections 1997

A catalogue record for this book is available from the British Library

Library of Congress Cataloguing in Publication data

Biskamp, Dieter.
Nonlinear magnetohydrodynamics/Dieter Biskamp.
    p.   cm. — (Cambridge monographs on plasma physics)
Includes index.
ISBN 0 521 40206 9
1. Magnetohydrodynamics. 2. Nonlinear theories. 3. Solar flares.
4. Plasma astrophysics.   I. Title. II. Series.
QC718.5.M36B56   1993
538'.6–dc20   93-3237 CIP

ISBN 0 521 40206 9 hardback
ISBN 0 521 59918 0 paperback

Transferred to digital printing 2004

To Eleonore, Stefan, and Susanne

Contents

ix

Preface

Magnetohydrodynamics (MHD) is the macroscopic theory of electrically conducting fluids, providing a powerful and practical theoretical framework for describing both laboratory and astrophysical plasmas. Most textbooks and monographs on the topic, however, concentrate on two particular aspects, magnetostatic equilibria and linear stability theory, while nonlinear effects, i.e. real magnetohydro*dynamics*, are considered only briefly if at all. I have therefore felt the need for a book with a special focus on the nonlinear aspects of the theory for some time.

In contrast to linear theory which, in particular in the limit of ideal MHD, rests on mathematically solid ground, nonlinear theory means adventures in a, mathematically speaking, hostile world, where few things can be proved rigorously. While in linear stability analysis numerical calculations are mainly quantitative evaluations, they obtain a different character in the study of nonlinear phenomena, which are often even qualitatively unknown. Hence this book frequently refers to results from numerical simulations, as a glance at the various illustrations reveals, but consideration is focused on the physics rather than the numerics.

In spite of the numerous references to the literature the book is essentially self-contained. Even the individual chapters can be studied quite independently as introductions to or current overviews of their particular topics. The main theoretical part, chapters 5–7, is written without special emphasis on either laboratory or astrophysical plasmas, while the applications part, chapters 8–10, includes examples from both fields. I hope that the book will thus be interesting for fusion and astro-plasma physicists, though I know from personal experience that both tend to have quite different ways of thinking in spite of the similarities between the subjects. Apart from some elementary knowledge of plasma physics and fluid dynamics there are no special requirements; the book is suitable for both researchers and graduate students.

I gratefully acknowledge the numerous stimulating and clarifying discussions I enjoyed over the years with many colleagues from the international community of plasma physicists, and in particular from the Max Planck Institutes for Plasma Physics, Astrophysics and Extraterrestrial Physics, so conveniently located on the same campus. My special thanks go to Karl Schindler for encouraging me to write this book in the new Cambridge series of monographs in plasma physics and for several useful comments. Most importantly, the book would never have been completed without the invaluable technical help of my collaborators for many years, Marianne Walter and Helmut Welter. I also acknowledge the permission granted by the authors and the copyright holders to reproduce many of the figures included in the book.

Garching, October 1992 *Dieter Biskamp*

1

Introduction

Magnetohydrodynamics (MHD) describes the macroscopic behavior of electrically conducting fluids, notably of plasmas. However, in contrast to what the name seems to indicate, work in MHD has usually little to do with dynamics, or at least has had so in the past. In fact, most MHD studies of plasmas deal with magnetostatic configurations. This is not only a question of convenience – powerful mathematical methods have been developed in magnetostatic equilibrium theory – but is also based on fundamental properties of magnetized plasmas. While in hydrodynamics of nonconducting fluids static configurations are boringly simple and interesting phenomena are in general only caused by sufficiently rapid fluid motions, conducting fluids are often confined by strong magnetic fields for times which are long compared with typical flow decay times, so that the effects of fluid dynamics are weak, giving rise to quasistatic magnetic field configurations. Such configurations may appear in a bewildering variety of shapes generated by the particular boundary conditions, e.g. the external coils in laboratory experiments or the "foot point" flux distributions in the solar photosphere, and their study is both necessary and rewarding.

In addition to finding the appropriate equilibrium solutions one must also determine their stability properties, since in the real world only stable equilibria exist. Stability theory, however, often predicts instability for equilibrium solutions which appear to describe experimentally observed configurations quite well. What happens to these solutions if a weak perturbation is applied, do they merely relax into a neighboring equilibrium or slightly oscillating state, thus effectively enlarging the class of realizable equilibrium configurations? Such questions cannot be answered by linear stability theory.

A further aspect is connected with the various types of disruptive processes which are observed in laboratory and astrophysical plasmas to

1

occur occasionally after a period of quiescent plasma behavior. According
to the conventional picture the configuration evolving because of slow
changes of the boundary conditions becomes unstable at a certain point. A
little reflection, however, shows that such an explanation is unsatisfactory
and insufficient. Instabilities are usually weak for conditions close to the
stability threshold, or marginal point, giving rise to a slow growth of
the unstable perturbation, which completely misses the rapid explosive
character of the observed process the instability is intended to explain. In
addition, as mentioned above, linear instability theory does not allow an
estimate of the final extent of the unstable dynamics. In particular, rapid
linear growth rates do not guarantee that a large amount of energy is
released. A somewhat more adequate approach to the problem of explosive
processes appears to be equilibrium bifurcation theory. In particular a
loss of equilibrium, called catastrophe, is often associated with the onset
of rapid dynamics. However, a catastrophe usually occurs only within
a certain equilibrium class, such that the system may still escape into
a neighboring equilibrium state belonging to a more general class, for
instance by introducing an X-type neutral point.

Hence it is necessary to leave the framework of equilibrium and stability
theory and consider nonlinear dynamics explicitly. Mathematically this
means leaving safe ground and embarking on unknown, perilous waters.
In addition some price has in general to be paid for practical tractabil-
ity. While equilibrium and stability theory can deal quantitatively with
geometrically complicated systems, nonlinear MHD studies are usually
restricted to obtain a qualitative picture in the simplest possible geometry.
The pioneering papers date back to the early seventies, when theorists
became aware of the importance of nonlinear effects. Among such papers
are, notably, Rutherford's theory of the tearing mode evolution, Kadomtsev and Pogutse's theory of vacuum bubbles, the nonlinear theories of the
ideal kink mode by Rosenbluth et al. and of the resistive kink mode by
Kadomtsev, Syrovatskii's theory of current sheet formation and Taylor's
theory of relaxed states, all of which have since been very influential. However, these nonlinear theories do not deal with truly dynamic processes but
consider slowly evolving equilibrium states or asymptotic states of systems
relaxed from some initial state under certain physical constraints. By contrast genuine dynamics is considered in a different line of approach, that
of fully developed MHD turbulence. In the case of turbulence, nonlinear
theory becomes tractable by applying statistical averaging together with
some closure assumption. Here work started with Kraichnan's paper on
the Alfvén effect in the sixties and a number of fundamental contributions,
notably by Frisch, Montgomery and Pouquet, in the seventies.

For more general processes, however, numerical computations become
the major tool, a trend observed in many branches of physics. In fact

the computational approach has reached a new dimension, which can be called computational theory. Fifty years ago a problem in fluid dynamics was considered as solved if the result could be expressed in terms of tabulated functions, twenty-five years ago if it could be reduced to an ordinary differential equation. With present-day supercomputers the partial differential equations for many two-dimensional fluid dynamic problems can be solved "exactly" for interesting Reynolds numbers and arbitrary boundary conditions. Moreover by performing a series of computer runs with different values of the externally given parameters scaling laws can be obtained. It appears that a problem should be considered solved if "exact" numerical solutions are available to provide scaling laws in a certain parameter regime, and when a basic physical mechanism is found, i.e. when the behavior of the system is "understood". This is the realm of computational theory. It might seem little in view of the beauty of exact analytical results; it should, however, be compared with the actually achievable, highly approximate analytical approaches often encountered. Since such approximations tend to be made more on grounds of convenience and feasibility than of mathematical rigor, they should be guided by the "exact" results obtained by relevant numerical simulation. Incidentally, the latter term, frequently used, is somewhat misleading. In fact one should distinguish between "physics simulations" and "real-world simulations". While the former simply provide an exact solution of some usually time-dependent model partial differential equations (exact within known and controllable discretization errors), the latter often incorporate many different, possibly equally important effects (e.g. in tokamak transport simulations), or complicated geometry (e.g. in stellerator development). In this book the term 'simulation' normally refers to the first category.

In a field such as nonlinear MHD, where no unifying methodical framework exists, the selection of topics is necessarily somewhat arbitrary, biased by personal taste. I have tried to concentrate more on the basic effects rather than include a broad scale of individual investigations. The book consists of three major parts, an introductory part, chapters 2–4, the main part treating three different aspects of nonlinear MHD, chapters 5–7, and an applications part, chapters 8–10.

Chapter 2 introduces the MHD equations in a macroscopic way, without recourse to concepts of kinetic theory. The ideal invariants play a crucial role, in particular magnetic helicity. We also derive a simplified set of equations for strongly magnetized plasmas, called reduced MHD, which has proved to be very convenient for nonlinear MHD studies. Finally the important dissipative effects are discussed, the magnitudes of which are measured by the corresponding Reynolds and Lundquist numbers.

In order to make the book self-contained, chapters 3 and 4 give an outline of the classical topics of MHD theory, equilibrium and stability theory. Chapter 3 derives the general two-dimensional equilibrium equation, discusses some exact two-dimensional solutions often used as paradigms, and gives a brief overview of numerical methods to compute general equilibria in 2-D and 3-D.

The linear stability problem is addressed in chapter 4. Since the normal mode spectrum can analytically only be obtained for 1-D configurations, the energy principle has become the main tool for a qualitative stability analysis of more complicated systems. We first discuss the ideal MHD stability of a linear pinch and the modifications which arise in the toroidal case. Allowing for finite electrical resistivity significantly broadens the class of possibly unstable plasma motions. Two prototypes of resistive instabilities are treated in more detail, the tearing instability and the resistive kink instability, corresponding respectively to ideally stable and marginal unstable modes.

After these preliminaries the reader is prepared to enter the world of nonlinear processes. Throughout the book the emphasis is on relatively slow, essentially incompressible processes, excluding fast shock phenomena. Chapter 5 considers the laminar nonlinear evolution of MHD instabilities, where the system remains within the geometry defined by the most unstable mode, e.g. helical symmetry in the case of a kink mode in a cylindrical pinch. First the quasi-linear approximation is introduced, which provides a practical estimate of the instability saturation level in cases of a nonsingular final state. Then two rare examples of analytically solvable models are presented, the theory of vacuum bubbles resulting from external kink modes in an unsheared plasma cylinder, and the theory of the saturation of the ideal internal kink mode. These ideal MHD models are, however, of limited practical significance, since dynamical processes in magnetized plasma are strongly affected by the presence of finite resistivity even if the latter is numerically very small. Here the nonlinear tearing mode with its different variants is probably the most important individual MHD process, which is hence discussed in some detail, in particular the universal small-amplitude phase, called the Rutherford regime, and the saturation properties depending on the geometry of the configuration, the current distribution and the transport properties of the resistivity.

Magnetic reconnection, which in a stricter sense means the *fast* dynamic decoupling of plasma and magnetic field, can be called the essence of nonlinear MHD. Also in the case of ideal instability reconnection usually determines the nonlinear evolution. In magnetized plasmas reconnection takes place in current sheets. Chapter 6 first introduces the Sweet-Parker model, which incorporates the basic properties of dynamic current sheets. Reconnection theory has long been dominated by two schools of thought, Petschek's slow shock model and Syrovatskii's theory of current sheet generation. While the former has, however, been shown in recent years to be invalid in the limit of high conductivity, for which it had been devised, the latter has been verified in detail by numerical simulations and in fact describes in a simple, elegant way a fundamental effect in the dynamics of highly conducting magnetized fluids. Though dynamic current sheets are significantly more stable than static ones they become tearing-unstable at sufficiently high Reynolds number, above which no stationary reconnection configurations appear to exist. As examples three well-known systems involving reconnection are discussed, the coalescence of magnetic islands, the nonlinear evolution of the resistive kink instability and the dynamics of plasmoids. While most of the chapter is restricted to two-dimensional systems, the final two sections discuss some aspects of three-dimensional reconnection.

Chapter 7 deals with fully developed turbulence, the most probable dynamical state at high Reynolds numbers. The absolute equilibrium distributions of truncated nondissipative systems provide insight into the properties of nonlinear mode interactions which determine the cascade directions in dissipative turbulence. In contrast to the hydrodynamic (Navier-Stokes) case MHD turbulence exhibits strong self-organization processes, which are connected with the existence of inverse cascades. These processes are selective decay leading to large-scale, static, force-free magnetic states, and dynamic alignment of velocity and magnetic fields. Also, spectral properties in MHD turbulence are different from those of hydrodynamic turbulence. The only quantitative approach for MHD turbulence theory developed to date is based on closure theory, a tractable version being the eddy-damped, quasi-normal Markovian approximation. Turbulent energy dissipation is discussed, in particular in 2-D MHD systems, which differs fundamentally from the behavior in 2-D hydrodynamics. An important aspect of modern turbulence theory is intermittency. Several intermittency models developed for Navier-Stokes turbulence are introduced and some results for MHD turbulence are given. Finally the magnetoconvection in primarily unmagnetized fluid turbulence is discussed, which is intimately connected with the turbulent dynamo effect.

The remaining chapters are devoted to three important applications in laboratory and astrophysical plasmas. Chapter 8 discusses the MHD properties of disruptive processes observed in tokamaks. The sawtooth oscillation is a periodic relaxation process restricted to the central region of the plasma column. Observations show beyond reasonable doubt that the sawtooth collapse is connected with the $m = 1, n = 1$ kink mode. However, Kadomtsev's model assuming full reconnection of the helical magnetic flux by the resistive kink mode, which had long been the generally accepted sawtooth model, does not seem to apply to present-day large-diameter hot tokamak plasmas, characterized by very large values

of the Lundquist number S. Observed time scales seem to be too fast
to allow full reconnection. In fact, measurements of the central safety
factor indicate that full reconnection does not take place and that the
thermal energy release is caused by some effect which is different from the
convective process in Kadomtsev's model. Numerical simulations, being
still confined to rather low S-values, have not been able to elucidate the
high S-value behavior. In addition the fast onset of the collapse is still
poorly understood. The major disruption in a tokamak occurs accidentally,
when plasma parameters, in particular current and pressure, exceed certain
operational limits, the density limit being effectively a current limit due
to transport processes. The disruption proper appears to be a fast MHD
process caused by the nonlinear instability of a large amplitude $m = 2$
tearing mode, which leads to a turbulent state. It can be described as an
interaction of modes of different helicity, notably $(m, n) = (2, 1), (1, 1), (3, 2)$,
which gives rise to an anomalous resistivity. The third type of disruptive
process is again a quasi-periodic relaxation oscillation, affecting the outer
plasma region, hence the name 'edge-localized mode' (ELM). It occurs
primarily in divertor plasmas in the high-confinement (H-) regime, where
owing to a local transport barrier steep pressure gradients are generated
at the plasma edge. The ELM can be associated with high-m ballooning
modes. A common feature observed in all three types of disruptive events
is that of a two-stage process, consisting of a coherent precursor and a
more rapid turbulent relaxation phase.

The reversed-field pinch (RFP) considered in chapter 9 is particularly
rich in MHD effects. Because of the low value of the safety factor
the RFP is prone to instability in contrast to a tokamak plasma and
hence tends to relax to the minimum energy configuration with a reversed
toroidal field predicted by Taylor's theory, which we discuss in some
detail. RFP plasmas are usually maintained in a quasi-stationary state,
where turbulent relaxation to the minimum energy state is counteracted
by resistive diffusion. The process of transforming the externally supplied
poloidal field into internal toroidal field is called the RFP dynamo effect.
The simplest theoretical model is a helical ohmic state, where a helical
stationary flow balances resistive diffusion. Such stationary dynamo states
are not forbidden by Cowling's antidynamo theorem, which applies only
to the more restricted case of axisymmetry. In general, however, such
stationary states are unstable with respect to modes with different helicities.
The resulting turbulent behavior can only be investigated by numerical
simulation.

Finally, chapter 10 deals with solar flares, which are among the most
spectacular explosive events observed in astrophysical plasmas. Flares
result from a sudden release of magnetic energy and are hence MHD processes, though a treatment of the different channels of energy dissipation

which give rise to the wealth of observed phenomena is outside the scope
of this book. We first discuss the process of magnetic field generation in
the solar convection zone and the typical magnetic configurations emerging into the corona. Flares appear over a wide energy range, from the
very weak microflares or even nanoflares forming a background noise responsible for the continuous coronal heating, to the macroflares occuring
sporadically. The latter are loosely classified as simple loop or compact
flares and two-ribbon flares, the largest events. The current MHD models
are presented for flares of both types.

A few remarks concerning the notation used in the book may be in
place. I have tried as much as possible to stay within the notations used
in the current literature. Occasionally conflicts arise, for instance for the
energy which is usually denoted by W in stability theory, but by E in
turbulence theory. I follow such conventions, but give an explicit warning.
The equations are written in SI units, which seem to be the most practical
for macroscopic plasma physics. However, to simplify notation I will set
the vacuum permeability $\mu_0 = 1$. If necessary μ_0 can be reintroduced at any
stage by a simple dimensionality consideration. In addition, since most
of the MHD processes considered are incompressible, a homogeneous
density distribution ρ is often assumed, setting $\rho = 1$. In these units the
magnetic field B has the dimension of a velocity, the corresponding Alfvén
velocity $v_A = B/\sqrt{\mu_0\rho}$, and v_A and B will be used interchangeably.

2

Basic properties of
magnetohydrodynamics

This chapter gives an introduction to the general properties of MHD theory. In section 2.1 a brief heuristic derivation of the MHD equations is presented, followed in section 2.2 by a discussion of the corresponding conservation laws. The magnetic helicity is considered in more detail in section 2.3. In section 2.4 a set of reduced MHD equations is introduced, which is a very convenient model for strongly magnetized plasmas. Finally, section 2.5 discusses the validity limits of the MHD approximation and introduces dissipation effects.

2.1   The MHD equations

For a plasma the MHD approximation can be derived in a systematic manner starting from the kinetic equations for ions and electrons, as done for instance by Braginskii (1965). However, we refrain from introducing microscopic concepts of plasma theory. Instead we derive the MHD equations in a purely macroscopic heuristic way, valid for any electrically conducting fluid, the specific properties of which appear only in the equation of state. Consider the force balance for a fluid element of volume δV and mass $\rho \, \delta V$, where ρ is the mass density. The inertial term is

$$\rho \delta V \frac{d\mathbf{v}}{dt} \equiv \rho \delta V \left(\partial_t \mathbf{v} + \mathbf{v} \cdot \nabla \mathbf{v} \right) . \tag{2.1}$$

The acceleration is caused by the total force on the fluid element consisting of the following parts:

(a) The thermal pressure force. Assuming conditions close to local thermodynamic equilibrium, the pressure tensor is isotropic, exerting the

8

force

$$- \oint p \, d\mathbf{F} = -\delta V \nabla p \, , \tag{2.2}$$

where the integral is over the surface of the volume element and $d\mathbf{F} = \mathbf{n} \, dF$ is the surface element.

(b) The magnetic force. The force on a particle of charge q_i is the Lorentz force $q_i (\mathbf{E} + \mathbf{v} \times \mathbf{B})$. Hence the force on a macroscopic fluid element is the sum of the forces acting on its individual particles $\delta q \mathbf{E} + \delta \mathbf{j} \times \mathbf{B}$, where δq is the net charge and $\delta \mathbf{j}$ the electric current carried by the fluid element. Since in dense fluids electrostatic fields enforce charge neutrality over macroscopic distances, $\delta q \simeq 0$ (which is called quasi-neutrality and does *not* imply a vanishing of the electrostatic field), the magnetic force is the macroscopic Lorentz force

$$\delta V \mathbf{j} \times \mathbf{B} \, . \tag{2.3}$$

(c) The gravitational force

$$\delta V \rho \mathbf{g} = -\delta V \rho \nabla \phi_g \, , \tag{2.4}$$

where ϕ_g is the gravitational potential. While this force is negligible in laboratory plasmas, it may play an important role in astrophysical systems. Thus the force balance becomes

$$\rho \left(\partial_t \mathbf{v} + \mathbf{v} \cdot \nabla \mathbf{v} \right) = -\nabla p + \mathbf{j} \times \mathbf{B} - \rho \nabla \phi_g \, . \tag{2.5}$$

In the following we usually omit the gravitational force.

The relation between the magnetic field and the current density is obtained from Ampère's law:

$$\nabla \times \mathbf{B} = \mathbf{j} \, . \tag{2.6}$$

(Throughout the book we simplify notation by setting $\mu_0 = 1$. When necessary μ_0 can be reintroduced at any point.) Compared with the full Maxwell's equation the displacement current is missing. This reflects the fact that in nonrelativistic magnetohydrodynamics, to which this treatise is restricted, space charges can be neglected, which does not exclude, however, phase velocities of waves reaching or exceeding the velocity of light.

The dynamics of the magnetic field follows from Faraday's law:

$$\partial_t \mathbf{B} = -\nabla \times \mathbf{E} \, . \tag{2.7}$$

To express the electric field by known quantities, we perform a Galilean transformation

$$\mathbf{x}' = \mathbf{x} - \mathbf{V}t \, , \quad t' = t \, ,$$

the nonrelativistic limit of a Lorentz transformation. The equations transform in the following way:

$$\nabla' \times \mathbf{B}' = \mathbf{j}' , \quad \partial_{t'}\mathbf{B}' = -\nabla' \times \mathbf{E}' , \tag{2.8}$$

with

$$\mathbf{B}' = \mathbf{B} , \quad \mathbf{E}' = \mathbf{E} + \mathbf{V} \times \mathbf{B} . \tag{2.9}$$

In a fluid element with a velocity $\mathbf{V}$ at time t the velocity vanishes in the transformed system. For a fluid at rest, however, Ohm's law, in the simplest case $\mathbf{E}' = \eta\mathbf{j}'$, indicates that the electric field vanishes, $\mathbf{E}' = 0$, neglecting dissipative effects for the moment. Hence in the laboratory frame we have $\mathbf{E} = -\mathbf{V} \times \mathbf{B}$. Performing this transformation for each fluid element we may replace the (uniform) transformation velocity by the actual fluid velocity $\mathbf{v}(\mathbf{x}, t)$. Hence we find the relation called Ohm's law for an infinitely conducting fluid:

$$\mathbf{E} + \mathbf{v} \times \mathbf{B} = 0 . \tag{2.10}$$

Eliminating the electric field, Faraday's law (2.7) becomes

$$\partial_t\mathbf{B} = \nabla \times (\mathbf{v} \times \mathbf{B}) . \tag{2.11}$$

To close the system of equations we still have to determine the mass density $\rho(\mathbf{x}, t)$ and the pressure $p(\mathbf{x}, t)$. The former obeys the continuity equation

$$\partial_t\rho + \nabla \cdot \rho\mathbf{v} = 0 , \tag{2.12}$$

which follows from mass conservation. Note that no assumption about the relation between mass and charge is implied, concerning the kind and number of charge carriers (degree of ionization in a plasma) except for the requirement that positive and negative charges balance within each macroscopic fluid element. The pressure contains the thermodynamic properties of the fluid. A plasma follows approximately the ideal gas law, $p = (n_i + n_e)k_B T = (R/\mu)\rho T$, where $n_{i,e}$ are the number densities of ions and electrons, k_B is Boltzmann's constant, $R = c_p - c_v$, the difference of specific heats, and μ the mean atomic weight, $\mu \simeq 1/2$ in a hydrogen plasma. In the MHD approximation variations of the thermodynamic state are assumed to be sufficiently fast and on sufficiently large spatial scales that dissipation effects, in particular heat conduction, are negligible. Hence changes of state are adiabatic, which can be written as

$$d\left(p\rho^{-\gamma}\right)/dt = 0 , \tag{2.13}$$

or by use of the continuity equation (2.12)

$$\partial_t p + \mathbf{v} \cdot \nabla p + \gamma p\nabla \cdot \mathbf{v} = 0 , \tag{2.14}$$

where $\gamma = c_p/c_v$ is the ratio of the specific heats.

A very useful limiting case valid for fluid velocities slow compared with the propagation speed of compressional waves is that of incompressibility,

$$\nabla \cdot \mathbf{v} = 0 . \tag{2.15}$$

In this case the density of a fluid element remains constant. We can therefore assume, without much loss of generality, a homogeneous density distribution which can be normalized such that $\rho = 1$. Condition (2.15) does not, however, imply that the pressure, too, is only advected, as eq. (2.14) might suggest, since incompressibility formally corresponds to $\gamma \rightarrow \infty$. In fact the pressure is no longer an independent dynamic variable, but is determined by the nonlinear terms in eq. (2.5), the divergence of which gives Poisson's equation for p. The incompressibility limit is particularly suitable for high-density fluids such as liquid metals, where the equation of state differs from the ideal gas law, but since the sound velocity is usually much higher than the flow speeds, these properties do not enter the flow dynamics. In the following we will consider either the case of fully compressible fluids with $\gamma = \frac{5}{3}$ or the case of incompressibility, assuming homogeneous density in the latter.

2.2   Conservation laws in ideal MHD

Equations (2.5), (2.11), (2.12) and (2.14) or (2.15) constitute a closed set of equations called ideal (i.e. nondissipative) MHD. The equations give rise to a number of global conservation laws. From the continuity equation (2.12) it follows that for constant volume V

$$\frac{d}{dt} \int_V \rho d^3 x = \int_V \partial_t \rho d^3 x = - \oint_S \rho \mathbf{v} \cdot d\mathbf{F} , \tag{2.16}$$

i.e. the mass enclosed in a volume V is constant if the normal velocity at the boundary S vanishes, $v_n = 0$. Equation (2.16) implies that the mass M contained in a volume $V(t)$ moving with the fluid is conserved,

$$\frac{dM}{dt} = \int_V \partial_t \rho d^3 x + \int_{\partial_t V} \rho d^3 x = 0 , \tag{2.17}$$

using eq. (2.16) and

$$\int_{dV} \rho d^3 x = \oint \rho \mathbf{v} \cdot d\mathbf{F} \, dt . \tag{2.18}$$

The equation of motion (2.5) gives the global momentum balance relation (again for constant V)

$$\frac{d}{dt} \int_V \rho \mathbf{v} d^3 x = - \oint_S \left(\rho \mathbf{v} \mathbf{v} + \left(p + \frac{B^2}{2} \right) \mathbf{I} - \mathbf{B} \mathbf{B} \right) \cdot d\mathbf{F} , \tag{2.19}$$

i.e. the momentum of the plasma volume V is conserved if $v_n = B_n = 0$ and $p + B^2/2 = $ const. on the boundary. The change of momentum $\mathbf{P}$ in a volume $V(t)$ moving with the fluid is due to the stresses exerted on its surface,

$$\frac{d\mathbf{P}}{dt} = -\oint_S \mathscr{T} \cdot d\mathbf{F} , \qquad (2.20)$$

where

$$\mathscr{T} = \left(p + \frac{B^2}{2}\right)\mathbf{I} - \mathbf{BB} \qquad (2.21)$$

is the stress tensor, and relation (2.18) has been used. From equations (2.5), (2.11), (2.12), (2.14) follows the conservation law of the total energy

$$\frac{d}{dt}\int_V \left(\rho\frac{v^2}{2} + \frac{B^2}{2} + \frac{p}{\gamma-1}\right)d^3x$$
$$= -\oint_S \left[\left(\rho\frac{v^2}{2} + \frac{p}{\gamma-1}\right)\mathbf{v} + p\mathbf{v} + \mathbf{E}\times\mathbf{B}\right]\cdot d\mathbf{F} . \qquad (2.22)$$

The particular form of the r.h.s. has been chosen to point out the physical meaning of the individual contributions, the kinetic and thermal energy flux, the work done by the pressure force and the Poynting flux, where Ohm's law (2.10) has been used. The energy is conserved if either $v_n = B_n = 0$ or $\mathbf{v} = 0$ at the boundary. The change of the energy W of the fluid contained in a comoving volume $V(t)$ is obtained from eq. (2.22)

$$\frac{dW}{dt} = -\oint_S (\mathscr{T}\cdot\mathbf{v})\cdot d\mathbf{F} . \qquad (2.23)$$

An important quantity in MHD theory is the magnetic helicity

$$H = \int_V \mathbf{A}\cdot\mathbf{B}\,d^3x . \qquad (2.24)$$

One finds

$$\frac{dH}{dt} = \int_V (\partial_t\mathbf{A}\cdot\mathbf{B} + \mathbf{A}\cdot\partial_t\mathbf{B})\,d^3x$$
$$= \int_V \mathbf{A}\cdot\nabla\times(\mathbf{v}\times\mathbf{B})\,d^3x$$
$$= \oint_S (\mathbf{A}\cdot\mathbf{v}\mathbf{B} - \mathbf{A}\cdot\mathbf{B}\mathbf{v})\cdot d\mathbf{F} , \qquad (2.25)$$

which vanishes if $B_n = v_n = 0$. Here we have chosen the gauge such that the scalar potential vanishes, $\partial_t\mathbf{A} = -\mathbf{E}$. The condition $B_n = 0$ is required to make H gauge-invariant. Performing a gauge transformation

$\mathbf{A}' = \mathbf{A} + \nabla \chi$ in eq. (2.24), we find

$$H' - H = \int_V \nabla \chi \cdot \mathbf{B} \, d^3 x$$

$$= \oint_S \chi \mathbf{B} \cdot d\mathbf{F} , \tag{2.26}$$

which vanishes only if $B_n = 0$, since χ is arbitrary. In the general case $B_n \neq 0$, where field lines cross the boundary surface, the magnetic helicity is not defined. A generalized helicity expression is given in section 6.7, eq. (6.78). In multiply connected systems gauge invariance imposes further conditions, as will be discussed in section 9.1.

In the case of incompressibility and homogeneous density $\rho = 1$ there is a further conserved quantity, the cross-helicity

$$K = \int_V \mathbf{v} \cdot \mathbf{B} \, d^3 x . \tag{2.27}$$

One finds

$$\frac{dK}{dt} = - \oint_S \left(\mathbf{v} \cdot \mathbf{B} \mathbf{v} - \frac{v^2}{2} \mathbf{B} + p \mathbf{B} \right) \cdot d\mathbf{F} \tag{2.28}$$

which vanishes for $v_n = B_n = 0$ at the boundary. The conservation of the cross-helicity replaces that of the kinetic helicity $H^V = \int \mathbf{v} \cdot \boldsymbol{\omega} \, d^3 x$, $\boldsymbol{\omega} = \nabla \times \mathbf{v}$, in incompressible nonmagnetic hydrodynamics described by the Navier-Stokes equation. (The conservation of H^V in the latter case can easily be seen from the conservation of H, eq. (2.25), by noting that the equation for the vorticity $\boldsymbol{\omega}$ is identical with eq. (2.11) for $\mathbf{B}$.)

A particularly simple situation arises if periodic boundary conditions are assumed, which is often done in turbulence theory (chapter 7). Since in this case the boundary terms vanish, there are three quadratic invariants, total energy, magnetic helicity and cross-helicity.

A fundamental concept in MHD theory is that of a magnetic flux tube. It is based on the conservation of the magnetic flux

$$\phi = \int_F \mathbf{B} \cdot d\mathbf{F} \tag{2.29}$$

through an arbitrary surface $F(t)$ bounded by a curve which moves with the fluid as illustrated in Fig. 2.1.

Taking the surface integral of eq. (2.11) one obtains

$$\int_F \partial_t \mathbf{B} \cdot d\mathbf{F} = \oint (\mathbf{v} \times \mathbf{B}) \cdot d\mathbf{l} = - \oint \mathbf{B} \cdot (\mathbf{v} \times d\mathbf{l}) ,$$

and hence

$$\frac{d\phi}{dt} = \int \partial_t \mathbf{B} \cdot d\mathbf{F} + \oint \mathbf{B} \cdot (\mathbf{v} \times d\mathbf{l}) = 0 \tag{2.30}$$

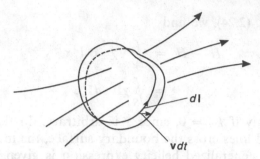

Fig. 2.1. Conservation of magnetic flux.

using

$$\int_{dF} \mathbf{B} \cdot d\mathbf{F} = \oint \mathbf{B} \cdot (\mathbf{v} \times d\mathbf{l}) dt \ .$$

Sweeping the boundary curve along the field lines defines a magnetic flux tube. Because of flux conservation the picture of field lines frozen to the fluid has a well-defined physical meaning as flux tubes of infinitesimal diameter.

Let us consider the magnetic helicity H_ε and, in the incompressible case, also the cross-helicity K_ε of a flux tube V_ε. Conservation of H_ε, the Woltjer invariant (Woltjer, 1958), follows from eq. (2.25),

$$\int_{V_\varepsilon} \partial_t (\mathbf{A} \cdot \mathbf{B}) \, d^3x = - \oint_{S_\varepsilon} \mathbf{A} \cdot \mathbf{B} \mathbf{v} \cdot d\mathbf{F} \ ,$$

since $B_n = 0$, the subscript ε indicating the flux tube. Using eq. (2.18) we find

$$\frac{dH_\varepsilon}{dt} = \int_{V_\varepsilon} \partial_t (\mathbf{A} \cdot \mathbf{B}) \, d^3x + \oint_{S_\varepsilon} \mathbf{A} \cdot \mathbf{B} \mathbf{v} \cdot d\mathbf{F} = 0 \ . \qquad (2.31)$$

On the same lines one obtains from eq. (2.28) that

$$\frac{dK_\varepsilon}{dt} = 0 \ . \qquad (2.32)$$

2.3   Magnetic helicity

The infinite set of magnetic helicities H_ε of the flux tubes in a magnetic configuration provide a measure of its complexity. In general the helicity has two contributions (Berger & Field, 1984; Moffat & Ricca, 1992): the internal helicity connected with twists and kinks of the individual tubes and the external one due to tube interlinkage. Consider a configuration consisting of a single closed untwisted tube (Fig. 2.2a). Using $\mathbf{B}d^3x = \phi d\mathbf{l}$, where ϕ is the magnetic flux in the tube, we have

$$H_\varepsilon = \phi \oint_{C_\varepsilon} \mathbf{A} \cdot d\mathbf{l} = 0 , \tag{2.33}$$

since the magnetic flux across the surface bounded by the curve C_ε is zero. If the field lines in the flux tube are twisted, one finds $H_\varepsilon = T\phi^2$, where T is the twist number = rotational transform ι, see below. Field line twist $T = \pm 1$ is equivalent to a kinking of the tube, which determines the helicity of a knotted tube, for instance $H_\varepsilon = \pm 3\phi^2$ for a (nontwisted) trefoil knot (Fig. 2.2b) (see Berger & Field). Now consider two linked flux tubes (Fig. 2.2c). For nontwisted tubes the total helicity H is the sum of the individual contributions $H_{\varepsilon 1}$, $H_{\varepsilon 2}$,

$$H_{\varepsilon 1} = \phi_1 \oint_{C_1} \mathbf{A} \cdot d\mathbf{l} = \phi_1\phi_2 = H_{\varepsilon 2},$$
$$H = 2\phi_1\phi_2 . \tag{2.34}$$

In the general case of N twisted and linked flux tubes the total helicity becomes

$$H = \sum_i T_i\phi_i^2 + \sum_{i,j} L_{ij}\phi_i\phi_j, \tag{2.35}$$

where L_{ij} is the number of linkages of tubes i and j (Moffat, 1969). In a complicated system the total number of linkages is $O(N^2)$ and hence the internal helicity contribution, the first term in eq. (2.35), becomes negligible compared with the external helicity, the interlinkage sum.

It is interesting to note that magnetic configurations with nonzero helicity cannot be represented *globally* by two scalar functions $\alpha(x, y, z)$, $\beta(x, y, z)$, the Euler potentials,

$$\mathbf{B} = \nabla\alpha \times \nabla\beta . \tag{2.36}$$

(This representation is a very convenient way of describing individual field lines by the intersection of two surfaces $\alpha = c_1, \beta = c_2$.) The most general form of the vector potential giving rise to the field (2.36) is

$$\mathbf{A} = \alpha\nabla\beta + \nabla\chi . \tag{2.37}$$

If χ is single-valued then the corresponding magnetic helicity vanishes,

$$\begin{aligned}
H &= \int_V \nabla\chi \cdot \nabla\alpha \times \nabla\beta \, d^3x \\
&= \int_V \nabla \cdot (\chi\nabla\alpha \times \nabla\beta) \, d^3x \\
&= \oint_S \chi\mathbf{B} \cdot d\mathbf{F} = 0 ,
\end{aligned} \tag{2.38}$$

where $B_n = 0$ on the boundary of a simply connected volume V is assumed. If the system is multiply connected, as for instance in a torus,

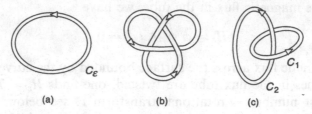

Fig. 2.2. Closed flux tubes. (a) single unknotted tube, (b) trefoil knot, (c) two interlinked tubes.

where χ may not be single-valued, H is not gauge-invariant and has to be generalized (see section 9.1).

The concept of magnetic helicity plays an important role in characterizing the field line topology in a toroidal equilibrium configuration, which will be treated in more detail in chapter 3. Such equilibria are usually determined by magnetic surfaces ψ nested around a magnetic axis. A field line running along a surface ψ either closes onto itself after m rotations around the major axis of the torus involving n rotations around the magnetic axis (or minor torus axis), or continues indefinitely filling the entire surface ergodically. In the first case ψ is called a rational surface characterized by the rotational transform $\iota = n/m$, in the second case ψ is called irrational with $\iota = \lim_{m\to\infty} n/m$.

In general $\iota(\psi)$ is a smoothly varying function, which makes the concept of flux tubes somewhat ill-defined. Consider a small, circular vicinity of a field line at a particular point to define the cross-section of a flux tube at this point. As we follow this flux tube on its long way around and around the torus the cross-section becomes more and more distorted since the field lines on the outer side at $\psi + d\psi$ rotate around the magnetic axis at a rate slightly different from that of those on the inner side at $\psi - d\psi$. To overcome this conceptual difficulty let us assume that the configuration is very slightly perturbed so that instead of smooth surfaces rational surfaces are found to be broken up into a chain of narrow islands, where the island width (see eq. (4.84)) is smaller than the distance between neighboring rational surfaces to prevent field line stochastization. (More details concerning the process of breaking of flux surfaces are found in section 3.5.) These islands define flux tubes, the cross-sections of which are approximately constant all along their entire extent. Consider such a flux tube characterized by $\iota = n/m$ and calculate the helicity H_ε. Since the field line encircles the torus m times the long way and n times the short way we have

$$H_\varepsilon = n\chi(\psi) + m\psi \,, \tag{2.39}$$

where ψ is defined as the poloidal flux through the inside bore of the torus up to the surface labeled ψ, and $\chi(\psi)$ is the toroidal flux enclosed by the surface ψ. Defining a helicity per toroidal turn $\hat{H}_\varepsilon = H_\varepsilon/m$, we find

$$\hat{H}_\varepsilon = \iota(\psi)\chi(\psi) + \psi \ . \tag{2.40}$$

Hence a toroidal configuration is characterized by the values of the helicity $\hat{H}_\varepsilon(\psi)$ on each surface.

In this section we have concentrated on the helicity of individual flux tubes H_ε. As will be seen in chapters 7 and 9, invariance of H_ε is easily destroyed by finite resistivity, leaving only the global helicity H of the entire plasma volume V as a robust invariant.

2.4 Reduced MHD equations

The MHD equations derived in section 2.1 contain seven independent dynamic variables, three velocity components, two magnetic field components (because of $\nabla \cdot \mathbf{B} = 0$), density and pressure. In the limiting case of incompressibility and homogeneous density these reduce to four, two velocity components (because of $\nabla \cdot \mathbf{v} = 0$) and two magnetic field components, since the pressure now is a function of $\mathbf{v}$ and $\mathbf{B}$ and the density can be assumed constant. For a strongly magnetized plasma a further approximation called reduced MHD is very useful (Kadomtsev & Pogutse, 1974; Strauss, 1976). A Hamiltonian formulation of the reduced equations has been developed (Morrison & Hazeltine, 1984), underlining the significance of this approximation.

Assume a plasma column immersed in a strong, nearly constant magnetic field, the main field B_z in the axial direction. By strong, we mean that thermal and kinetic plasma energies are small compared with the main field energy, $\rho v^2 \sim p \ll B_z^2$. In this case the plasma motions are highly anisotropic. Small spatial scales associated with strong local gradients such as current sheets can be generated only in the plane perpendicular to B_z, the poloidal plane, since spatial variations along B_z corresponding to a bending of field lines are smooth. We may hence make the ordering ansatz

$$B_z \sim \nabla_\perp \sim 1 \ , \quad B_\perp \sim \partial_z \sim \varepsilon \ . \tag{2.41}$$

One can assume $\rho = 1$, since the flow turns out to be incompressible. Dynamics will lead to equipartition (order of magnitude-wise) $v_\perp^2 \sim p \sim B_\perp^2$, hence one has

$$v_\perp \sim \varepsilon \ , \quad p \sim \varepsilon^2 \ , \quad \partial_t \sim \mathbf{v} \cdot \nabla_\perp \sim \varepsilon \ . \tag{2.42}$$

Since the plasma pressure is balanced by a variation of the axial field

$$\delta B_z^2 = 2 \, \delta B_z \, B_z \sim p \,,$$

we find

$$\delta B_z \sim \varepsilon^2 \,, \qquad (2.43)$$

hence to lowest order

$$B_z = \text{const.} \qquad (2.44)$$

The poloidal field is conveniently written in terms of the axial component of the vector potential A_z, usually denoted by ψ, $\psi = -A_z$,

$$\mathbf{B} = \mathbf{e}_z \times \nabla\psi + B_z \mathbf{e}_z \,, \qquad (2.45)$$

which is divergence-free neglecting terms of order ε^3, $\nabla \cdot \mathbf{B} = \partial_z B_z \sim \varepsilon^3$. The axial forces are small,

$$\partial_z p \sim j_\perp B_\perp \sim \varepsilon^3 \,,$$

hence the axial velocity is small, $v_z \sim \varepsilon^2 \ll v_\perp$, and is neglected to lowest order,

$$v_z = 0 \,. \qquad (2.46)$$

Integration of Faraday's law eq. (2.11) gives

$$\partial_t \mathbf{A} = \mathbf{v} \times \mathbf{B} - \nabla\chi \,, \qquad (2.47)$$

where $\mathbf{B} = \nabla \times \mathbf{A}$ and χ is the scalar potential. From $B_z = (\nabla_\perp \times \mathbf{A}_\perp) \cdot \mathbf{e}_z$ and eqs (2.42), (2.43) we find that

$$\partial_t \mathbf{A}_\perp \sim \varepsilon^3 \,,$$

such that in the poloidal part of eq. (2.47) the l.h.s. is negligible. This gives in lowest order

$$(\mathbf{v} \times \mathbf{B})_\perp = \mathbf{v} \times B_z \mathbf{e}_z = \nabla_\perp \chi$$

and hence the poloidal velocity

$$\mathbf{v}_\perp = \mathbf{e}_z \times \nabla\phi \qquad (2.48)$$

introducing the stream function $\phi = \chi/B_z$. Equation (2.48) implies that the poloidal flow is incompressible. From the z-component of eq. (2.47) we obtain the equation for the flux function to leading order

$$\partial_t \psi + \mathbf{v} \cdot \nabla\psi = B_z \partial_z \phi \,, \qquad (2.49)$$

which can also be written in the form

$$\partial_t \psi - \mathbf{B} \cdot \nabla\phi = 0 \,. \qquad (2.50)$$

The equation for the stream function ϕ is obtained by considering the axial component of the curl of the equation of motion (2.5). Since

$$\mathbf{e}_z \cdot \nabla \times (\mathbf{j} \times \mathbf{B}) = \mathbf{B} \cdot \nabla j_z - \mathbf{j} \cdot \nabla B_z \simeq \mathbf{B} \cdot \nabla j_z$$

to order ε^2, one obtains

$$\partial_t \omega + \mathbf{v} \cdot \nabla \omega = \mathbf{B} \cdot \nabla j \tag{2.51}$$

where

$$\omega = \mathbf{e}_z \cdot \nabla \times \mathbf{v} = \nabla_\perp^2 \phi \,,$$
$$j = j_z = \nabla_\perp^2 \psi \,.$$

Equations (2.49) (or (2.50)) and (2.51) for the two scalar quantities ψ, ϕ together with the definitions (2.45) and (2.48) constitute a closed set, which are called the (lowest order) reduced MHD equations.

It is interesting to note that the reduced equations are almost identical with the incompressible two-dimensional MHD equations; the only coupling in the z-direction enters through the *linear* terms $B_z \partial_z j$, $B_z \partial_z \phi$ (since $B_z = \text{const.}$). This indicates that in strongly magnetized systems where the main processes occur in the plane perpendicular to the axial field two- and three-dimensional systems are dynamically rather similar. We shall come back to this point in chapter 7.

Multiplying eq. (2.49) by j, eq. (2.51) by ϕ and integrating over the volume V we obtain the energy conservation law of reduced MHD

$$\frac{d}{dt} \int_V \frac{1}{2} \left(v^2 + B^2 \right) d^3 x$$
$$= \oint_S \phi \left[\omega \mathbf{v} - j\mathbf{B} + \partial_t \nabla \phi \right] \cdot d\mathbf{F} + \oint_S \partial_t \psi \nabla \psi \cdot d\mathbf{F} \,. \tag{2.52}$$

The r.h.s. of this equation has an unusual form depending explicitly on ϕ. However, it is easy to see that the expression is invariant with respect to a scale transformation $\phi \to \phi + \phi_0$, where ϕ_0 is a constant. In fact

$$\oint (\omega \mathbf{v} - j\mathbf{B} + \partial_t \nabla \phi) \cdot d\mathbf{F}$$
$$= \int [\nabla \cdot (j\mathbf{B} - \omega \mathbf{v}) - \partial_t \omega] d^3 x = 0 \,. \tag{2.53}$$

The physical meaning of the last term in eq. (2.52) becomes clear when considering the special case of a two-dimensional stationary system. Here $\partial_t \psi = E_z = \text{const.}$ is the inductively applied electric field and $\oint \nabla \psi \cdot d\mathbf{F} = \oint \mathbf{B} \cdot d\mathbf{l} dz = IL$, where I is the total current in the z-direction and L the length of the system. With $U = E_z L$ this term becomes

$$\oint \partial_t \psi \nabla \psi \cdot d\mathbf{F} = UI \,, \tag{2.54}$$

the inductively injected power. The result (2.52) is generalized in section 5.2 to a plasma surrounded by a vacuum region.

The magnetic helicity degenerates to the linear expression

$$H = \int_V \psi \, d^3x \,, \tag{2.55}$$

which is conserved for a flux tube,

$$\frac{d}{dt} \int_{V_\varepsilon} \psi \, d^3x = \oint_{S_\varepsilon} \phi \mathbf{B} \cdot d\mathbf{F} = 0 \,, \tag{2.56}$$

just as in the general MHD case eq. (2.31). Also the cross-helicity $K = \int \nabla\phi \cdot \nabla\psi d^3x = -\int \omega\psi d^3x$ is conserved in reduced MHD.

Because of their simplicity and the fact that fast, compressible time scales are eliminated, the reduced equations have been used extensively in numerical simulations of various MHD processes in strongly magnetized plasmas, notably for tokamaks. Since this model, which is the lowest-order theory in an expansion in the inverse aspect ratio of a toroidal system (see section 3.2), contains several obviously crude approximations regarding both dynamics and geometry, higher-order approximations have been developed. These are notably: first-order explicit toroidal corrections (Carreras et al., 1981a), high plasma pressure $p \sim \varepsilon$ (compared with $p \sim \varepsilon^2$) in order to account for ballooning effects (see section 4.6) which introduces the pressure as a third dynamic variable (Strauss, 1977), and equations including both effects (Schmalz, 1981). Still higher-order reduced equations have been derived (Izzo et al., 1983).

However, the complexity of these equations rapidly increases, which soon seems to offset the advantage gained by using a reduced dynamical model as compared with the full MHD equations. This is especially true since the advent of efficient semi-implicit numerical schemes (Harned & Kerner, 1985; Harned & Schnack, 1986; Lerbinger & Luciani, 1991), which allow the numerical solution of the full compressible MHD equations using time steps Δt similar to those for the reduced equations. It therefore appears that if physics requires to go beyond the lowest-order equations (2.49), (2.51), one should use the full MHD equations directly, since in addition to their complexity the physical implications involved in the higher-order reduced models are only poorly known, in particular those regarding the nonlinear behavior.

2.5 Validity limits of ideal MHD and dissipation effects

Ideal fluid equations are, strictly speaking, a contradiction in itself, since the fluid approximation is based on the assumption that the system is locally close to thermodynamic equilibrium, which requires a certain rate of collisions and hence dissipation. Let us first discuss the conditions for the validity of a fluid description in general. Conventional theory requires that the mean free path λ_c be short compared with typical gradient scales, $\lambda_c |\nabla f| \ll f$. While this is usually satisfied for liquids and neutral gases, the mean free path in hot plasmas becomes very long, $\lambda_c \propto T^2$, where T is the temperature, such that formally the condition for a fluid approximation may easily be violated. This is true even for some astrophysical plasmas of large extent, for instance the solar wind, and much more so for hot laboratory plasmas such as in tokamaks.

The formal argument is, however, somewhat misleading. Firstly the plasma behavior is in general strongly anisotropic owing to the presence of a magnetic field. Even in a collisionless plasma the effective mean free path in the direction perpendicular to the field is the gyroradius ρ_c, which is usually very small, $\rho_c = v_t / \Omega_c$, where $v_t = \sqrt{2 k_B T / m}$ is the thermal velocity of a particular particle species and $\Omega_c = eB/m$ its gyrofrequency, m and e are the particle mass and charge. On the other hand it turns out that in a magnetized plasma gradients parallel to the field where the mean free path is long tend to be much weaker than in the perpendicular direction $|\nabla_\parallel f| \ll |\nabla_\perp f|$. Secondly a collisionless plasma is not dissipationless. Even if Coulomb collisions are absent, small-scale plasma turbulence usually gives rise to stochastic particle orbits and phase mixing in velocity space and hence to efficient dissipation, though velocity distribution functions do not in general relax to a Maxwellian. Thus most plasma particles feel a rather short effective mean free path, so that for large-scale plasma motions a fluid approximation may be well justified, even in a collisionless plasma.

The question of the validity of the *ideal* fluid approximation needs a more differentiated argumentation. Dissipation effects are represented by various kinds of diffusion processes and are called weak if the diffusion coefficients normalized to some average spatial scale are small. The dissipation rates, however, depend on the local spatial scales, which are usually determined by the internal dynamics. For smooth, quasi-static plasma configurations dissipation rates are usually negligible, so that they may be treated in the framework of ideal MHD. This is also true in the case of nonsingular eigenmodes, for which ideal stability theory has a solid foundation. Singular equilibria containing certain discontinuities, such as current sheets, and singular eigenmodes may be strongly influenced by dissipation effects and inherently require a nonideal theory. For strongly nonlinear dynamic processes dissipation is in general important, because, apart from special cases, large-scale motions rapidly build up small-scale structures, corresponding to singularities in the framework of ideal theory. It is known from hydrodynamic turbulence that independently of the smallness of the viscosity coefficient energy dissipation rates are finite. Reducing the viscosity automatically leads to the excitation of finer spatial scales. As we see in section 7.6, this is also true in MHD. Hence for the remainder of this book we include dissipation, though usually in the limit of almost vanishing dissipation coefficients.

Let us now discuss the important dissipation effects, which are primarily corrections to Ohm's law (2.10). Assuming that these depend mainly on the current density a Taylor expansion gives

$$\mathbf{E} + \mathbf{v} \times \mathbf{B} = \eta_1 \mathbf{j} - \eta_2 \nabla^2 \mathbf{j} + \cdots , \qquad (2.57)$$

since only even-order differential operators correspond to dissipation. The coefficients are in general tensors accounting for the plasma anisotropy caused by the magnetic field. For η_1, however, the difference between $\eta_\perp$ and $\eta_\parallel$ is only a factor of 2, so that the assumption of a scalar resistivity is justified in the present context. Though we call $\eta = \eta_1$ resistivity, it will be used in the sense of a magnetic diffusivity, $\eta_1 = (c/\omega_{pe})^2 \nu_e$, where ν_e is the electron–ion collision frequency, ω_{pe} the electron plasma frequency, $\omega_{pe}^2 = 4\pi n e^2/m_e$ in Gaussian units, and c the velocity of light.

The coefficient η_2 of the next-order term is often called hyperresistivity. In a plasma it is associated with the perpendicular electron viscosity μ_e, $\eta_2 \simeq (c/\omega_{pe})^2 \mu_e$. It is true that the collisional value of μ_e is very small, $\mu_e \sim \rho_e^2 \nu_e$, ρ_e = electron Larmor radius. But it has been argued (Furth et al., 1973; Kaw et al., 1979) that microscale field line stochasticity ("braiding") caused by weak magnetic fluctuations gives rise to an effective perpendicular momentum transport and hence to an anomalous electron viscosity much larger than the collisional one, such that the η_2-term in eq. (2.57) may be important. Inserting eq. (2.57) into Faraday's law and neglecting spatial variations of the transport coefficients gives

$$\partial_t \mathbf{B} - \nabla \times (\mathbf{v} \times \mathbf{B}) = \eta_1 \nabla^2 \mathbf{B} - \eta_2 \nabla_\perp^{(4)} \mathbf{B} . \qquad (2.58)$$

In most applications where dissipation in Ohm's law is considered explicitly we confine ourselves to the resistivity, but occasionally also discuss the effect of hyperresistivity, for instance for the internal kink mode. In the context of magnetic reconnection in the sawtooth collapse a further, nondissipative correction in Ohm's law (2.57) is also discussed, the electron inertia term $\propto d\mathbf{j}/dt$ (see section 8.1.5).

Dissipation must also be considered in the equation of motion (2.5). In a plasma this enters through the ion viscous force, the divergence of the stress tensor, which is highly anisotropic. However, this term is usually replaced for simplicity by a diffusion term with a scalar kinematic viscosity ν,

$$\rho \left(\partial_t + \mathbf{v} \cdot \nabla \right) \mathbf{v} = -\nabla p + \mathbf{j} \times \mathbf{B} + \rho \nu \nabla^2 \mathbf{v} . \qquad (2.59)$$

The justification for this crude approximation is that the dynamic behavior is often found to be rather insensitive to the magnitude and form of the viscous term, the more important effect being the magnetic dissipation in eq. (2.58). On the hydrodynamic level viscosity is usually not expected to mimic real transport processes but to provide an efficient energy sink at small scales in order to avoid the formation of flow singularities. It is believed, though not rigorously proved, that the introduction of scalar resistivity and viscosity ensures regularity of the solutions of the MHD equations for all times.

A convenient measure of the magnitude of the dissipation coefficients is obtained by the ratios of the average of the dissipation and flow terms in eqs (2.58), (2.59)

$$\frac{\langle |\eta \nabla^2 \mathbf{B}| \rangle}{\langle |\nabla \times (\mathbf{v} \times \mathbf{B})| \rangle} \sim \frac{\eta}{vL} = R_m^{-1} , \qquad (2.60)$$

$$\frac{\langle |\nu \nabla^2 \mathbf{v}| \rangle}{\langle |\mathbf{v} \cdot \nabla \mathbf{v}| \rangle} \sim \frac{\nu}{vL} = Re^{-1} , \qquad (2.61)$$

where R_m and Re are the magnetic and kinetic Reynolds numbers, v being a typical flow velocity and L a global scale length.

Alternatively to R_m, the Lundquist number S is often used for magnetized plasmas,

$$S = \frac{v_A L}{\eta} , \qquad (2.62)$$

where the Alfvén velocity replaces the fluid velocity. S is a useful measure of the magnitude of resistive effects in static configurations, in particular for growth rates of resistive stabilities, but for strongly nonlinear, often turbulent systems R_m is the more significant quantity.

3

Magnetostatic equilibria

Slowly varying magnetically confined plasma configurations are of fundamental interest. They are usually well approximated by magnetostatic equilibria. In section 3.1 we introduce two one-dimensional configurations, the sheet pinch and the cylindrical pinch. These highly idealized systems are the archetypical plasma configurations, which exhibit many features of real systems. In section 3.2 we derive the equation for the magnetic flux function ψ in the most general two-dimensional case, that of helical symmetry. In section 3.3 several exact analytical solutions are discussed, which are often used in theoretical and numerical studies. Section 3.4 gives a brief introduction to the numerical methods of computing general 2-D equilibria. The general nonsymmetric case is discussed in section 3.5.

3.1 One-dimensional configurations

In magnetostatic plasma configurations the magnetic and pressure forces in eq. (2.5) balance (neglecting the gravitational term)

$$\nabla p = \mathbf{j} \times \mathbf{B} \equiv \mathbf{B} \cdot \nabla \mathbf{B} - \nabla B^2/2 . \tag{3.1}$$

The formal simplicity of this equation is deceptive, however. The existence of strict solutions has only been shown for symmetric systems, while in the general nonsymmetric toroidal case equilibria seem to exist only in an approximate sense.

Apart from the trivial homogeneous case the simplest systems are those with the highest degree of symmetry, viz. one-dimensional equilibria, where a coordinate system ξ, η, ζ exists such that all physical quantities depend only on ξ. In principle there are three symmetry classes with these properties: the plane or slab symmetry, where in cartesian coordinates x, y, z physical quantities depend only on x, the rotationally symmetric cylinder, where in cylindrical coordinates r, θ, z dependence is only on

24

r, and spherical symmetry. Only in the first two cases do nontrivial MHD configurations exist, since for spherical symmetry $\nabla \cdot \mathbf{B} = 0$ requires $\mathbf{B} = 0$. The corresponding configurations, the (plane) sheet pinch and the (circular) cylindrical pinch, represent the most fundamental types of magnetically confined plasmas. Though real plasmas are not symmetric such as the geomagnetic tail, a solar magnetic loop or the various types of laboratory pinch plasma columns, they often resemble the symmetric configurations. Even very irregular cosmical plasmas have sheet- or tube-like local structures.

One-dimensional plasmas are necessarily open, i.e. either extend to infinity or are bounded by some kind of electrodes (in which case they are no longer strictly one-dimensional). To avoid such undesirable conceptual features one often assumes these systems to be topologically equivalent to closed toroidal systems, which amounts to imposing periodic boundary conditions with certain periodicity lengths.

In the case of a *sheet pinch* the equilibrium is determined by

$$p = p(x) , \quad \mathbf{B} = \mathbf{B}(x) . \tag{3.2}$$

For completeness we also specify the mass density $\rho = \rho(x)$, which does, however, not enter the equilibrium equation (3.1) since we are not considering gravitational forces. From $\nabla \cdot \mathbf{B} = dB_x/dx = 0$ it follows that the B_x is constant and hence vanishes if perfectly conducting boundary conditions at $x = \pm L_x$ are assumed. Thus $\mathbf{B} = (0, B_y(x), B_z(x))$, which implies that field lines are straight lines on surfaces $x = $ const., with a direction characterized by the ratio $q(x) = B_z(x)/B_y(x)$, the analog of the safety factor in a cylindrical pinch, as introduced below.

From eq. (3.1) one obtains

$$\frac{d}{dx}\left(p + \frac{B^2}{2}\right) = 0 , \tag{3.3}$$

since $\mathbf{B} \cdot \nabla \mathbf{B} = 0$. In general both B_y and B_z vary across the sheet, but two special cases are of particular interest: (a) The field line direction is constant. Without loss of generality we may choose $B_z = 0$. Prescribing $p(x)$ and $B_y(-L_x)$ determines $B_y(x)$. In theoretical investigations one often assumes a symmetric behavior $p(x) = p(-x)$ and $B_y(x) = -B_y(-x)$, i.e. $x = 0$ is a neutral surface, or neutral line in projection, where B_y reverses sign. A well-known example is the geomagnetic tail. (b) The plasma pressure is negligible, $p = 0$, hence $B^2 = $ const. from eq. (3.3). This is a simple example of a force-free configuration defined by $\mathbf{j} \times \mathbf{B} = 0$. The sheet is characterized by a region where $\mathbf{j} \neq 0$, joining two current-free regions of different field direction. In the limit of vanishing sheet thickness this is called a rotational discontinuity.

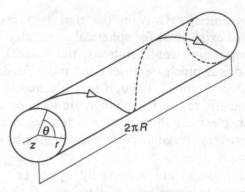

Fig. 3.1. Schematic drawing of a cylindrical pinch configuration.

Since in real plasma configurations sheet-like structures are often narrow, they require some external driving mechanism to prevent rapid expansion due to resistive diffusion. The most important effect is plasma flow toward the sheet balancing resistive broadening. Since, however, the inflowing plasma is ejected at a high velocity along the sheet, conditions are strictly speaking neither static nor one-dimensional. These dynamic current sheets (Sweet-Parker sheets) play a fundamental role in the theory of magnetic reconnection (chapter 6).

In the *cylindrical pinch* (Fig. 3.1) the equilibrium quantitites are

$$p(r) \, , \quad \mathbf{B} = (0, B_\theta(r), B_z(r)) \, , \tag{3.4}$$

following the equation

$$\frac{dp}{dr} = j_\theta B_z - j_z B_\theta \equiv -\frac{d}{dr}\frac{B_z^2}{2} - B_\theta \frac{1}{r}\frac{d}{dr} r B_\theta \, . \tag{3.5}$$

Hence the plasma confinement is in general due to both poloidal and axial currents. Special cases, interesting mainly for historical reasons, are $B_\theta = j_z = 0$, called θ-pinch, since the current flows in the θ-direction, and $B_z = j_\theta = 0$, called z-pinch, since the current flows in the z-direction. Multiplying eq. (3.5) by r^2 and integrating over the plasma cross-section, one obtains

$$\langle p \rangle = \frac{1}{2} \left[B_\theta^2(a) + B_z^2(a) - \langle B_z^2 \rangle \right] \, , \tag{3.6}$$

where $r = a$ is the boundary of the plasma column, defined by the vanishing of the pressure, and $\langle \ldots \rangle = 2 \int_0^a \ldots r dr / a^2$. A convenient measure of the plasma pressure is given by the plasma beta, the ratio

$$\beta = 2 \frac{\langle p \rangle}{\langle B^2 \rangle} \, . \tag{3.7}$$

If the axial magnetic field is reduced inside the plasma compared with its constant outside value, $\langle B_z^2 \rangle < B_z^2(a)$, the poloidal current j_θ is called diamagnetic, while it is called paramagnetic when B_z is enhanced. In the paramagnetic case the current density lines encircle the axis in the same sense as the magnetic field lines, while they do so in the opposite sense in the diamagnetic case. The distinction can be expressed by the parameter β_p, the poloidal β,

$$\beta_p = 2\frac{\langle p \rangle}{B_\theta^2(a)} \quad \begin{cases} < 1 & \text{paramagnetic} \\ > 1 & \text{diamagnetic} . \end{cases} \tag{3.8}$$

A very useful quantity characterizing the magnetic field line twist is the rotational transform ι (introduced in section 2.3), or its inverse, called the safety factor q, commonly used in tokamak physics:

$$q(r) = \iota^{-1}(r) = \frac{rB_z(r)}{RB_\theta(r)} . \tag{3.9}$$

ι is the angle by which a field line is rotated in poloidal direction when advancing by a distance R in the axial direction. The dependence of ι or q on r indicates how the field line twist varies between different surfaces $r = \text{const.}$, a quantitative measure being the shear parameter

$$s(r) = \frac{d \ln q}{d \ln r} . \tag{3.10}$$

In section 3.2 $q = q(\psi)$ is defined for more general configurations with magnetic surfaces ψ nested around a magnetic axis.

The equilibrium determined by eq. (3.5) contains two free functions, which are usually chosen as $p(r)$ and $j_z(r)$ or $q(r)$. In practice, however, the radial profiles are strongly constrained by stability requirements and transport effects. In most applications configurations are assumed to exist for periods which are long enough for transport processes to determine the radial profiles $p(r), j_z(r)$, making them essentially bell-shaped.

3.2 The two-dimensional equilibrium equation

To obtain geometrically more general equilibria than a one-dimensional pinch, it appears that all we have to do is to deform this configuration into the desired shape, for instance bending a cylindrical pinch into a torus, squeezing it into a strongly noncircular cross-section or applying certain axial corrugations, all of which seem to imply purely quantitative changes. This is, however, not true. Apart from the problem of practical evaluation which requires rather sophisticated numerical techniques, the transition to higher-dimensional configurations in general introduces qualitatively new features which in the three-dimensional case lead to a basic existence problem. While the latter case is addressed in section 3.5, we here consider the most general two-dimensional case, where the existence of equilibria can still be shown rigorously.

From eq. (3.1) follows immediately that p must be constant along magnetic field lines as well as current density lines

$$\mathbf{B} \cdot \nabla p = 0 \,, \quad \mathbf{j} \cdot \nabla p = 0 \,. \tag{3.11}$$

Since for a finite pressure gradient $\mathbf{B}$ and $\mathbf{j}$ are not parallel, one can locally construct a surface $\psi(x, y, z) = \text{const.}$, where p is constant, $p = p(\psi)$. In an equilibrium configuration this property must, however, hold globally, i.e. there must be surfaces of constant pressure, called magnetic surfaces, on which a field line runs on forever, a condition which turns out to be rather stringent for toroidal configurations with nonclosed field lines.

That infinitely long field lines span smooth magnetic surfaces has only been shown for configurations with a continuous symmetry, implying the existence of a coordinate system ξ, η, ζ such that all physical quantities including the elements of the metric tensor depend only on two coordinates ξ, η. If this condition holds then the equilibrium equation (3.1) can be reduced to a quasilinear elliptic differential equation for a scalar function $\psi(\xi, \eta)$ such that surfaces $\psi = \text{const.}$ have the desired properties of magnetic surfaces (Edenstrasser, 1980a).

The question now arises: what are the different classes of continuous symmetries? It turns out that the requirements of the coordinates ξ, η, ζ given above are rather restrictive. In fact it can be shown (Edenstrasser, 1980b) that the most general case is that of helical symmetry. It corresponds to the invariance with respect to a general rigid motion in space which is the combination of a rotation around an axis and a translation along this axis. If we introduce cylindrical coordinates r, φ, z with z along the rotation axis, then helical symmetry implies that any physical quantity depends only on r and $u = l\varphi + kz$. Though a helix is characterized by just one parameter, for instance $\alpha = k/l$, it is convenient to keep both l and k in order to discuss the two limiting cases, $l = 0$, corresponding to rotational or axisymmetry, and $k = 0$, corresponding to translational or plane symmetry. Let us define the vector

$$\mathbf{h} = \frac{l\nabla z - kr^2 \nabla \varphi}{l^2 + k^2 r^2} = \frac{r}{l^2 + k^2 r^2} \nabla r \times \nabla u \,, \tag{3.12}$$

which is tangent to the helix $r = \text{const.}$, $u = \text{const.}$, where $\nabla u = l\nabla\varphi + k\nabla z = (l/r)\mathbf{e}_\varphi + k\mathbf{e}_z$. It follows from eq. (3.12) that $\mathbf{h} \cdot \nabla F = 0$ for any function $F(r, u)$. As can easily be checked, $\mathbf{h}$ has the properties

$$\nabla \cdot \mathbf{h} = 0 \,, \quad \nabla \times \mathbf{h} = \frac{-2kl}{l^2 + k^2 r^2} \mathbf{h} \,. \tag{3.13}$$

The general solution of $\nabla \cdot \mathbf{B} = 0$ in helical symmetry can be written in the form

$$\mathbf{B} = \mathbf{h} \times \nabla\psi(r,u) + \mathbf{h}f(r,u) . \tag{3.14}$$

Here ψ is called the helical flux function, since $\psi(r,u) - \psi(0,u)$ is the magnetic flux through a helical ribbon $u = $ const. between the axis $r = 0$ and the value r of a cylinder of unit length in z. ψ is essentially the component of the vector potential in the direction of the ignorable coordinate

$$\psi = -\mathbf{A} \cdot \mathbf{h}/h^2 = -(lA_z - krA_\varphi) \tag{3.15}$$

and f is called the helical magnetic field[*],

$$f = \mathbf{B} \cdot \mathbf{h}/h^2 = lB_z - krB_\varphi . \tag{3.16}$$

The current density $\mathbf{j} = \nabla \times \mathbf{B}$ can be written in a form analogous to $\mathbf{B}$. Using relations (3.13) one derives in a straightforward way that

$$\mathbf{j} = -\mathbf{h} \times \nabla f(r,u) + \mathbf{h}\left(\mathcal{L}\psi - \frac{2kl}{l^2 + k^2r^2}f\right) , \tag{3.17}$$

where $\mathcal{L}$ is the generalized Laplacian

$$\mathcal{L} \equiv \frac{l^2 + k^2r^2}{r}\left(\partial_r \frac{r}{l^2 + k^2r^2}\partial_r + \frac{1}{r}\partial_{uu}\right) . \tag{3.18}$$

Let us now proceed to the equilibrium equation. Using the representations (3.14) and (3.17) one obtains

$$\nabla p = -\frac{1}{l^2 + k^2r^2}\left[f\nabla f + \left(\mathcal{L}\psi - \frac{2kl}{l^2 + k^2r^2}f\right)\nabla\psi\right]$$

$$- (\mathbf{h} \times \nabla f) \times (\mathbf{h} \times \nabla\psi) , \tag{3.19}$$

where the last term can be written in the form $(\nabla\psi \times \nabla f) \cdot \mathbf{h}\,\mathbf{h}$. Since $\mathbf{h} \cdot \nabla p = \mathbf{h} \cdot \nabla f = \mathbf{h} \cdot \nabla\psi = 0$ this term must vanish, $\nabla f \times \nabla\psi = 0$. It then follows from eq. (3.19) that also $\nabla p \times \nabla\psi = 0$. Hence f and p are flux functions

$$f = f(\psi) , \quad p = p(\psi) . \tag{3.20}$$

With these results the equilibrium equation reduces to a scalar equation for ψ

$$\mathcal{L}\psi = \frac{2kl}{l^2 + k^2r^2}f - ff' - (l^2 + k^2r^2)p' , \tag{3.21}$$

where the prime indicates the derivative with respect to the argument ψ. Surfaces $\psi = $ const. obviously have the properties of global magnetic

---

[*] In tokamak theory the first term on the right side of eq. (3.14) is usually called helical field $\mathbf{B}_*$, see eq. (5.12).

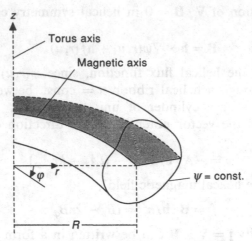

Fig. 3.2.   Axisymmetric toroidal configuration; shaded region = toroidal ribbon between magnetic axis and flux surface ψ.

surfaces.   Equation (3.21) is a quasilinear elliptic equation, containing two free profile functions $f(\psi), p(\psi)$. However, to determine a solution completely requires in general not only the specification of these functions and of the boundary conditions, but also some additional constraint. These issues are discussed in section 3.4.

For the special case of axisymmetry $l = 0, k = 1$ eq. (3.21) becomes

$$(r\partial_r r^{-1}\partial_r + \partial_{zz})\psi \equiv \Delta^*\psi = -ff' - r^2 p' . \tag{3.22}$$

This equation is usually called the Grad-Shafranov equation, though the historically correct but impractical name should be Grad-Lüst-Rubin-Schlüter-Shafranov equation, since it has independently been derived by Lüst & Schlüter (1957), Shafranov (1958), and Grad & Rubin (1958). From eqs (3.14) and (3.17) we obtain in the case of axisymmetry

$$\mathbf{B} = \nabla\varphi \times \nabla\psi + f\nabla\varphi = \mathbf{B}_p + B_\varphi \mathbf{e}_\varphi \ ,$$
$$\mathbf{j} = -\nabla\varphi \times \nabla f + \Delta^*\psi\nabla\varphi = \mathbf{j}_p + j_\varphi \mathbf{e}_\varphi \ .$$

It can easily be seen that $f(\psi) - f(0) = I_p(\psi)$ is the total current through a toroidal ribbon between the magnetic axis $\psi = 0$ and the flux surface ψ (Fig. 3.2).

For the case of plane symmetry $l = 1, k = 0$ eq. (3.21) reduces to

$$(r^{-1}\partial_r r\partial_r + r^{-2}\partial_{\varphi\varphi})\psi \equiv \nabla^2\psi = -ff' - p' \tag{3.23}$$

and the magnetic field and current density become

$$\mathbf{B} = \nabla z \times \nabla\psi + f\nabla z = \mathbf{B}_p + B_z \mathbf{e}_z \ ,$$
$$\mathbf{j} = -\nabla z \times \nabla f + \nabla^2 f\nabla z = \mathbf{j}_p + j_z \mathbf{e}_z \ .$$

A different transition to plane symmetry is often used as the lowest-order approximation of slender axisymmetric toroidal configurations characterized by a small inverse aspect ratio $\varepsilon = a/R \ll 1$, where R is the radius of the magnetic axis, called the major radius, and a is the (suitably defined) radius of the plasma column, called the minor radius. Introducing cartesian coordinates x, y in the poloidal plane by

$$r = R + x,$$
$$z = y,$$

eq. (3.22) reduces to the case of plane geometry eq. (3.23) in the neighborhood of the magnetic axis $x, y \ll R$. This approximation forms the geometrical basis of the reduced MHD equations derived in section 2.4.

In the two-dimensional case the rotational transform ι, or the safety factor q, cannot be written in terms of the local magnetic field components in contrast to the one-dimensional expression (3.9). As has already been discussed in section 2.3, a field line in a toroidal configuration is characterized by the ratio $\iota = \lim_{m \to \infty} n/m$, where n is the number of turns of a field line around the magnetic axis after m turns around the major torus axis. (If there is more than one magnetic axis then the axis must be specified, as the value of ι depends on its choice.) In general the rotational transform becomes a function of the flux surface ψ, since a field line fills its surface ergodically (apart from closed field lines with rational ι; in a system with continuously varying ι we can either use rational or irrational surfaces, whichever properties we need in the course of the argument, since both are densely distributed).

Let us derive an explicit expression for $\iota(\psi)$ for the case of axisymmetry, which is of particular practical interest. If ψ is normalized such that $\psi = 0$ on the magnetic axis then the magnetic flux through a toroidal ribbon between the axis and the surface ψ (see Fig. 3.2) is called the poloidal flux Ψ,

$$\Psi = \int_{\psi=0}^{\psi} \mathbf{B}_p \cdot d\mathbf{F} = \int \nabla \varphi \times \nabla \psi \cdot d\mathbf{F} = 2\pi\psi , \qquad (3.24)$$

while the flux through a meridional cross-section $\varphi = \text{const.}$ is called the toroidal flux χ,

$$\chi = \int^{\psi} \mathbf{B}_t \cdot d\mathbf{F} = \int^{\psi} f \nabla \varphi \cdot d\mathbf{F} = \int^{\psi} \frac{f}{r} \, dr dz . \qquad (3.25)$$

Consider the ratio of the differential fluxes,

$$\frac{\delta\chi}{\delta\Psi} = \frac{1}{2\pi} \oint \frac{B_t}{|\nabla\psi|} dl = \oint \frac{B_t}{B_p} \frac{dl}{2\pi r} = q(\psi) , \qquad (3.26)$$

where the line integral is along the closed curve $\psi = \text{const.}$ in the poloidal plane $\varphi = 0$. $q(\psi)$ is called the safety factor in tokamak physics, since it must be sufficiently large for stability, as is discussed in chapter 4. Since the ratio B_p/B_t is a measure of the local inclination of a field line with respect to the φ-axis, $B_p/B_t = dl/rd\varphi$, such that the field line advances by $rd\varphi$ in φ-direction when proceeding by dl in poloidal direction, we find that

$$\oint \frac{B_t}{B_p} \frac{dl}{2\pi r} = \int \frac{d\varphi}{2\pi}$$

equals the (in general fractional) number of turns of a field line the long way around the torus per one turn the short way, which is just the inverse of ι. Hence

$$\frac{\delta\chi}{\delta\Psi} = q(\psi) = \iota^{-1}(\psi) \,.$$

It can be shown (Kruskal & Kulsrud, 1958) that this relation is also valid for general nonsymmetric toroidal configurations with nested magnetic surfaces.

3.3   Some exact two-dimensional equilibrium solutions

In most cases of practical interest the equilibrium equation (3.21) must be solved numerically, as will be addressed in the next section. Closed analytical solutions can only be found under very restrictive conditions. The equilibrium equation has to be separable in some coordinate system and the boundaries must be coordinate surfaces (if boundary conditions are imposed explicitly), so that the partial differential equation reduces to a set of ordinary ones. Nevertheless, analytic solutions have significant practical value both as paradigms illustrating general features and as reference points to check the accuracy of numerical methods. In this section we therefore present several well-known examples, some of which also play a role in subsequent chapters.

3.3.1   *Solov'ev's solution*

In the case of axisymmetry a rather simple exact solution is obtained, if the r.h.s. of the Grad-Shafranov equation (3.23) is assumed to be independent of ψ (Solov'ev, 1968). We write

$$p = -|p'|\psi + p_0 \,, \quad f^2 = f_0^2 + \frac{2\gamma|p'|}{1+\alpha^2}\psi \,, \tag{3.27}$$

where p_0 is the pressure and f_0/R the toroidal field on the magnetic axis $r = R$, $z = 0$, $\psi = 0$. Hence the equilibrium equation (3.23) becomes

$$\Delta^*\psi = |p'|\left(r^2 - \frac{\gamma}{1+\alpha^2}\right) \,. \tag{3.28}$$

Since $\Delta^* \psi = r j_\varphi$, j_φ depends only on r, which implies that for a large toroidal aspect ratio R/a the toroidal current profile is almost flat. The parameter γ characterizes the modification of the toroidal field B_φ due to poloidal currents, $\gamma < 0$ corresponding to the paramagnetic, $\gamma > 0$ to the diamagnetic case.

A polynomial solution of eq. (3.28) is

$$\psi = \frac{|p'|}{2\,(1+\alpha^2)} \left(\left(r^2 - \gamma\right) z^2 + \frac{\alpha^2}{4} \left(r^2 - R^2\right)^2 \right) , \tag{3.29}$$

as can easily be verified. First consider the region close to the magnetic axis. Writing $r - R = x \ll R$ and expanding eq. (3.29) in x, we find that the magnetic surfaces have approximately elliptical cross-sections

$$(x + \delta(x,z))^2 + \frac{R^2 - \gamma}{\alpha^2 R^2} z^2 = x_0^2 , \tag{3.30}$$

where for given γ the parameter α determines the ellipticity, $\alpha^2 = 1 - \gamma/R^2$, corresponding to circular surfaces. The centers of the surface cross-sections are shifted from the magnetic axis toward the inside of the torus with a shift δ,

$$\delta(x,z) = \frac{1}{2R} \left(x^2 + \frac{2}{\alpha^2} z^2 \right) = O\left(\frac{x_0^2}{R} \right) , \tag{3.31}$$

increasing with the "radius" x_0 of the flux surface. These properties of the magnetic surfaces in the vicinity of the axis are characteristic for axisymmetric equilibria. The shift δ, called the Shafranov shift, has the effect of moving the magnetic axis toward the torus outside relative to the plasma boundary, thus compressing magnetic surfaces on the outside and diluting them on the inside. This effect is due to the so-called hoop force. Since a torus has a larger surface on the outside than on the inside, while the pressure is constant on magnetic surfaces, a net outward force results, as in a tire, which has to be balanced by a corresponding asymmetry of the magnetic force. From eq. (3.27) we can easily compute the necessary Lorentz force in the plane $z = 0$,

$$(\mathbf{j} \times \mathbf{B})_r = j_\varphi B_z - j_z B_\varphi = -|p'| \, \partial_r \psi ,$$

which is larger in magnitude on the outside of the torus, where flux surfaces are compressed, than on the inside corresponding to larger $\partial_r \psi$. The Solov'ev equilibrium has the peculiarity that the poloidal β is independent of p' and γ, $\beta_p \sim p_0/B_\theta^2(a) \sim 1$, $p_0 = |p'|\psi_0$, if $p = 0$ on the plasma boundary $\psi = \psi_0$. For more general tokamak equilibria, where β_p is a free parameter, the Shafranov shift is proportional to β_p, $\delta(a)/a = O(\beta_p a/R)$, $a =$ plasma radius.

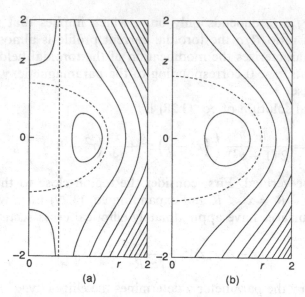

Fig. 3.3. Solov'ev's solution (3.29), $\alpha = \sqrt{2}$; (a) $\gamma = 0.25$, (b) $\gamma = -0.25$.

Globally the solution (3.29) has the general character shown in Fig. 3.3. There is a surface $\psi = \psi_s$ indicated by the dashed line which separates closed surfaces nested around the axis from open surfaces extending to infinity and is therefore called the separatrix. For $\gamma > 0$ there are two stagnation or X-type neutral points given by $B_p = 0$, i.e. $\partial_r \psi = \partial_z \psi = 0$,

$$r_X^2 = \gamma , \quad z_X^2 = \frac{\alpha^2}{2}(R^2 - \gamma) , \tag{3.32}$$

which move to the main axis $r = 0$ for $\gamma \to 0$, so that for $\gamma < 0$ the vertical branch of the separatrix is missing. However, one should realize that the solution (3.29) is only valid in the plasma, the boundary $\psi = \psi_0$ of which is located in the region of closed magnetic surfaces. Outside the plasma boundary $p' = 0$ and the solution is *not* given by eq. (3.29) (there need not even be a separatrix at all), but has to be obtained from the homogeneous equation

$$\Delta^* \psi = 0 .$$

The matching conditions at the plasma boundary are both $\psi = \psi_0$ and the continuity of the normal derivatives $\partial_n \psi$ in order to avoid a surface current contribution. Contrary to its formal simplicity the problem is quite complicated. Since both ψ and $\partial_n \psi$ are prescribed on a boundary curve for an elliptic differential equation, which constitutes an incorrect boundary value problem, regular solutions exist at most within a limited region outside the plasma (see also section 3.4).

For $\gamma > 0$ the solution (3.29) can be valid up to $\psi = \psi_s$, i.e. the plasma can fill the entire volume inside the separatrix, which constitutes a natural plasma limit. Such magnetically limited plasmas are of fundamental importance in nuclear fusion research. Let us discuss the behavior of the safety factor $q(\psi)$ in the vicinity of the separatrix. From eq. (3.26) we see that the integrand becomes singular for $\psi \to \psi_s$ such that the integral can be approximated by considering only the vicinity of an X-point, eq. (3.32). Introducing the coordinates $x = r - r_X, y = z - z_X$, the flux function becomes approximately

$$\psi - \psi_s = \psi_s'' xy \ .$$

Hence for $|\psi - \psi_s| \ll \psi_s$ we find

$$
\begin{aligned}
q(\psi) &\simeq \frac{B_t}{2\pi} \int \frac{dl}{|\nabla\psi|} \\
&= \frac{B_t}{2\pi\psi_s''} \int_{x_{min}}^{x_0} \frac{2dx}{x} \\
&\simeq -\frac{B_t}{\pi\psi_s''} \ln(\psi_s - \psi)
\end{aligned}
\tag{3.33}
$$

using $dl = dx\sqrt{x^2 + y^2}/x$, $|\nabla\psi| = \psi_s''\sqrt{x^2 + y^2}$, $x_{min} = (\psi_s - \psi)/\psi_s'' z_X$, and x_0 is some rather arbitrary upper integration limit. Hence $q(\psi) \to \infty$ for $\psi \to \psi_s$. However, because of the logarithmic nature of the singularity the approximation (3.33) is only valid if ψ is very close to the separatrix. For $\psi = 0.95\psi_s$, for instance, the contribution from the X-point region is still negligibly small.

3.3.2 Elliptical plasma cylinder

The plasma vacuum boundary value problem can be solved in the case of plane symmetry, corresponding to Solov'ev's solution for a large aspect ratio, $\psi_0 \ll \psi_s$, where the plasma boundary cross-section is an ellipse and toroidal curvature is negligible, $\Delta^* \to \nabla^2$, and the plasma current density is constant. Plane problems of this kind can be treated by the method of conformal mapping, which we will apply in a subsequent example in section 3.3.4. For the present problem, that of homogeneous current density j_0 in an elliptical plasma cylinder, we follow the direct approach by Gajewski (1972). We introduce elliptical coordinates ξ, θ (see e.g. Abramowitz & Stegun, section 21.1)

$$
\begin{aligned}
x &= \alpha \cosh \xi \cos \theta \ , \\
y &= \alpha \sinh \xi \sin \theta \ ,
\end{aligned}
\tag{3.34}
$$

in which Laplace's equation is separable and the plasma boundary given by

$$(x/a)^2 + (y/b)^2 = 1 , \quad a > b , \tag{3.35}$$

is a coordinate line $\xi = \xi_0$. The parameter α is related to a, b, ξ_0 by

$$a = \alpha \cosh \xi_0 , \quad b = \alpha \sinh \xi_0 ,$$

and the ellipticity ε is given by

$$\varepsilon = \frac{a^2 - b^2}{a^2 + b^2} = \frac{1}{\cosh 2\xi_0} . \tag{3.36}$$

The solution of the internal equation

$$\nabla^2 \psi = j_0$$

with $\psi = \psi_0$ on the boundary is

$$\begin{aligned}
\psi_{int} &= \psi_0 [(x/a)^2 + (y/b)^2] \\
&= \psi_0 \left[\frac{\cosh^2 \xi}{\cosh^2 \xi_0} \cos^2 \theta + \frac{\sinh^2 \xi}{\sinh^2 \xi_0} \sin^2 \theta \right]
\end{aligned} \tag{3.37}$$

for $\xi \leq \xi_0$, where

$$\psi_0 = j_0 a^2 b^2 / 2(a^2 + b^2) .$$

Solution (3.37) implies that magnetic surfaces inside the plasma are concentric ellipses. The external solution satisfies Laplace's equation, which in elliptical coordinates has the same form as in cartesian, namely

$$(\partial_{\xi\xi} + \partial_{\theta\theta}) \psi = 0 . \tag{3.38}$$

The general solution of this equation symmetric in both x and y, which implies $f(\xi, \theta) = f(\xi, -\theta)$, $f(\xi, \theta) = f(\xi, \pi - \theta)$, has the form

$$\begin{aligned}
\psi = c_0 + d(\xi - \xi_0) + \sum_{n=1}^{\infty} [c_n \cosh 2n(\xi - \xi_0) \\
+ s_n \sinh 2n(\xi - \xi_0)] \cos 2n\theta
\end{aligned} \tag{3.39}$$

for $\xi \geq \xi_0$. Using the boundary condition

$$\psi(\xi_0, \theta) = \psi_0$$

one finds $c_0 = \psi_0$, $c_n = 0, n \geq 1$. Continuity of the normal derivatives $\partial_\xi \psi_{int} = \partial_\xi \psi_{ext}$ at $\xi = \xi_0$ gives

$$2\psi_0 (1 - \varepsilon \cos 2\theta) \Big/ \sqrt{1 - \varepsilon^2} = d + \sum_{n=1}^{\infty} 2n s_n \cos 2n\theta , \tag{3.40}$$

which determines the remaining coefficients d, s_n. Hence the external solution becomes

$$\psi_{ext} = \psi_0 + \frac{\psi_0}{\sqrt{1 - \varepsilon^2}} \left[2(\xi - \xi_0) - \varepsilon \sinh 2(\xi - \xi_0) \cos 2\theta \right] . \qquad (3.41)$$

The complete solution is shown in Fig. 3.4 for $a/b = 2$. There is a separatrix $\psi = \psi_s$ between closed and open magnetic surfaces connecting two X-points where the poloidal field vanishes, $\partial_\xi \psi = \partial_\theta \psi = 0$, with the coordinates

$$\xi = 2\xi_0 , \quad \theta = 0, \pi . \qquad (3.42)$$

It follows from eq. (3.36) that for $a/b \to 1$ the neutral points move to infinity. Hence the separatrix disappears in the limit of a circular pinch, which is a further indication of the special character of one-dimensional configurations. The external solution can be considered as the superposition of the magnetic field generated by the plasma current $I = j_0 \pi a b$ and a quadrupole field generated at infinity, which dominates at large distances. For $\xi \gg \xi_0$ we obtain from eq. (3.41) the asymptotic behavior, which we write in two slightly different forms

$$\begin{aligned} \psi_{ext}|_{\xi \gg \xi_0} &= \frac{j_0}{4} \frac{\lambda(\lambda - 1)}{(\lambda^2 + 1)(\lambda + 1)} (y^2 - x^2) \\ &= \frac{I}{4\pi a^2} \frac{1 - \lambda'}{(1 + \lambda'^2)(1 + \lambda')} (y^2 - x^2) , \end{aligned} \qquad (3.43)$$

where $\lambda = a/b$, $\lambda' = b/a$. Hence for any ellipticity $\lambda > 1$ or $\lambda' < 1$, which implies a finite value of ξ_0, an external quadrupole field is required. Let us consider the inverse problem, where a quadrupole field

$$\psi_\infty = C(y^2 - x^2)$$

of amplitude C is superimposed on a plasma column with uniform current density. In the case corresponding to the upper form of ψ_{ext} in eq. (3.43) we fix the current density j_0 and the minor axis b of the column and determine λ (Zakharov & Shafranov, 1986):

$$\lambda(\lambda - 1) - C'(\lambda^2 + 1)(\lambda + 1) = 0$$

with $C' = 4C/j_0$. Solutions exist only if $0 < C' \leq C_{max} \cong 0.15$. For $C' < 0.15$ there are two states $\lambda_{1,2}$, $\lambda_1 < \lambda_0$, $\lambda_2 > \lambda_0$, and $\lambda_0 \cong 2.89$ is the value of λ where $C' = C'_{max}$. In the limit $C' \to 0$ $\lambda_1 \to 1$ (circular pinch), $\lambda_2 \to \infty$ (sheet pinch). Hence the lower and the higher λ-branches correspond to perturbations of a circular and a plane configuration, respectively.

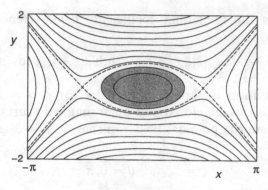

Fig. 3.4. Solution $\psi(x, y)$, eqs (3.37) and (3.41), for an elliptical plasma cylinder of constant current density (shaded area).

In the second case corresponding to the lower form of ψ_{ext} in eq. (3.43) the total plasma current I and the major axis are fixed, which gives

$$1 - \lambda' - C''(1 + \lambda'^2)(1 + \lambda') = 0 ,$$

with $C'' = 4\pi a^2 C/I$. Here only a single solution branch exists for $0 \leq C'' < 1$, the lower limit corresponding to a circular pinch $\lambda' = 1$, while for $C'' \to 1$ the plasma configuration approaches a singular current sheet $\lambda' \to 0$, i.e. $j_0 \to \infty$. The line current density is

$$J_0(x) = \int j_0 \, dy = \frac{2I}{\pi a^2} \sqrt{a^2 - x^2} , \qquad (3.44)$$

which is a special case of Syrovatski's current sheet solution (section 6.2).

3.3.3   Constant-μ, force-free equilibria

If the plasma pressure is negligible, the Lorentz force must vanish in equilibrium. Hence the plasma current $\mathbf{j}$ is parallel to $\mathbf{B}$,

$$\nabla \times \mathbf{B} = \mu \mathbf{B} . \qquad (3.45)$$

Taking the divergence of this equation gives the condition that μ is constant along field lines,

$$\mathbf{B} \cdot \nabla \mu = 0 . \qquad (3.46)$$

The case $\mu = $ const. throughout is of particular importance, as it corresponds to a preferred state to which a plasma tends to relax (sections 7.3.1 and 9.1). In this case exact analytical solutions can be given. By taking the curl of eq. (3.45) for constant μ we arrive at the following equivalent equations:

$$(\nabla^2 + \mu^2)\mathbf{B} = 0 , \quad \nabla \cdot \mathbf{B} = 0 . \qquad (3.47)$$

For the case of a helical force-free pinch the solution can easily be obtained from (3.47), which is much simpler than solving the corresponding equation (3.21) for the helical flux function ψ. Assuming $B_z \propto \cos(m\varphi + kz)$, the z-component of eq. (3.47) becomes

$$\left(\partial_{rr} + \frac{1}{r}\partial_r + \alpha^2 - \frac{m^2}{r^2}\right)B_z = 0 , \quad \alpha^2 = \mu^2 - k^2 ,$$

which has the solution

$$B_z^{m,k} = J_m(\alpha r)\cos(m\varphi + kz) . \tag{3.48a}$$

The other components follow from the r- and φ-components of eq. (3.45):

$$B_r^{m,k} = -\left[\frac{k}{\alpha}J_m'(\alpha r) + \frac{m\mu}{r\alpha^2}J_m(\alpha r)\right]\sin(m\varphi + kz) , \tag{3.48b}$$

$$B_\varphi^{m,k} = -\left[\frac{\mu}{\alpha}J_m'(\alpha r) + \frac{mk}{r\alpha^2}J_m(\alpha r)\right]\cos(m\varphi + kz) . \tag{3.48c}$$

Because of the linearity of eq. (3.47) the general solution is a linear superposition of $\mathbf{B}^{m,k}$:

$$\mathbf{B} = \sum_{m,k} \alpha_{m,k}\,\mathbf{B}^{m,k} . \tag{3.49}$$

In the special case of cylindrical symmetry $m = k = 0$ the solution is

$$B_z = J_0(\mu r) , \quad B_r = 0 , \quad B_\varphi = J_1(\mu r) . \tag{3.50}$$

The solution (3.50) is plotted in Fig. 3.5. The configuration is characterized by a reversal of the toroidal field B_z at $\mu r \simeq 2.4$, the first root of the Bessel function J_0, and is fundamental to the theory of the reversed-field pinch (chapter 9).

In plane geometry a particular constant-μ, force-free solution is used to model magnetic arcade configurations in the solar atmosphere. Since $j_z = \mu B_z$, eq. (3.23) gives $-ff' = -B_z B_z' = \mu B_z$, whence $B_z = -\mu\psi$, and ψ obeys the equation

$$(\nabla^2 + \mu^2)\psi(x, y) = 0 . \tag{3.51}$$

Choosing $y = 0$ to represent the solar surface, separation of variables gives the solution which is periodic in x and decays exponentially in y

$$\psi = e^{-y\sqrt{k^2 - \mu^2}}\cos kx , \tag{3.52}$$

plotted in Fig. 3.6. The general solution of eq. (3.51) with the boundary conditions $\psi(x, 0) = u(x)$ and $\lim_{y\to\infty}\psi = 0$ is obtained by a linear superposition of solutions (3.52).

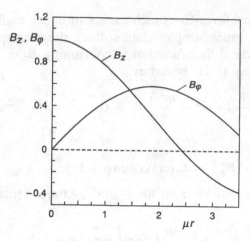

Fig. 3.5.   The Bessel function force-free equilibrium solution (3.50).

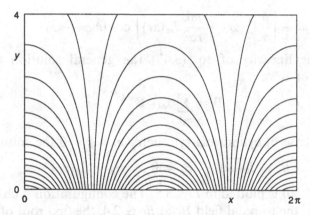

Fig. 3.6.   Force-free magnetic arcade eq. (3.52) for $k = 1$, $\mu = 1/\sqrt{2}$.

3.3.4   Nonlinear equilibria with $j_z = e^{-\psi}$

In the preceding examples the r.h.s. of the equilibrium equation was a linear function of ψ. Hence if a solution to the boundary value problem exists, it is unique. This is no longer valid for a nonlinear ψ-dependence of the r.h.s. of the equilibrium equation, where in general multiple solutions exist and specification of an additional parameter such as the total current is necessary to determine the solution uniquely. This problem will be addressed in section 3.4. Here we do not consider the boundary problem but give some explicit solutions. While in the more general cases of axial or helical symmetry the nonlinear equilibrium equation can only be solved numerically, in plane geometry the method of conformal mapping provides a large class of solutions in closed form. Exponential ψ-dependence is of particular physical relevance. It naturally appears for a weakly collisional plasma with nearly Maxwellian velocity distributions. We consider the equation

$$\nabla^2 \psi = k^2 e^{-\psi} . \tag{3.53}$$

Since the factor k^2 can be incorporated in the exponential by an appropriate gauge transformation $\psi - \psi_0 \to \psi$, $e^{\psi_0} = k^2$, it will be omitted for simplicity. The solution of (3.53) can be written in the form

$$e^{-\psi} = 8 \frac{|w'|^2}{\left(1 + |w|^2\right)^2} = j_z(x, y) , \tag{3.54}$$

where $w(z)$ is an arbitrary analytical solution, $z = x + iy$ (Fadeev et al., 1965). The result can rather easily be checked by writing $w(z) = u(x, y) + iv(x, y)$, $w'(z) = \partial_x u + i\partial_x v$ and using the Cauchy-Riemann relations $\partial_x u = \partial_y v$, $\partial_y u = -\partial_x v$. Note that $\nabla^2 \log |w| = 0$ for any analytical function with $|w| \neq 0$.

The function $w(z)$ is chosen according to the symmetry of the problem. Two simple classes of functions are $w = z^m$ for cylindrical symmetry since $|z^m| = r^m$, and $w = e^{\alpha z}$ for plane or slab symmetry since $|e^{\alpha z}| = e^{\alpha x}$. Let us consider some special cases which have been used in various applications. For $w = cz$ one obtains from eq. (3.54)

$$j_z = \frac{8c^2}{\left(1 + c^2 r^2\right)^2} , \quad B_\theta = \frac{4c^2 r}{1 + c^2 r^2} . \tag{3.55}$$

This is the Bennet pinch solution (Bennet, 1934), which is also called the "peaked current profile" (Furth et al., 1973). More fundamentally it is predicted to correspond to the relaxed state in a current-carrying plasma, in particular in a tokamak (Biskamp, 1986b; Kadomtsev, 1987; Taylor, 1990). Higher powers $w = cz^m, m > 1$, yield hollow current profiles with zero current density on the axis, which may account for transient states caused by rapid induction of the plasma current (skin current) or steady configurations with severe radiation cooling and hence enhanced resistivity in the column center.

Choosing $w = e^{\alpha z}$ gives

$$j_z = \frac{2\alpha^2}{\cosh^2 \alpha x} , \quad B_y = 2\alpha \tanh \alpha x , \tag{3.56}$$

the well-known plane sheet pinch, originally studied in collisionless plasma theory. Since a sheet pinch tends to break up into a chain of nearly circular pinches or magnetic islands, the appropriate equilibrium configuration is a corrugated sheet pinch, for which a simple expression is obtained by the

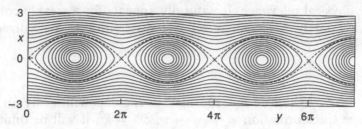

Fig. 3.7. Corrugated sheet pinch solution eq. (3.57) for $\alpha = c = 1$.

choice $w = c + (1 + c^2)^{1/2}e^{\alpha z}$:

$$j_z = \frac{2\alpha^2}{1 + c^2} \left(\cosh \alpha x + \frac{c}{(1 + c^2)^{1/2}} \cos \alpha y \right)^{-2} . \qquad (3.57)$$

In Fig. 3.7 the configuration is plotted for $\alpha = c = 1$. Two limiting cases are interesting. For $c \to 0$ the corrugation, i.e. the y-dependence, vanishes and expression (3.57) becomes identical with the slab configuration (3.56), while for $c \to \infty$ the current density is concentrated at the O-points, representing a periodic chain of line currents.

3.4 Numerical solution of the Grad-Shafranov equation

In this section we briefly outline methods of computing the general solution of the two-dimensional equilibrium equation for given profile functions and boundary conditions. We concentrate on axisymmetric configurations which are of particular interest in laboratory plasmas, especially in tokamaks and reversed-field pinch devices. The methods can be modified for helical symmetry, which has some practical importance in the approximate modelling of three-dimensional toroidal configurations.

Specifying the free profile functions $p(\psi), f(\psi)$, the Grad-Shafranov equation (3.22) can be written in the form

$$\Delta^* \psi = -(r^2 f_1(\psi) + f_2(\psi)) \equiv F(r, \psi) , \qquad (3.58)$$

where F is in general a nonlinear functional of ψ. In the usual case the plasma is surrounded by a vacuum region. The following situations can be distinguished:

(a) The vacuum region is bounded by a conducting wall (copper shell) Γ_0 (Fig. 3.8a). We may choose the boundary condition $\psi = 0$ at the wall. The plasma boundary Γ_p is not known, but we can fix the value $\psi_p = \psi(r_p, z_p)$ at one point on Γ_p corresponding to the position of a limiter, or fix the ratio ψ_p/ψ_{max}, where ψ_{max} is the value on

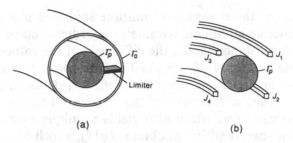

Fig. 3.8. Schematic drawing of a tokamak plasma column (a) enclosed in a metallic casing, (b) shaped and positioned by the poloidal field generated by a number of external toroidal currents J_i.

the magnetic axis. In the absence of a surface current the boundary conditions on Γ_p are

$$[\psi] = [\partial_n \psi] = 0 .$$

(b) In most applications the position and shape of the plasma column are determined by a set of external poloidal field coils with currents J_i (Fig. 3.8b). In this case the metallic casing serves only to maintain the vacuum conditions for the plasma discharge, but is penetrated by the field. Hence the outer boundary conditions are taken at infinity.

(c) It is often desirable, for instance for stability reasons, to *prescribe* the plasma shape and calculate the necessary external currents. This corresponds to the inverse of problem (b).

Cases (a) and (b) are free-boundary problems, which are inherently nonlinear even for a linear r.h.s. in eq. (3.58), since the plasma boundary depends on the solution. Only in the special case of $\Gamma_0 = \Gamma_p$, when the conducting wall is very close to the plasma surface or when the latter is known a priori (for instance experimentally), a simple fixed-boundary problem results. Case (c) corresponds to an ill-posed boundary value problem, since prescription of both ψ and $\partial_n \psi$ is only suitable for a hyperbolic and not an elliptic system. As a result only approximate "optimal" solutions can be obtained (Lackner, 1976; Degtyarev & Drozdov, 1985; see also the review by Takeda & Tokuda, 1991).

The nonlinearity of the problem requires iterative methods of solution, using for instance the Picard algorithm,

$$\Delta^* \psi^{(n+1)} = F\left(r, \psi^{(n)}\right) \tag{3.59}$$

or the Newton-Raphson algorithm,

$$\left(\Delta^* - F'\left(\psi^{(n)}\right)\right) \psi^{(n+1)} = F\left(\psi^{(n)}\right) - \psi^{(n)} F'\left(\psi^{(n)}\right) . \tag{3.60}$$

However, because of the existence of multiple solutions it is usually necessary to introduce an additional parameter in order to make the solution unique. We can for instance choose the difference of the values on the axis and at the plasma boundary $\psi_0 - \psi_p$, or the total toroidal plasma current I. A very convenient way is to specify the profiles $f_{1,2}(s)$ in eq. (3.58) along a ray across the plasma as functions of the normalized distance s from the magnetic axis instead of ψ, which also yields a unique solution. In each iteration cycle one inverts $\psi^{(n)}(s)$ to obtain $s(\psi^{(n)})$, which inserted into the radial profiles $f_{1,2}(s)$ gives the r.h.s. of (3.59) as a function of space

$$\Delta^* \psi^{(n+1)} = -r^2 f_1 \left[s \left(\psi^{(n)}(r,z) \right) \right] - f_2 \left[s \left(\psi^{(n)}(r,z) \right) \right] . \tag{3.61}$$

The solution gives the equilibrium configuration for experimentally measured or inferred radial profiles of, say, $p(s)$ and $q(s)$.

The linear equation arising at each iteration step, such as eq. (3.59), can be solved in different ways. Discretization on a fixed (Eulerian) grid r_i, z_i and application of a fast direct solver such as the double cyclic reduction method by Buneman (1969) is a flexible and robust method in particular for free-boundary problems of type (b), where a separatrix may occur (a review of different methods is given in Hockney & Eastwood (1988), chapter 6). The drawback is that for stability investigations equilibrium quantities are needed on flux surfaces, which have to be constructed from the solution $\psi(r_i, z_i)$ by some mapping procedure.

For problems with nested surfaces, essentially of type (a) with a sufficiently small vacuum region, very efficient methods have been developed to obtain the equilibrium directly in flux coordinates. In the inverse variable technique (Degtyarev & Drozdov, 1985) one performs a transformation $(r, z) \rightarrow (\rho, \theta)$, where ρ is a radial coordinate with $\rho = $ const. coinciding with flux surfaces, $\psi = \psi(\rho)$, and θ is a poloidal angle coordinate, which can be chosen in a variety of ways. This results in a set of equations consisting of the transformed Grad-Shafranov equation for $\psi(\rho)$ and equations for the transformation functions $r(\rho, \theta), z(\rho, \theta)$. In a variant called the variational moment method (Lao et al., 1981; Lao, 1984) the θ-dependence of $r(\rho, \theta), z(\rho, \theta)$ is expanded in a finite Fourier series and the expansion coefficients are determined from a variational principle. This method has been extended to compute three-dimensional equilibria very efficiently; see section 3.5.2.

3.5   Three-dimensional equilibria

3.5.1   The general equilibrium problem

Real plasma configurations are not, or at most only approximately, symmetric. For nonsymmetric equilibria the equilibrium equation (3.1) cannot be reduced to a scalar equation for a flux function and there is no proof of the existence of three-dimensional magnetic surfaces and hence of such equilibria, at least not in a rigorous sense. In fact one can give reasons for the contrary to be true. Applying methods of Hamiltonian mechanics it can be shown that the (nonsymmetric) perturbation $\varepsilon \mathbf{B}_1$ of a symmetric magnetic field $\mathbf{B}_0$, even for $\varepsilon \ll 1$, leads to drastic changes in the topology of the magnetic surfaces. Since there is no closed expression $\psi(x, y, z) =$ const. describing a flux surface, one must follow individual field lines encircling the torus many times in order to find out whether they span smooth surfaces. Because of $d\mathbf{x} \parallel \mathbf{B}$ along a field line the differential equations of a field line are

$$\frac{dx}{B_x} = \frac{dy}{B_y} = \frac{dz}{B_z}, \qquad (3.62)$$

which must in general be integrated numerically. The field line behavior is visualized by considering the points where a field line pierces a fixed poloidal plane, called a Poincaré plot (see for instance Lichtenberg & Lieberman (1983), which gives a general introduction to nonlinear mechanics and chaos theory). The field line is said to span a flux surface, if the points form a smooth curve around the magnetic axis. In the special case of a rational surface, which closes onto itself after m turns around the torus, the Poincaré plot of an individual field line consists of m isolated points, and the corresponding magnetic surface carries infinitely many such closed field lines. Perturbing the configuration by a magnetic field $\mathbf{B}_1$ containing a contribution which is constant along a rational field line, called a resonant perturbation, the rational surface and the surfaces close to it are strongly affected, forming a chain of m islands of width proportional to the square root of the resonant perturbation amplitude. Since a realistic perturbation in general contains all kinds of resonant contributions, all resonant surfaces are expected to be more or less strongly disrupted. If the perturbations on neighboring surfaces become so large that the corresponding islands, taken individually, would overlap, magnetic surfaces no longer exist in this region. Instead field lines fill a finite space volume ergodically in a stochastic way. Dynamically evolving magnetic fields, for instance during disruptive events in tokamak plasmas, usually exhibit extended stochastic regions (see Fig. 8.23).

If resonant perturbations are small the situation is described by the KAM (Kolmogorov-Arnol'd-Moser) theorem. It says that for finite but sufficiently small ε most surfaces are conserved, namely those corresponding to irrational values of the rotational transform, while rational surfaces with $\iota = n/m$ are destroyed but the size of the corresponding islands decreases faster than the distance between rational surfaces, when m and n increase. As a result there are no stochastic regions and flux surfaces appear to be smooth for practical purposes (in particular in view of the physically imposed limit of resolution due to finite gyroradii of the plasma particles), apart from a few low-m-number rational surfaces which may be destroyed on macroscopic scales.

This behavior is confirmed by the numerical computation of 3-D equilibria which exhibit seemingly smooth, nested magnetic surfaces as seen in the corresponding Poincaré plots. Hence assuming the existence of magnetic surfaces as done in analytical studies and also in most numerical calculations of 3-D equilibria seems to be justified and can always be checked a posteriori.

The most widely studied non-symmetric equilibria are toroidal configurations of the stellarator type, which term comprises stellarators in a stricter sense, heliotrons and torsatrons and which in nuclear fusion research are considered as alternatives to the tokamak configuration. They are characterized by a large vacuum field with a dominant toroidal and a smaller helical component (which may both be generated by the same external coil system) and a minor component due to plasma currents. These are mainly diamagnetic currents $\mathbf{j} \simeq \mathbf{j}_\perp = \mathbf{B} \times \nabla p / B^2$, while the parallel current density, at least the net toroidal current, should be small. In these systems the vacuum field already exhibits nested flux surfaces, so that the plasma can in principle be well confined even at very low density, in contrast to tokamaks, where a finite plasma current is required which leads to strong confinement deterioration for densities below a certain threshold.

Prior to the advent of present-day supercomputers and the development of efficient numerical codes, stellarator equilibrium studies were primarily analytic or semi-numerical, using expansion techniques such as the "stellarator expansion" by Greene & Johnson (1961). The smallness parameter is the ratio of the helical to the toroidal field

$$\delta = B_{hel}/B_t \ll 1$$

and the other characteristic quantities are assumed to scale as

$$a/R \sim \beta \sim N^{-1} \sim \delta^2 \, ,$$
$$\iota \sim 1 \, ,$$

where N is the number of helical periods around the torus. Evidently the helical pitch $\propto N$ is widely different from the average field line pitch $\propto \iota = O(1)$, such that undesirable resonance effects by the helical geometry on the magnetic surfaces are weak. We will not dwell on these expansion studies but refer for instance to the comprehensive presentation by Freidberg (1987, section 7.5). The main advantage of such investigations is to obtain a survey of the different types of equilibria in terms of a relatively small number of parameters, which is useful in equilibrium optimization studies.

3.5.2 *Numerical equilibrium computations*

The actual computation of equilibria for finite aspect ratio is nowadays performed entirely numerically. The starting point is the variational principle of ideal MHD, namely that the potential energy

$$W = \int \left(\frac{B^2}{2} + \frac{p}{\gamma - 1} \right) d^3x \qquad (3.63)$$

is stationary for an equilibrium state, $\delta W = 0$, subject to the conservation of mass, magnetic flux and entropy, the latter giving rise to the relation $p = \rho^\gamma$ between pressure p and density ρ. If, in addition, the equilibrium corresponds to a (local) minimum it is (linearly) stable. The computational procedure consists in successively decreasing W from some suitably chosen initial distribution $p(\mathbf{x})$ and $\mathbf{B}(\mathbf{x})$ until the residual force $\delta \mathbf{F} = \mathbf{j} \times \mathbf{B} - \nabla p$ falls below some desired limit. Both Eulerian (Chodura & Schlüter, 1981) and Lagrangian (Bauer et al., 1978) spatial discretization schemes have been applied. While the former is more flexible in allowing multiple magnetic axes, the latter is considerably more accurate and is mostly used in practical computations. The magnetic field is represented in terms of two flux functions $\psi(\rho)$, $\chi(\rho)$:

$$\mathbf{B} = \nabla \zeta \times \nabla \psi - \nabla \theta \times \nabla \chi , \qquad (3.64)$$

generalizing the corresponding representation for a two-dimensional configuration. Here ρ (do not confuse with the plasma density) labels a family of nested toroidal surfaces $\rho = \text{const.}$, with $\rho = 0$ on the magnetic axis and $\rho = 1$ on the plasma boundary, θ and ζ are angular coordinates on the surface in poloidal and toroidal direction, respectively, and $\psi(\rho)$ and $\chi(\rho)$ are the poloidal and toroidal fluxes, respectively, as defined in eqs (3.24), (3.25) for axisymmetry. There is considerable freedom in the choice of angular coordinates θ, ζ. One often uses $\zeta = \varphi$, the toroidal angle in cylindrical coordinates r, φ, z, while continuously readjusting θ during the relaxation procedure in order to optimize convergence.

The energy is expressed in flux coordinates ρ, θ, ζ using the invariant flux quantities $\psi(\rho)$, $\chi(\rho)$, and $M(\rho)$, the total mass enclosed by the surface $\rho = \text{const.}$ The variational problem is essentially reduced to computing the shape of the magnetic surfaces $\rho(\mathbf{x})$ for which W is minimal, and thus becomes a purely geometrical problem. Variation of W is performed with respect to $x_i(\rho, \theta, \xi)$, the cartesian coordinates as functions of the flux coordinates, in a way analogous to the inverse variable technique in the two-dimensional case discussed in the preceding section. To perform the variation an artificial time parameter t is introduced such that

$$\frac{dW}{dt} = -\int \delta \mathbf{F} \cdot \dot{\mathbf{x}} \, d^3x ,$$

where the path $x_j(t)$ is chosen in such a way as to make $dW/dt < 0$, until a minimum energy state is reached where $dW/dt \rightarrow 0$, implying $\delta F_j \rightarrow 0$. The "time" evolution is discretized, i.e. divided into a number of "time" steps or iterations. In practice typically 10^3 iterations are required to reduce the residual force $\langle \delta F^2 \rangle$ by a factor of 10^{-10}.

The spatial discretization plays an important role for the efficiency of the numerical computation. Fourier representation of x_j in the angular coordinates θ, ζ has been shown to be a particularly efficient numerical method (Hirshman & Whitson, 1983), a generalization of the variational moment method for 2-D equilibria (Lao et al., 1981). Continued refinement of this method has led to highly optimized codes which allow computation of a typical stellarator equilibrium within minutes of supercomputer time, as compared with many hours only ten years ago.

The power and efficiency of these computations tend to make us forget that the existence of smooth nested magnetic surfaces, assumed in these codes, is not guaranteed. In fact, Hayashi et al. (1990) have shown, using a sophisticated Eulerian numerical code, that by increasing the plasma pressure β in a stellarator equilibrium above some critical value of the order of a few per cent, magnetic surfaces are broken up by the formation of islands which first appear at the major rational surfaces close to the plasma boundary. As β is further increased the range of disrupted surfaces extends inward, thus more and more reducing the central region of intact surfaces. Simultaneously the islands grow in size until island overlapping creates a belt of stochastic field lines around the central plasma. The effect is mainly due to the poloidal field generated by field aligned currents, which can be reduced but not completely avoided, and appears to constitute a limitation of the achievable β, an important figure of merit of any potential fusion plasma configuration.

4
Normal modes and instability

Plasma physics has sometimes been called the science of instabilities. In fact during the last three decades of plasma research, stability theory was probably the most intensively studied field. The reason for this widespread activity is the empirical finding that in general plasmas, especially those generated in laboratory devices, are not quiescent but spontaneously develop rapid dynamics which often tend to terminate the plasma discharge. MHD instabilities are considered as particularly dangerous because they usually involve large-scale motions and short time scales. Though a realistic picture of dynamic plasma processes requires a nonlinear theory, the knowledge of the basic linear instability is usually a very helpful starting point, in particular since linear theory has a solid mathematical foundation.

The organization of the chapter is as follows. Section 4.1 presents the linearized MHD equations. In section 4.2 we consider the simplest case of linear eigenmodes, waves in a homogeneous plasma. The energy principle is introduced in section 4.3. In section 4.4 we then derive in some detail the theory of eigenmodes in a circular cylindrical pinch, which contains many qualitative features of geometrically more complicated configurations. In section 4.5 this theory is applied to the cylindrical tokamak model. The influence of toroidicity, which most severely affects the $n = 1$ mode, is discussed briefly in section 4.6. Finally section 4.7 deals with the effect of finite resistivity, giving rise to new types of instabilities, notably the tearing mode and the resistive kink mode.

The presentation in this chapter is necessarily rather concise and selective, serving only as an introduction to the main topic of the book. For a more detailed study of the linear eigenmode problem the reader is referred to several recent publications, for instance by Bateman (1980) and Freidberg (1987) for a general physical discussion, by Lifshitz (1989) for the mathematical aspects, and by Gruber & Rappaz (1985) for numerical methods.

4.1   The normal mode problem in MHD

To study the behavior of small perturbations of an equilibrium state, we write $f = f_0 + \varepsilon f_1$, $f = \mathbf{v}, \mathbf{B}, p, \rho$, and linearize the MHD equations about the equilibrium f_0, assuming $\mathbf{v}_0 = 0$. The resulting equations for the perturbations f_1 are

$$\rho_0 \partial_t \mathbf{v}_1 = -\nabla p_1 + (\nabla \times \mathbf{B}_1) \times \mathbf{B}_0 + (\nabla \times \mathbf{B}_0) \times \mathbf{B}_1 + \nu \nabla^2 \mathbf{v}_1 , \quad (4.1)$$

$$\partial_t \mathbf{B}_1 = \nabla \times (\mathbf{v}_1 \times \mathbf{B}_0) + \eta \nabla^2 \mathbf{B}_1 , \quad (4.2)$$

$$\partial_t p_1 = -\mathbf{v}_1 \cdot \nabla p_0 - \gamma p_0 \nabla \cdot \mathbf{v}_1 , \quad (4.3)$$

$$\partial_t \rho_1 = -\nabla \cdot (\mathbf{v}_1 \rho_0) . \quad (4.4)$$

In the ideal case $\eta, \nu \to 0$ these equations can be reduced to a single equation for the displacement vector $\boldsymbol{\xi}$ defined by $\mathbf{v}_1 = \partial_t \boldsymbol{\xi} = \dot{\boldsymbol{\xi}}$. Integration of eqs (4.2)–(4.4) gives $\mathbf{B}_1, p_1, \rho_1$ in terms of $\boldsymbol{\xi}$,

$$\mathbf{B}_1 = \nabla \times (\boldsymbol{\xi} \times \mathbf{B}_0) , \quad (4.5)$$

$$p_1 = -\boldsymbol{\xi} \cdot \nabla p_0 - \gamma p_0 \nabla \cdot \boldsymbol{\xi} , \quad (4.6)$$

$$\rho_1 = -\nabla \cdot (\boldsymbol{\xi} \rho_0) , \quad (4.7)$$

which can be inserted into eq. (4.1) :

$$\rho \ddot{\boldsymbol{\xi}} = \mathbf{F}(\boldsymbol{\xi}) \quad (4.8)$$
$$= (\nabla \times \mathbf{B}_1) \times \mathbf{B} + (\nabla \times \mathbf{B}) \times \mathbf{B}_1 + \nabla (\boldsymbol{\xi} \cdot \nabla p + \gamma p \nabla \cdot \boldsymbol{\xi}) ,$$

where we have dropped the subscript zero denoting the equilibrium quantities. Note that eq. (4.8) is independent of the density gradient. In sections 4.2–4.6 we restrict ourselves to ideal modes described by eq. (4.8), while in the last section, 4.7, the effect of finite resistivity will be investigated, where the equation of motion (4.1) and Faraday's law (4.2) have to be considered simultaneously.

Equation (4.8) gives the time evolution of a perturbation applied at $t = 0$, determined by the initial values $\boldsymbol{\xi}(\mathbf{x}, 0)$, $\dot{\boldsymbol{\xi}}(\mathbf{x}, 0)$ and appropriate boundary conditions. However, instead of solving this initial value problem for all different perturbations it is more practical to consider the equivalent problem of determining the normal modes of the system. Linearity of the equation and stationarity of the equilibrium quantities allow us to assume a time dependence $\boldsymbol{\xi} \propto \exp\{-i\omega t\}$. Equation (4.8) then takes the form

$$-\rho \omega^2 \boldsymbol{\xi} = \mathbf{F}(\boldsymbol{\xi}) . \quad (4.9)$$

Because of the imposition of boundary conditions for $\boldsymbol{\xi}$ there is in general no solution for arbitrary values of ω^2. Instead, eq. (4.9) constitutes an eigenvalue problem for the eigenvalue ω^2 and the eigenfunctions $\boldsymbol{\xi}$. The set of all admissible eigenvalues is called the spectrum, which is a function of the equilibrium state and the boundary conditions. Knowledge of the spectrum and the corresponding eigenfunctions then allows us to solve the initial value problem by expanding the initial perturbations in terms of eigenfunctions.

4.2 Waves in a homogeneous plasma

The simplest equilibrium state is a homogeneous plasma. Since the equilibrium quantities are constant in space we can make the ansatz $\xi = \xi_0 \exp(i\mathbf{k} \cdot \mathbf{x})$. In an infinitely extended system the components of $\mathbf{k}$ can assume any real value, while for instance in the case of a homogeneous sheet of thickness L_x with periodic boundary conditions the values of k_x are quantized, $k_x = 2\pi n/L_x$, $n =$ integer. Choosing a coordinate system such that

$$\mathbf{B} = B_0 \mathbf{e}_z , \quad \mathbf{k} = k_\perp \mathbf{e}_y + k_\parallel \mathbf{e}_z , \tag{4.10}$$

eq. (4.8) can be written in the matrix form

$$\begin{pmatrix} \omega^2 - k_\parallel^2 v_A^2 & 0 & 0 \\ 0 & \omega^2 - k_\perp^2 v_s^2 - k^2 v_A^2 & -k_\perp k_\parallel v_s^2 \\ 0 & -k_\perp k_\parallel v_s^2 & \omega^2 - k_\parallel^2 v_s^2 \end{pmatrix} \begin{pmatrix} \xi_x \\ \xi_y \\ \xi_z \end{pmatrix} = 0 , \tag{4.11}$$

where $v_A = B_0/\sqrt{\rho}$ is the Alfvén speed and $v_s = \sqrt{\gamma p/\rho}$ the sound speed, and $k^2 = k_\perp^2 + k_\parallel^2$. Solutions of this homogeneous system of equations exist only if the determinant vanishes, which gives the dispersion relations for three types of modes,

$$\omega_1^2 = k_\parallel^2 v_A^2 , \tag{4.12}$$

$$\omega_{2,3}^2 = \frac{k^2}{2} \left[\left(v_A^2 + v_s^2\right) \pm \sqrt{(v_A^2 + v_s^2)^2 - 4v_A^2 v_s^2 k_\parallel^2/k^2} \right] . \tag{4.13}$$

Since only ω^2 appears in these expressions there are always two modes for each $\mathbf{k}$, propagating in opposite directions. The dispersion curves, $\omega_{1,2,3}(\theta)$, $k_\parallel/k = \cos\theta$, are plotted in Fig. 4.1.

The first mode, eq. (4.12), is called the shear Alfvén wave. The corresponding plasma motion $\xi = (\xi_x, 0, 0)$ is perpendicular to both $\mathbf{B}$ and $\mathbf{k}$. Also, the magnetic field perturbation is perpendicular to $\mathbf{B}$, since from eq. (4.5) $\mathbf{B}_1 = i\mathbf{k} \cdot \mathbf{B}\xi$, which causes a bending of the field lines. Hence the shear Alfvén wave is a purely transverse, incompressible mode generating no density or pressure perturbation. Since $\omega = \pm \mathbf{k} \cdot \mathbf{B}$, choosing $\rho = 1$, one has $\mathbf{B}_1 = \pm \mathbf{v}_1$. In a dynamic plasma state fluctuating about an average field $\mathbf{B}$, the fields $\mathbf{z}_1^+$ and $\mathbf{z}_1^-$ (Elsässer, 1950),

$$\mathbf{z}_1^\pm = \mathbf{v}_1 \pm \mathbf{B}_1 , \tag{4.14}$$

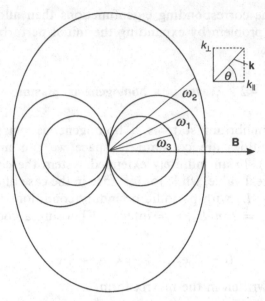

Fig. 4.1.   Dispersion curves $\omega_{1,2,3}(\theta)$, eqs (4.12), (4.13), for $v_A^2/v_s^2 = 2$.

describe Alfvén waves moving in the direction of and oppositely to **B**, respectively. Since ω is independent of $k_\perp$, the mode may be strongly localized to some field line. This property is the origin of the singular Alfvén modes in an inhomogeneous sheared configuration which form the Alfvén continuum (see section 4.4).

The second and third modes, eq. (4.13), are in general compressible. The second mode, corresponding to the plus sign on the r.h.s., is called the fast magnetosonic wave, since the phase velocity is large, $v_A^2 + v_s^2 \geq (\omega/k)^2 \geq \max\{v_A^2, v_s^2\}$, where the former equality sign is attained for perpendicular propagation, $k_\parallel = 0$, and the latter for parallel propagation, $k_\perp = 0$. In the first case the mode is purely longitudinal, $\boldsymbol{\xi} \parallel \mathbf{k}$, and the field perturbation $\mathbf{B}_1 = -i\mathbf{k} \cdot \boldsymbol{\xi}\mathbf{B}$ is parallel to **B**, corresponding to pure field compression. Since both B^2 and p are simultaneously compressed, the restoring force and hence the frequency are particularly high, $\omega_2^2 = k^2 \left(v_A^2 + v_s^2\right)$. In the case of parallel propagation we have

$$\omega_2^2 = \frac{k^2}{2}(v_A^2 + v_s^2 + |v_A^2 - v_s^2|) \ . \tag{4.15}$$

For $v_A^2 > v_s^2$, where $\omega_2^2 = k^2 v_A^2$, the mode is purely transverse and coincides with the shear Alfvén wave, while for $v_s^2 > v_A^2$, where $\omega_2^2 = k^2 v_s^2$, it is the purely longitudinal nonmagnetic sound wave. In the general oblique case the polarization is mixed.

The third mode, corresponding to the minus sign in eq. (4.13), is called the slow magnetosonic mode, since the phase velocity is small. ω_3 vanishes in the limit of perpendicular propagation, where the mode represents a quasi-static change of equilibrium, $\left(B^2/2\right)_1 \equiv \mathbf{B} \cdot \mathbf{B}_1 = -p_1$, i.e. opposite in phase. Hence there is no restoring force, which leads to the somewhat unusual situation that $\lim_{k_\parallel \to 0} \mathbf{k} \cdot \mathbf{v}_1 = 0$, although the mode is purely longitudinal, $\mathbf{k} \times \boldsymbol{\xi} = 0$. In the case of parallel propagation the dispersion relation becomes

$$\omega_3^2 = \frac{k^2}{2}(v_A^2 + v_s^2 - |v_A^2 - v_s^2|) . \tag{4.16}$$

Hence we have $\omega_3^2 = k^2 v_s^2$ and longitudinal polarisation for $v_A^2 > v_s^2$, $\omega_3^2 = k^2 v_A^2$ and transverse polarisation for $v_A^2 < v_s^2$, which is just the opposite behavior to that in the fast magnetosonic mode.

Consider two limiting cases of ω_3^2 in eq. (4.13). For $v_s^2 \gg v_A^2$, corresponding to high β, the plasma motion is incompressible, $\mathbf{k} \cdot \boldsymbol{\xi} = 0$, and the dispersion relation becomes, independently of $k_\perp / k_\parallel$,

$$\omega_3^2 = k_\parallel^2 v_A^2 . \tag{4.17}$$

This mode plays a special role in the theory of magnetic reconnection (section 6.2.2). In the opposite case, $v_s^2 \ll v_A^2$, corresponding to low β, the plasma motion is essentially parallel to the field, $\xi_z \gg \xi_y$, and the dispersion relation is, again independently of $k_\perp / k_\parallel$,

$$\omega_3^2 = k_\parallel^2 v_s^2 . \tag{4.18}$$

Thus we find that in these limits the slow magnetosonic mode, just as the shear Alfvén wave, allows strong localization perpendicular to the mean magnetic field. This behavior is the origin of the slow mode continuum in the spectrum of a sheared system (see section 4.4).

Equations (4.12), (4.13) indicate that $\omega^2 \geq 0$. Hence frequencies are real and eigenmode amplitudes $\boldsymbol{\xi}$ are constant in time. Adding dissipation will only introduce wave damping, $\omega \to \omega + i\gamma, \gamma < 0$. There are no unstable MHD modes in a homogeneous plasma, which represents global thermodynamic equilibrium, the lowest energy state a system may reach. Instability arises only for sufficiently strong spatial inhomogeneity.

4.3 The energy principle

In contrast to the homogeneous plasma, where the eigenmode problem could be reduced to a purely algebraic form, more interesting inhomogeneous systems lead to differential equations, which in general do not allow exact analytical solutions but require a numerical treatment. In many cases, however, one is interested not in the complete spectrum but only in the unstable part. The main practical problem is to determine stability thresholds in terms of characteristic equilibrium parameters and, if these are exceeded, to understand the properties of the unstable modes. To achieve this goal the energy principle provides us with a very useful tool. First note that the linearized force operator $\mathbf{F}(\xi)$ in eq. (4.8) is self-adjoint,

$$\int \boldsymbol{\eta} \cdot \mathbf{F}(\xi)\, d^3x = \int \boldsymbol{\xi} \cdot \mathbf{F}(\boldsymbol{\eta})\, d^3x \ . \tag{4.19}$$

Relation (4.19) can be proved by direct though somewhat tedious calculation. As a consequence, eigenvalues ω^2 in eq. (4.9) are real. Multiplication of the latter by ξ^* and integration gives

$$-\omega^2 \int \rho\, |\xi|^2\, d^3x = \int \xi^* \mathbf{F}(\xi) d^3x \ . \tag{4.20}$$

Subtraction of the complex conjugate and use of eq. (4.19) then yields immediately $\omega^{*2} = \omega^2$. Hence eigenmodes are either purely oscillating, $\omega^2 > 0$, i.e. ω is real, or purely growing or decaying, $\omega^2 < 0$, i.e. ω is imaginary. Transition from stability to instability occurs through $\omega^2 = 0$, which implies that marginal stable perturbations ξ correspond to neighboring equilibria. (There are some mathematical subtleties which are connected with the existence of continuous spectra; see e.g. Lifshitz, 1989). Hence modes are the more stable the larger ω^2, i.e. the faster they are oscillating, while unstable modes $\gamma > 0$, where $\gamma^2 = -\omega^2$, always have a damped counterpart $\gamma < 0$. However, the latter does not manifest itself, since any real perturbation contains a contribution of the growing mode, which will soon dominate.

Let us now introduce the energy principle. It states that the equilibrium is stable if and only if

$$\delta W\,(\xi,\xi) \equiv -\tfrac{1}{2} \int \xi \cdot \mathbf{F}(\xi) d^3x \geq 0 \tag{4.21}$$

for all square-integrable displacement functions $\xi(\mathbf{x})$ (Bernstein et al., 1958; Hain et al., 1957).

4.3.1 Proof of the energy principle

An elegant, almost elementary proof has been given by Laval et al. (1965). Multiplying eq. (4.8) by $\dot{\xi}$ and integrating over space gives

$$\int \dot{\xi} \cdot (\rho\ddot{\xi} - \mathbf{F}(\xi)) d^3x = 0 \ . \tag{4.22}$$

Time integration gives

$$K(\dot{\xi},\dot{\xi}) + \delta W(\xi,\xi) = H = \text{const.} \ , \tag{4.23}$$

where

$$K(\xi,\xi) = \tfrac{1}{2} \int \rho|\xi|^2 d^3x \; .$$

H can be interpreted as the total energy associated with the perturbation ξ, the value of which is determined by the initial conditions $\xi(0), \dot{\xi}(0)$. Equation (4.23) shows immediately the sufficiency of condition (4.21), since for positive-definite δW the mode cannot grow beyond its initial level. To show the necessity assume that there is a displacement η which makes δW negative, which can be written in the form

$$\delta W(\eta,\eta) = -\gamma^2 K(\eta,\eta) \; , \tag{4.24}$$

defining γ, which can be taken $\gamma > 0$. Choosing initial conditions $\xi(0) = \eta$, $\dot{\xi}(0) = \gamma\eta$ gives $H = 0$, hence $\xi(t)$ obeys the equation

$$K(\dot{\xi},\dot{\xi}) + \delta W(\xi,\xi) = 0 \; . \tag{4.25}$$

By Schwartz's inequality one has

$$\left[\frac{d}{dt}K(\xi,\xi)\right]^2 = 4\left[K(\dot{\xi},\xi)\right]^2 \le 4K(\xi,\xi)\,K(\dot{\xi},\dot{\xi}) \; . \tag{4.26}$$

Multiplying eq. (4.8) by ξ and integrating over space, one derives the relation

$$\frac{1}{2}\frac{d^2}{dt^2}K(\xi,\xi) = K(\dot{\xi},\dot{\xi}) - \delta W(\xi,\xi) = 2K(\dot{\xi},\dot{\xi}) \; , \tag{4.27}$$

which allows us to write Schwartz's inequality (4.26) in the form

$$\frac{d^2}{dt^2}K(\xi,\xi) \ge \left[\frac{d}{dt}K(\xi,\xi)\right]^2 / K(\xi,\xi) \; . \tag{4.28}$$

With the new variable $y(t) = \ln[K(\xi,\xi)/K(\eta,\eta)]$ satisfying the initial conditions $y(0) = 0, dy(0)/dt = 2\gamma$, eq. (4.28) becomes

$$\frac{d^2y}{dt^2} \ge 0 \; . \tag{4.29}$$

Integration gives $dy/dt \ge 2\gamma$, whence

$$y \ge 2\gamma t \; . \tag{4.30}$$

Using the definition (4.24) and the inequality (4.26) yields the estimate

$$K(\dot{\xi},\dot{\xi}) \ge |\delta W(\eta,\eta)|e^{2\gamma t} \; , \tag{4.31}$$

i.e. $\xi(t)$ grows at least exponentially. The result can easily be understood in the normal mode picture. The equality sign in relation (4.31) is valid for a pure normal mode with growth rate γ, while for an initial mixture of normal modes γ is the average initial growth rate and the perturbation $\xi(t)$ grows faster than $e^{\gamma t}$, as the most unstable mode gradually dominates.

The energy principle can be used to obtain a qualitative picture of the stability properties of a plasma configuration, which is considerably less demanding than the full solution of the normal mode problem and has therefore been the standard tool in stability theory in the first two decades of plasma research. Guessing a trial function ξ with $\delta W(\xi, \xi) < 0$ proves that the equilibrium is unstable. It provides a lower limit γ of the maximum growth rate by use of eq. (4.24); it also gives a rough picture of the expected unstable plasma motion. The procedure can be refined by minimization within certain classes of trial functions, which yields necessary stability criteria, some of which will be discussed in the subsequent sections.

4.3.2   *Different forms of the energy integral*

After integration by parts and some algebra δW can be written in the form

$$\delta W = \delta W_P + \delta W_S + \delta W_V . \tag{4.32}$$

δW_P is the contribution from the plasma volume Ω_P

$$\delta W_P = \tfrac{1}{2} \int_{\Omega_P} d^3x (|\mathbf{B}_1|^2 + \gamma p |\nabla \cdot \xi|^2 - \xi^* \cdot (\mathbf{j} \times \mathbf{B}_1) + \xi_\perp \cdot \nabla p \nabla \cdot \xi_\perp^*) , \tag{4.33}$$

where $\mathbf{j} = \nabla \times \mathbf{B}$ is the equilibrium current density and the subscripts $\perp, \|$ refer to the equilibrium field $\mathbf{B}$. A general complex function ξ has been admitted, which is convenient when complex Fourier representation is used. δW_P is real because of the self-adjointness of $\mathbf{F}(\xi)$.

The remaining terms δW_S, δW_V in eq. (4.32) are surface contributions, which arise from the integration by parts. δW_S can be written in the form

$$\delta W_S = \tfrac{1}{2} \oint_S d\mathbf{F} \cdot \left[\nabla \left(p + \frac{B^2}{2} \right) \right]_S |\mathbf{n} \cdot \xi|^2 , \tag{4.34}$$

where $[f]_S = f_V - f_P$ is the jump in f at the plasma boundary S. In the absence of equilibrium surface currents, to which case the remainder of this chapter is restricted, the integrand in eq. (4.34) vanishes, since in this case $\mathbf{B}_V = \mathbf{B}_P$ on the surface and hence

$$\left[\nabla \left(p + \frac{B^2}{2} \right) \right]_S = \mathbf{B}_V \cdot \nabla \mathbf{B}_V - \mathbf{B}_P \cdot \nabla \mathbf{B}_P = 0 .$$

The last term δW_V can be written as a volume integral over the vacuum region Ω_V

$$\delta W_V = \tfrac{1}{2} \int_{\Omega_V} d^3x |B_{1V}|^2 , \tag{4.35}$$

the vacuum field energy perturbation.

In the derivation of eq. (4.32) boundary conditions are required for the vacuum field perturbation $\mathbf{B}_{1V}$ at the plasma boundary and the conducting wall. While these are simple at the latter,

$$\mathbf{n} \cdot \mathbf{B}_{1V}|_w = 0 \,, \tag{4.36}$$

they need some more reflection at the former. First we derive the expression of the normal component $\mathbf{n} \cdot \mathbf{B}_{1V}$. We start by observing that the field remains tangent to the surface after applying the infinitesimal perturbation $\boldsymbol{\xi}$,

$$\delta \left(\mathbf{n} \cdot \mathbf{B}_V\right) = \delta\mathbf{n} \cdot \mathbf{B}_V + \mathbf{n} \cdot \delta\mathbf{B}_V = 0 \,. \tag{4.37}$$

Let us compute the change $\delta\mathbf{n}$ of the surface normal vector $\mathbf{n}$. The surface element is

$$d\mathbf{F} = \mathbf{n}dF = d\mathbf{x} \times d\mathbf{x}' \,.$$

With $\delta d\mathbf{x} = d\mathbf{x} \cdot \nabla\boldsymbol{\xi}$, $\delta d\mathbf{x}' = d\mathbf{x}' \cdot \nabla\boldsymbol{\xi}$ we can write $\delta d\mathbf{F}$ in the form

$$\begin{aligned}
\delta d\mathbf{F} &= \delta d\mathbf{x} \times d\mathbf{x}' + d\mathbf{x} \times \delta d\mathbf{x}' \\
&= -\left[(d\mathbf{x} \times d\mathbf{x}') \times \nabla\right] \times \boldsymbol{\xi} = -dF \left(\mathbf{n} \times \nabla\right) \times \boldsymbol{\xi} \,.
\end{aligned}$$

On the other hand we have

$$\delta d\mathbf{F} = \delta\mathbf{n}dF + \mathbf{n}\delta dF \,.$$

Since $\mathbf{n}^2 = 1$, i.e. $\mathbf{n} \cdot \delta\mathbf{n} = 0$, $\delta\mathbf{n}$ is given by the component of $\delta d\mathbf{F}$ perpendicular to $\mathbf{n}$,

$$\begin{aligned}
\delta\mathbf{n} &= -\left[(\mathbf{n} \times \nabla) \times \boldsymbol{\xi} - \mathbf{n}\mathbf{n} \cdot (\mathbf{n} \times \nabla) \times \boldsymbol{\xi}\right] \\
&= -\left[(\nabla\boldsymbol{\xi}) \cdot \mathbf{n} - \mathbf{n}\mathbf{n} \cdot (\nabla\boldsymbol{\xi}) \cdot \mathbf{n}\right] \,. \tag{4.38}
\end{aligned}$$

Inserting the result in eq. (4.37) gives

$$\mathbf{n} \cdot \delta\mathbf{B}_V = \mathbf{B}_V \cdot (\nabla\boldsymbol{\xi}) \cdot \mathbf{n}$$

and, since $\delta\mathbf{B}_V$ is the total change of $\mathbf{B}_V$, $\delta\mathbf{B}_V = \mathbf{B}_{1V} + \boldsymbol{\xi} \cdot \nabla\mathbf{B}_V$ (note that the quantities carrying a subscript 1, eqs (4.5)–(4.7), refer to the change at a fixed point in space), we obtain the result

$$\mathbf{n} \cdot \mathbf{B}_{1V} = \mathbf{n} \cdot (\nabla \times (\boldsymbol{\xi} \times \mathbf{B}_V)) \,. \tag{4.39}$$

This expression is formally identical with the corresponding one in the plasma, but the derivation given above is necessary since $\boldsymbol{\xi}$ is not defined in the vacuum.

Next we consider the pressure balance at the plasma boundary. For the force on a plasma element to be finite, the total pressure $P = p + B^2/2$ must be continuous perpendicular to $\mathbf{B}$. Hence the unperturbed equilibrium pressure P is continuous at the unperturbed boundary S and so is the perturbed pressure $P + \delta P$ at the shifted boundary S', so that we have to lowest order

$$[P + \delta P]_{S'} - [P]_S = [P_1 + \xi \cdot \nabla P]_S$$

$$= \left[p_1 + \mathbf{B} \cdot \mathbf{B}_1 + \xi \cdot \nabla \left(p^2 + B^2/2 \right) \right]_S = 0 , \qquad (4.40)$$

which determines $\mathbf{B}_V \cdot \mathbf{B}_{1V}$, i.e. the parallel component of $\mathbf{B}_{1V}$. In the absence of an equilibrium surface current this simplifies to

$$p_1 + \mathbf{B} \cdot \mathbf{B}_1 = \mathbf{B} \cdot \mathbf{B}_{1V} . \qquad (4.41)$$

One also needs to express the tangential component of the vector potential $\mathbf{n} \times \mathbf{A}_1$, $\mathbf{B}_{1V} = \nabla \times \mathbf{A}_1$, in terms of ξ. It follows from Faraday's law that in the system moving with the plasma the tangential component of the electric field is continuous,

$$\left[\mathbf{n} \times \mathbf{E}' \right]_S = 0 . \qquad (4.42)$$

Since $\mathbf{E}' = 0$ in the plasma, this relation implies

$$\mathbf{n} \times (\mathbf{E}_{1V} + \mathbf{v}_1 \times \mathbf{B}_V) = 0 ,$$

where $\mathbf{v}_1 = \dot{\xi}$ is the velocity of the plasma boundary. Using the gauge such that $\mathbf{E}_{1V} = -\partial_t \mathbf{A}_1$, time integration gives

$$\mathbf{n} \times \mathbf{A}_1 = -\mathbf{n} \cdot \xi \mathbf{B}_V . \qquad (4.43)$$

Equations (4.39), (4.40) or (4.41), and (4.43) give the boundary conditions at the plasma–vacuum surface.

Finally, the plasma contribution δW_P can be cast into a form (Greene & Johnson, 1968) which makes the physical meaning of the different energy contributions particularly transparent:

$$\delta W_P = \tfrac{1}{2} \int d^3x [|\mathbf{B}_{1\perp}|^2 + |\mathbf{B}_{1\parallel} - (\xi_\perp \cdot \nabla p) \, \mathbf{b}/B|^2 + \gamma p \, |\nabla \cdot \xi|^2$$

$$- j_\parallel \left(\xi_\perp^* \times \mathbf{b} \right) \cdot \mathbf{B}_1 - 2 \left(\xi_\perp \cdot \nabla p \right) \left(\kappa \cdot \xi_\perp^* \right)] , \qquad (4.44)$$

where $\mathbf{b} = \mathbf{B}/B$, and $\kappa = \mathbf{b} \cdot \nabla \mathbf{b}$ is the field line curvature vector. The first term in eq. (4.44) is the potential energy of the shear Alfvén mode, the second term essentially that of the fast magnetosonic mode, and the third that of the nonmagnetic sound wave. The fourth term, proportional to the parallel current density $j_\parallel$, represents the free energy source for current-driven instabilities. The last term is responsible for pressure-driven instabilities. The latter arise if $\kappa \cdot \nabla p > 0$, i.e. the curvature vector must have a component along the pressure gradient, which is called unfavorable field line curvature.

For instability to occur the stabilizing terms in δW_P must be sufficiently small. First consider the term $\gamma p |\nabla \cdot \xi|^2$ arising from plasma compressibility.

Since the parallel displacement $\xi_\parallel$ enters only in this term, one can minimize δW_P by choosing $\xi_\parallel$ such that $\nabla \cdot \boldsymbol{\xi} = 0$. However, this does not mean that the actual unstable plasma motions are exactly incompressible. All we can infer is that compressibility effects are weak, vanishing only at marginal stability. If $\xi_\parallel$ is eliminated δW_P contains two independent functions, a displacement of the magnetic surface ξ_ψ ("radial") and a displacement within the surface ξ_θ ("poloidal"). A particular class of trial functions ξ_ψ are those localized radially in the vicinity of some rational magnetic surface. In this case, where the radial gradients of $\boldsymbol{\xi}$ are much larger than the equilibrium gradients, instability requires that $|\mathbf{B}_1|^2 \simeq |\mathbf{B} \cdot \nabla \xi|^2$ is sufficiently small, hence the displacement must be essentially constant along the field line to minimize field line bending. While in general current-driven unstable modes are radially extended depending on the global current distribution (kink modes), pressure-driven modes are rather sensitive to the damping effect of magnetic shear and hence tend to be radially localized (interchange modes, ballooning modes).

If the conducting wall is located at the plasma boundary one has $\mathbf{n} \cdot \boldsymbol{\xi}|_S = 0$ and the vacuum term vanishes. With this restriction the plasma is less unstable, in spite of the fact that in the case of free boundary modes $\mathbf{n} \cdot \boldsymbol{\xi}|_S \neq 0$ the vacuum contribution $\delta W_V \geq 0$ is formally stabilizing.

Finally, we derive the energy principle for the reduced MHD equations (2.49), (2.51). In the case of a two-dimensional equilibrium the equation $\mathbf{B} \cdot \nabla j = 0$ gives $j(x, y) = j(\psi(x, y))$. Introducing u_1 by the relation $\partial_t u_1 = \phi_1$, u_1 is the streamfunction of the displacement vector $\boldsymbol{\xi} = \mathbf{e}_z \times \nabla u_1$, and eq. (2.49) can be integrated to yield

$$\psi_1 = \mathbf{B} \cdot \nabla u_1 . \tag{4.45}$$

Using this result, the linearized equation of motion becomes

$$-\omega^2 \nabla_\perp^2 u_1 = F(u_1) ,$$

$$F(u_1) = \mathbf{B} \cdot \nabla \nabla_\perp^2 (\mathbf{B} \cdot \nabla u_1) - \frac{dj}{d\psi} \mathbf{B}_\perp \cdot \nabla (\mathbf{B} \cdot \nabla u_1) . \tag{4.46}$$

Multiplication by u_1, integration over the plasma volume and partial integration gives

$$\omega^2 \int_{\Omega_P} (\nabla_\perp u_1)^2 \, d^3x = \int_{\Omega_P} \left[(\nabla_\perp \mathbf{B} \cdot \nabla u_1)^2 + \frac{dj}{d\psi} (\mathbf{B}_\perp \cdot \nabla u_1)(\mathbf{B} \cdot \nabla u_1) \right] d^3x$$

$$+ \int_{\Omega_V} |\nabla \psi_V|^2 \, d^3x , \tag{4.47}$$

where $\nabla^2 \psi_V = 0$ has been used. We thus obtain the energy principle in reduced MHD: the equilibrium is stable if and only if the r.h.s. of eq. (4.47) is positive for all u_1, satisfying the condition that $|\nabla_\perp u_1|$ is square-integrable such that the kinetic energy remains finite. This energy principle is generalized in section 4.7 to the case of finite resistivity.

4.4 The cylindrical pinch

This section gives an introduction to the stability theory of the straight circular-cross-section pinch. We consider both the normal mode problem, which reduces to an ordinary differential equation, and the energy principle, demonstrating the efficiency of the latter. In the subsequent section 4.5 the theory is applied to the "cylindrical" tokamak model, while the discussion of the stability properties of the reversed-field pinch is postponed to chapter 9.

4.4.1 Normal modes in a cylindrical pinch

For a cylindrical pinch the equilibrium is determined by $B_\theta(r)$, $B_z(r)$, $p(r)$, $\rho(r)$, which follow the equilibrium equation (3.5). With the ansatz $\xi = \xi(r) \exp[im\theta + ik_z z]$ the normal mode equation (4.9) can be reduced to a second-order differential equation for $\xi_r(r)$, the Hain-Lüst equation (Hain & Lüst, 1958; Goedbloed & Hagebeuk, 1972)

$$\frac{d}{dr} \frac{f}{rh} \frac{d}{dr} (r\xi_r) - g\xi_r = 0 , \tag{4.48}$$

where f, g, h are functions of ω and the equilibrium quantities,

$$
\begin{aligned}
f &= \rho(v_A^2 + v_s^2)(\omega^2 - \omega_a^2)(\omega^2 - \omega_s^2) , \\
h &= (\omega^2 - \omega_2^2)(\omega^2 - \omega_3^2) ,
\end{aligned}
$$

$$
g = -\rho(\omega^2 - \omega_a^2) - \left(\frac{4k_z^2 B_\theta^2}{r^2}\right) \frac{\left(\omega^2 v_A^2 - \omega_a^2 v_s^2\right)}{h} \tag{4.49}
$$

$$
+ r\frac{d}{dr} \left[\frac{B_\theta^2}{r^2} - \left(\frac{2k_z B_\theta G}{r^2}\right) \frac{\left(v_A^2 + v_s^2\right)\left(\omega^2 - \omega_s^2\right)}{h} \right] ,
$$

with

$$
\begin{aligned}
\omega_a^2 &= F^2/\rho = k_\parallel^2(r)v_A^2 , \\
\omega_s^2 &= \omega_a^2 v_s^2/(v_A^2 + v_s^2) , \\
\omega_{2,3}^2 &= (k^2/2)[v_A^2 + v_s^2 \pm \sqrt{(v_A^2 + v_s^2)^2 - 4v_s^2 v_A^2 k_\parallel^2/k^2}\,] , \tag{4.50}
\end{aligned}
$$

$$
\begin{aligned}
F &= \mathbf{k} \cdot \mathbf{B} = mB_\theta/r + k_z B_z = k_\parallel B , \\
G &= (\mathbf{k} \times \mathbf{B})_r = mB_z/r - k_z B_\theta ,
\end{aligned}
$$

$$
v_A^2 = B^2/\rho , \quad v_s^2 = \gamma p/\rho , \quad k^2 = (m/r)^2 + k_z^2 .
$$

Boundary conditions are $r\xi_r = 0$ for $r \to 0$ and some condition at the plasma boundary, which will be given below. While for a homogeneous plasma column, $\rho, p, B_z = $ const. and $B_\theta = 0$, the dispersion relation essentially degenerates to that of a homogeneous system (4.12), (4.13) with three decoupled types of stable modes, inhomogeneity couples these modes, which may give rise to instability.

Equation (4.48) in general constitutes a non-Sturmian eigenvalue problem since f may become zero. For $f(r_0) = 0$ the equation admits singular solutions localized around $r = r_0$. The approximate solution in the vicinity of the singular radius r_0 can easily be obtained. Writing $f(r) \simeq xf'(r_0)$, $x = r - r_0$, we have

$$\frac{d}{dx} x \frac{d\xi_r}{dx} \simeq 0 \,,$$

which gives $\xi_r = \ln|x|$.[*] In this case the boundary conditions at $r = 0, a$ do not constrain the value of ω^2. Instead the eigenvalue is given by the condition $f = 0$ and hence may vary continuously. From the definition of f we see that there are two continuous spectra, the Alfvén continuum

$$\omega^2 = \omega_a^2 \qquad\qquad (4.51)$$

and the slow mode continuum

$$\omega^2 = \omega_s^2 \,, \qquad\qquad (4.52)$$

with a range between the maximum and the minimum values of $\omega_{a,s}^2(r)$. Evidently the continuum modes are stable, $\omega^2 > 0$. Because of their singular structure they are strongly affected by finite resistivity η. The spectrum of resistive modes has been obtained, which displays a finite damping rate even in the limit $\eta \to 0_+$ (Lortz & Spies, 1984). These resistive eigenmodes are, however, strongly nonorthogonal and ill-conditioned as a basis for expanding an arbitrary initial perturbation (Borba et al., 1994), such that for practical purposes, for instance the Alfvén wave heating problem, the continuum modes are the relevant ones.

No singularity of ξ_r and hence no continuous spectrum is associated with the vanishing of the denominator h, as pointed out by Appert et al. (1974). To avoid this pseudo-singularity eq. (4.48) is replaced by two first-order equations for ξ_r and the total pressure perturbation $P_1 \equiv p_1 + \mathbf{B} \cdot \mathbf{B}_1$,

$$f \frac{d}{dr} (r\xi_r) = c\, r\, \xi_r - h\, r\, P_1 \,,$$
$$f \frac{d}{dr} P_1 = d\, r\, \xi_r - c\, P_1 \,, \qquad\qquad (4.53)$$

---

[*] In general ξ_r exhibits also a finite jump at $r = r_0$, see e.g. Goedbloed (1975).

where

$$c = 2\frac{B_\theta}{r}\left[\omega^4 B_\theta - \frac{m}{r}F(v_A^2 + v_s^2)(\omega^2 - \omega_s^2)\right] ,$$

$$d = f\left[\rho(\omega^2 - \omega_a^2) + 2B_\theta\frac{d}{dr}\left(\frac{B_\theta}{r}\right)\right]$$

$$+4\frac{B_\theta^2}{r^2}[B_\theta^2\omega^4 - \rho\omega_a^2(v_A^2 + v_s^2)(\omega^2 - \omega_s^2)] .$$

Hence $h = 0$ does not give rise to a singularity but only to decoupling of ξ_r and P_1. In this form eq. (4.53) also allows a convenient representation of the boundary conditions in the case of a free plasma boundary at $r = a$. Writing the vacuum field perturbation in terms of a scalar potential $\mathbf{B}_{1V} = \nabla\phi_1$, eq. (4.39) becomes

$$\left(\frac{d\phi_1}{dr} - iF\xi_r\right)_{r=a} = 0 . \tag{4.54}$$

The pressure balance (4.41) gives $P_1 = \mathbf{B} \cdot \mathbf{B}_{1V}$ (in the absence of an equilibrium surface current), hence

$$(P_1 - iF\phi_1)_{r=a} = 0 . \tag{4.55}$$

The vacuum field solution, obeying Laplace's equation $\nabla^2\phi_1 = 0$, is readily obtained,

$$\phi_1 = \phi_0\left[I_m(k_z r) - \frac{I'_m(k_z b)}{K'_m(k_z b)}K_m(k_z r)\right] , \tag{4.56}$$

assuming a conducting wall at $r = b > a$ with $d\phi_1/dr|_{r=b} = 0$ according to eq. (4.36).

Equations (4.53) constitute a nonlinear eigenvalue problem for the eigenvalue ω^2, which can be solved numerically for given equilibrium profiles. The full normal mode spectrum thus obtained shows a considerable complexity; see for instance Gruber & Rappaz (1985). Since, however, the properties of stable linear modes seem to have little relevance for the nonlinear plasma behavior, we restrict ourselves to a qualitative discussion of the stability thresholds and the most important unstable motions of a plasma column. Here the energy principle provides the most appropriate tool.

4.4.2 *The energy principle for a cylindrical equilibrium*

For a one-dimensional equilibrium the potential energy can be minimized analytically with respect to the poloidal component ξ_θ, such that δW_P contains only ξ_r. The energy integral can then be written in the following convenient form, introducing the normalized quantities

$\delta \widehat{W} = \delta W / \left(2\pi^2 a^2 R B_0^2 \right)$ and $\xi = \xi_r / a$:

$$\delta \widehat{W} = \int_0^a dr \left(\widehat{f} \left(\frac{d\xi}{dr} \right)^2 + \widehat{g} \xi^2 \right)$$
$$+ \left(\frac{k_z^2 r^2 B_z^2 - m^2 B_\theta^2}{k^2 r^2 B_0^2} + \frac{r}{|m|} \widehat{f} \Gamma \right)_{r=a} \xi_a^2 , \tag{4.57}$$

where

$$\widehat{f} = r \frac{F^2}{k^2 B_0^2} , \tag{4.58}$$

$$\widehat{g} = \frac{k_z^2}{k^2} \frac{2}{B_0^2} \frac{dp}{dr} + \left(\frac{m^2 - 1}{r^2} + k_z^2 \right) \widehat{f}$$
$$+ \frac{2}{r} \frac{k_z^2}{k^4} \frac{1}{B_0^2} \left(k_z^2 B_z^2 - \left(\frac{m}{r} \right)^2 B_\theta^2 \right) , \tag{4.59}$$

$$\Gamma = -\frac{|m|}{k_z a} \frac{I_{ma} K'_{mb} - K_{ma} I'_{mb}}{I'_{ma} K'_{mb} - K'_{ma} I'_{mb}} > 0 , \tag{4.60}$$

$$\Gamma \simeq \frac{1 + (a/b)^{2|m|}}{1 - (a/b)^{2|m|}} \quad \text{for } k_z b \ll 1 ,$$

$$I_{ma} = I_m (k_z a) \text{ etc.}$$

The first of the two contributions in the surface term in eq. (4.57) arises from partial integrations performed on δW_P in the course of the minimization procedure, while the second term is the vacuum contribution δW_V. Variation of $\delta \widehat{W}$ gives the Euler-Lagrange equation for the minimum energy solution ξ,

$$\frac{d}{dr} \widehat{f} \frac{d\xi}{dr} - \widehat{g} \xi = 0 . \tag{4.61}$$

This equation is obviously much simpler than the normal mode equation (4.48). Only in the case of marginal stability $\omega = 0$ do both become identical (direct verification requires some algebra).

Exact minimization of $\delta \widehat{W}$ (i.e. solution of eq. (4.61)) can in general be performed only numerically. The main power of the energy principle, however, is the trial function approach. A particular type of trial function is a perturbation $\xi(r; m, k_z)$ localized to its mode rational surface $r = r_s$, defined by $\mathbf{k} \cdot \mathbf{B} \equiv F(r_s) = 0$. Such a function represents a local interchange of neighboring flux tubes and hence is called *interchange mode*, which is driven by the last term in expression (4.44). Expansion of the coefficients

$\widehat{f}(r), \widehat{g}(r)$ in eq. (4.61) about $r = r_s$ gives

$$\frac{d}{dx}\left(x^2\frac{d\xi}{dx}\right) + D\xi = 0, \quad x = r - r_s .\tag{4.62}$$

Here D depends on the equilibrium quantities at $r = r_s$,

$$D = -\frac{2r}{B_z^2}\frac{1}{s^2}\frac{dp}{dr}\bigg|_{r=r_s},\tag{4.63}$$

where $s = d\ln q/d\ln r$ is the shear parameter, eq. (3.10). The solution of eq. (4.62) is singular,

$$\xi_r = \alpha_1 x^{\mu_1} + \alpha_2 x^{\mu_2},$$
$$\mu_{1,2} = \tfrac{1}{2}\left(-1 \pm \sqrt{1-4D}\right),\tag{4.64}$$

and cannot in general be used as a trial function in the energy principle. We can, however, obtain a valid trial function by suitably truncating the solution (4.64) for $|x| < \varepsilon$ close to the resonant surface. While for $D < \frac{1}{4}$, ξ decreases monotonically away from r_s, it oscillates for $D > \frac{1}{4}$. Substitution of the regularized function into the energy integral gives $\delta W > 0$ in the former case, while $\delta W < 0$ in the latter, where the stabilizing term can be made to vanish with a suitable choice of ε. Hence one finds the stability criterion for localized modes (Suydam, 1958),

$$s^2 > -\frac{8r}{B_z^2}\frac{dp}{dr} \simeq \frac{8r^2}{B_\theta^2}\boldsymbol{\kappa}_c\cdot\nabla p,\tag{4.65}$$

where $\boldsymbol{\kappa}_c = -\mathbf{e}_r B_\theta^2/rB^2$ is the field line curvature vector in a cylindrical configuration. Equation (4.65) indicates that interchange perturbations can grow only if the pressure gradient is in the direction of the magnetic curvature, $dp/dr < 0$, and the shear is small enough. In a low-β plasma Suydam's criterion is usually satisfied if $s = O(1)$, since the r.h.s. is $O(\beta)$, except close to the magnetic axis, where $s \propto r^2$ and stability requires a flat pressure profile $p - p_0 \propto r^4$. The same argument applies to any surface $r = r_0$, where $q(r)$ has an extremum.

What is the significance of a violation of the criterion (4.65) over a finite radial range? The derivation using localized trial functions does not reveal the structure of the most unstable physical mode. It has been shown that this is in general a nonlocalized $m = 1$ mode (Goedbloed & Sakanaka, 1974). Hence contrary to the superficial impression that "Suydam modes" imply only some innocuous, small-scale fluctuations, instability arising from a radially extended Suydam-unstable region is expected to lead to large-scale dynamics.

4.5 The circular cylindrical tokamak

In application of the theory presented in the preceding section, we consider the stability of a slender circular plasma cylinder of radius a and length $L = 2\pi R$, satisfying the so-called tokamak ordering $q \sim 1$, $B_\theta/B_z \sim \varepsilon$, $B_z \simeq B_0 = $ const., where $\varepsilon = a/R \ll 1$, which is identical with the reduced MHD approximation (section 2.4). Periodic boundary conditions in the axial direction ensure toroidal topology. We expand $\delta\widehat{W}$, eq. (4.57), in powers of $k_z r \sim \varepsilon$, where the wavenumber k_z is quantized,

$$k_z = -n/R \,. \tag{4.66}$$

Hence the approximation is valid for $n \ll \varepsilon^{-1}$. The minus sign in eq. (4.66) is introduced for convenience, since interest is primarily in resonant modes, where $F = mB_\theta/r + k_z B_z = (B_\theta/r)(m - nq) = 0$. The cylindrical model provides a qualitative picture of tokamak stability with respect to global current-driven modes, so-called kink modes, except for the $n = 1$ internal kink mode. This mode as well as pressure-driven localized modes are more strongly affected by toroidal curvature and will be discussed in the next section 4.6. Kink modes are divided into internal and external modes, depending on whether the resonant surface r_s, defined by $q(r_s) = m/n$, is located within the plasma column, $r_s < a$, or outside in the surrounding vacuum region, $r_s > a$.

For *internal kink modes* one assumes $\xi(a) \equiv \xi_a = 0$, which is a good approximation even for a free plasma boundary, if r_s is not too close to this boundary. Hence the boundary term in eq. (4.57) vanishes. Expansion of the functions $\widehat{f}, \widehat{g}$ to second order in $k_z r$ gives the simple expression

$$\delta\widehat{W} = \frac{1}{R^2} \int_0^a \left(\frac{n}{m} - \frac{1}{q}\right)^2 \left[r^2 \left(\frac{d\xi}{dr}\right)^2 + (m^2 - 1)\,\xi^2\right] r\,dr$$
$$= O(\varepsilon^2)\,. \tag{4.67}$$

Expression (4.67) is identical with the first term on the r.h.s. in eq. (4.47) for the reduced MHD model as can easily be verified using $\mathbf{B}\cdot\nabla u_1 = r\mathbf{k}\cdot\mathbf{B}\xi/m$. Evidently in a slender plasma cylinder internal modes with $|m| \geq 2$ are stable.

The $m = 1$ internal mode is a special case in cylindrical geometry, since the wavenumber m/r becomes equal to the curvature of the cylinder. A similar behavior is found for the $n = 1$ mode in a toroidal system, as is discussed in section 4.6. Since the second-order contribution to $\widehat{g}$ vanishes, we have to include the fourth-order term. Hence for $m = 1$ we have

$$\delta \widehat{W} = \frac{1}{R^2} \int_0^a dr \left(n - \frac{1}{q} \right)^2 r^3 \left(\frac{d\xi}{dr} \right)^2 + \delta W_c$$

$$\delta W_c = \frac{n^2}{R^2} \int_0^a r dr \left[\frac{2r}{B_0^2} \frac{dp}{dr} + \frac{r^2}{R^2} \left(n - \frac{1}{q} \right) \left(3n + \frac{1}{q} \right) \right] \xi^2 = O(\varepsilon^4) .$$

(4.68)

If $q < 1/n$ for $r < r_s < a$, $q(r_s) = 1/n$, the field-line-bending first term in $\delta \widehat{W}$ can be made arbitrarily small by choosing a trial function which is constant for $r < r_s$ and falls to zero within a narrow interval around $r = r_s$. Instability arises for $\delta W_c < 0$, which may be due to either the pressure or the current profile. As indicated above the case $n = 1$, which is associated with the sawtooth phenomenon in a tokamak plasma, cannot be treated in the cylindrical model but requires consideration of toroidal geometry. It should also be mentioned that in a straight cylindrical pinch of *noncircular* cross-section the $m = 1$ mode (or predominantly $m = 1$, because of poloidal mode coupling) may already be unstable in the lowest order $O(\varepsilon^2)$ (Laval et al., 1974).

External kink modes exist and can be unstable, if the plasma column is surrounded by a vacuum region. For simplicity we consider only the most unstable situation where there is no conducting wall, $b \to \infty$, such that $\Gamma = 1$ in eq. (4.57). To second order in ε the energy integral becomes

$$\delta \widehat{W} = \int_0^a \frac{r dr}{R^2} \left(\frac{n}{m} - \frac{1}{q} \right)^2 \left(r^2 \left(\frac{d\xi}{dr} \right)^2 + (m^2 - 1)\xi^2 \right)$$

$$+ \frac{a^2}{R^2} \left(\frac{n}{m} - \frac{1}{q_a} \right) \left(\frac{n}{m} + \frac{1}{q_a} + |m| \left(\frac{n}{m} - \frac{1}{q_a} \right) \right) \xi_a^2 .$$

(4.69)

Instability of a mode (m, n) can arise only if $0 < n/m < q_a^{-1}$, which implies that for a q-profile increasing with r, the usual case in a tokamak, the mode rational surface must be outside the plasma, $r_s > a$, which defines an external mode.

For $m = 1$ the minimizing trial function is $\xi(r) = \xi_a$, a rigid shift of the entire plasma column. Equation (4.69) gives, independently of the plasma current distribution,

$$\delta \widehat{W} = \frac{a^2}{R^2} \xi_a^2 2n \left(n - \frac{1}{q_a} \right) .$$

(4.70)

$\delta \widehat{W} = O(\varepsilon^2)$ in contrast to the internal mode, eq. (4.68), where $\delta \widehat{W} = O(\varepsilon^4)$. Instability occurs for $q_a < 1/n$. The stability condition is most restrictive for $n = 1$,

$$q_a > 1 .$$

(4.71)

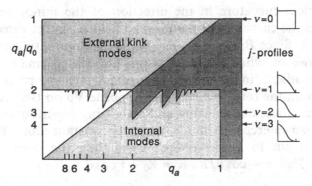

Fig. 4.2. Stability diagram for kink modes in a straight tokamak for current profiles $j(r) = j_0(1 - r^2/a^2)^\nu$. The stable region is unshaded (from Wesson, 1978).

This is the Kruskal–Shafranov criterion, which poses a strict limitation on tokamak operation. Note that for the external $(m, n) = (1, 1)$ mode toroidal effects being of order ε^4 are not important.

For $m \geq 2$ the stability depends on the current distribution and no general analytical threshold can be given. Instability of a mode (m, n) arises in the range

$$m - \alpha_{mn} \leq nq_a \leq m, \tag{4.72}$$

where α_{mn} is a functional of the current profile, $0 \leq \alpha_{mn} \leq 1$. For homogeneous current density, $q(r) = q_a$, one finds $\alpha_{mn} = 1$, i.e. the plasma is unstable with respect to external modes for all values of q_a (Shafranov, 1970). The instability range α_{mn} becomes smaller the more the current profile is peaked, i.e. the larger the ratio q_a/q_0, where $q_0 = q(0)$. If $q_a/q_0 \gtrsim 3.5$ (and $j(a) = 0$), all external modes are stable. These features are illustrated in the kink-mode stability diagram q_a/q_0 vs. q_a (Fig. 4.2) computed numerically for a rather general class of current distributions $j = j_0(1 - r^2/a^2)^\nu$, where j_0 and ν are determined by q_a and q_a/q_0.

4.6 Toroidal effects on ideal tokamak stability

4.6.1 Interchange and ballooning modes

It has been indicated in the preceding section that toroidal curvature is essential in determining the stability properties of pressure-driven modes. First consider *interchange modes* defined by $\mathbf{B} \cdot \nabla \xi \simeq 0$, the toroidal analog of the Suydam modes in the cylinder discussed in section 4.4.2. Toroidicity has in general a stabilizing effect, as can be estimated in the following way. Replace κ_c in Suydam's criterion, eq. (4.65), by the field line curvature in a toroidal configuration $\kappa \simeq \kappa_c + \kappa_t$, where $\kappa_c = -(B_\theta^2/rB_0^2)\mathbf{e}_r$

is the cylindrical curvature in the direction of the minor radius r and $\kappa_t = -\mathbf{e}_R/R$ is the curvature of the toroidal field B_φ in the direction of the major radius R. Since $B_\theta^2/B_0^2 \sim (r/R)^2$, κ_t is the dominating contribution, $\kappa_t \sim \kappa_c/\varepsilon$. However, while κ_c always points into the plasma column along the pressure gradient (in the normal case of a decreasing pressure profile), κ_t is along ∇p on the torus outside, and opposite ∇p on the inside. Hence, following a field line, destabilizing and stabilizing contributions alternate such that for an interchange mode, which is constant along $\mathbf{B}$, the effect of κ_t averages out to lowest order. Expanding the normal curvature $\kappa_{tn} = \kappa_t \cdot \nabla\psi/|\nabla\psi| = -\cos\theta/R$, $R = R_0 + r\cos\theta$,

$$\kappa_n \simeq -B_\theta^2/rB_0^2 - \frac{\cos\theta}{R_0}\left(1 - \frac{r}{R_0}\cos\theta + \cdots\right),$$

and averaging over a field line, assuming concentric circular flux surfaces $r = \text{const.}$, we obtain

$$\langle\kappa_n\rangle \simeq \kappa_c\left(1 - q^2/2\right).$$

A more rigorous calculation including the outward shift of the magnetic surfaces gives the factor $1-q^2$ instead of $1-q^2/2$ (Shafranov & Yurchenko, 1968). The stability criterion for interchange modes in a large-aspect-ratio, circular-cross-section tokamak with $\beta_p \sim 1$, the toroidal analog of Suydam's criterion (4.65), becomes

$$s^2 > -\frac{8r}{B_0^2}\frac{dp}{dr}(1 - q^2). \tag{4.73}$$

Hence for $q > 1$ interchange modes are stable in a tokamak. Equation (4.73) is a special form of Mercier's interchange criterion for general tokamak equilibria (Mercier, 1960).

Interchange modes are, however, not the most unstable localized modes in a toroidal configuration. Perturbations localized along a field line but concentrated on the outside of the torus will feel the full destabilizing effect of the toroidal curvature κ_t. We obtain an estimate of the stability threshold of such *ballooning modes* by balancing the destabilizing energy contribution due to the toroidal interchange effect, the last term in δW in eq. (4.44),

$$\delta W_\kappa \sim \frac{1}{R}\frac{dp}{dr}\xi_r^2,$$

with the stabilizing effect due to line bending, the first term,

$$\delta W_\parallel \sim (\mathbf{B} \cdot \nabla\xi_r)^2 \sim \frac{B_0^2}{q^2R^2}\xi_r^2.$$

This gives a limit of the plasma pressure,

$$\beta \sim \frac{\varepsilon}{q^2}, \qquad (4.74)$$

which is much lower than the value necessary to violate the interchange stability condition eq. (4.73), $\beta \sim 1$ for $s \sim 1$. A quantitative determination of the ballooning stability for arbitrary tokamak equlibria is obtained by numerically minimizing on each flux surface the one-dimensional form of the energy integral derived from the general three-dimensional expression (4.44) for localized perturbations (see for instance Gruber & Rappaz, 1985, pp. 142–5).

The stability threshold for ballooning modes can also be estimated in a different, more physical way. Pressure-driven instability arises, when the pressure gradient has a component in the direction of the magnetic curvature. The effect of the latter is equivalent to a gravitational force in a straight field line configuration (see also Fig. 5.12)

$$\mathbf{F}_g = -\kappa p, \qquad (4.75)$$

hence the pressure-driven instability in a curved magnetic field is essentially a Rayleigh-Taylor instability. When a fluid of mass density ρ is subjected to a gravitational force oppositely directed to the density gradient, it is convectively unstable with a growth rate which is maximal for small-scale perturbations $kl_\rho \gg 1$, $l_\rho = \rho/|\nabla\rho|$,

$$\gamma \simeq \sqrt{\frac{F_g}{\rho} \cdot \frac{1}{l_\rho}} = \sqrt{\frac{p}{\rho} \cdot \frac{1}{Rl_\rho}}, \qquad (4.76)$$

inserting eq. (4.75) and $\kappa = R^{-1}$. Since the region of unstable curvature on the torus outside is coupled to that of stable curvature on the inside, the mode is usually stabilized. However, the coupling becomes ineffective if the communication time between these regions, the Alfvén time $\tau_A \sim qR/v_A = qR\sqrt{\rho}/B$, is longer than the growth time γ^{-1} of a pressure perturbation localized on the outside, where γ is given by eq. (4.76). Hence instability occurs for

$$\beta > \frac{l_\rho}{Rq^2}, \qquad (4.77)$$

which agrees with condition (4.74) for smooth profiles $l_\rho \sim a$. (In a rigorous derivation the pressure gradient scale height $l_p = p/|\nabla p|$ appears instead of l_ρ.)

4.6.2 The toroidal internal kink mode

Besides localized pressure-driven modes the $n = 1$ *internal kink mode* is strongly affected by toroidal geometry. As mentioned in the preceding section, toroidal curvature has a similar effect on the $n = 1$ mode as cylindrical curvature has on $m = 1$. The fundamental theory has been developed by Bussac et al. (1975) by expanding δW up to $O(\varepsilon^4)$ for $\beta_p \sim s \sim 1$. In toroidal geometry the poloidal mode number m is no longer a quantum number, since different m couple, the most important coupling occurring between $m = 1$ and $m = 2$. The result of the minimization of the energy integral (the algebra is too involved to be presented here) amounts to making the replacement in eq. (4.68)

$$\delta W_c \rightarrow \left(1 - \frac{1}{n^2}\right)\delta W_c + \frac{1}{n^2}\delta W_t . \tag{4.78}$$

The cylindrical term δW_c dominates only for $n \geq 2$, while for $n = 1$, the internal kink mode properly speaking, only the toroidal term δW_t contributes. Assuming a current profile of the form $j(r) \propto 1 - (r/a)^\nu$, with $\nu < 4$ and $1 - q(0) \ll 1$, $r_1/r_2 \ll 1$, $q(r_j) = j$, one finds

$$\delta W_t \simeq 3\,\frac{r_1^4}{R^4}\,(1 - q(0))\left(\frac{13}{48}\frac{4 - \nu}{4 + \nu} - \tilde{\beta}_p^2\right)\xi_0^2 , \tag{4.79}$$

where $\tilde{\beta}_p$ is essentially the poloidal β_p,

$$\tilde{\beta}_p = -\frac{2}{B_\theta^2(r_1)}\int_0^{r_1}\frac{r^2}{r_1^2}\frac{dp}{dr}dr . \tag{4.80}$$

Since for sufficiently small $\tilde{\beta}_p$ one has $\delta W_t > 0$, there is a threshold value for instability, which depends on the shear for $r \leq r_1$ (determined by the profile parameter ν, $\tilde{\beta}_p \simeq 0.3$ for a parabolic profile $\nu = 2$), decreasing with decreasing shear. In fact it has recently been shown (Nave & Wesson, 1988) that the stabilizing effect vanishes for a current profile with $j = $ const. for $r \leq r_1$. This result can be understood by considering the effect of the coupling to the $m = 2$ mode. Since in straight geometry this mode is stable, it exerts a stabilizing effect on the $m = 1$ mode in a toroidal system, the coupling being proportional to the shear at $r = r_1$. The internal kink mode plays a fundamental role in the sawtooth phenomenon (section 8.1).

Bussac's theory has been confirmed by numerical studies of the internal kink mode (Kerner et al., 1980). Very efficient numerical codes have been developed to compute the full ideal spectrum for arbitrary two-dimensional equilibria. Instead of solving the linearized differential equation (4.9), it is numerically more convenient to consider an integral formulation of the eigenvalue problem: find solutions $\boldsymbol{\xi}(\mathbf{x})$ and values ω^2 such that for all $\boldsymbol{\eta}(\mathbf{x})$

$$\int \boldsymbol{\eta} \cdot F(\boldsymbol{\xi})\,d^3x = -\omega^2 \int \rho\boldsymbol{\eta} \cdot \boldsymbol{\xi}\,d^3x , \tag{4.81}$$

where $\boldsymbol{\xi}$ and $\boldsymbol{\eta}$ are square-integrable and satisfy appropriate boundary conditions. This problem is solved approximatively by a Ritz-Galerkin method, which consists of expanding ξ, η in a finite number of basis functions $\phi_1, \ldots, \phi_N$, $\xi(x) = \sum_{i=1}^{N} \xi_i \phi_i(x)$, thus reducing eq. (4.81) to a linear matrix eigenvalue problem

$$A_{ij} \cdot \xi_j^{(n)} = \lambda_n B_{ij} \xi_j^{(n)}$$

for N eigenvectors $\{\xi_i^{(n)}\}$ and eigenvalues $\lambda_n = -\omega_n^2$. Here the matrix elements are $A_{ij} = \int \phi_i F(\phi_j) d^3x$, $B_{ij} = \int \phi_i \rho \phi_j d^3x$. The integrands in eq. (4.81) are written in flux coordinates and the basis functions are usually chosen as finite elements in the ψ-(radial) direction combined with a Fourier representation in the θ-(poloidal) direction. Accumulation of radial mesh points at the main resonant surfaces also allows calculation of localized, quasi-singular eigenmodes. Convergence studies by varying the number of basis functions and extrapolation to the continuum limit provide very accurate values of the eigenvalues. For details see the book by Gruber & Rappaz (1985). By combining a spectral code for radially extended modes with the evaluation of the ballooning criterion the ideal stability properties of tokamak equilibria can now be determined routinely.

4.7 Resistive instabilities

In the previous sections we have only considered ideal MHD perturbations, neglecting the effect of omnipresent, though usually small, electrical resistivity. If the spatial structure of an eigenmode is smooth the effect is negligible. However, in the evaluation of the energy principle we have often used nonsmooth or even discontinuous trial functions, which correspond to marginally stable eigenfunctions, such as the internal kink mode. For such modes resistive effects are expected to be important. The same is true for certain stable ideal modes, which may be destabilized by finite resistivity.

The most important resistive effect is the violation of local magnetic flux conservation, which leads to a change of field line topology, a process called magnetic (field line) reconnection. While the general theory of reconnection will be treated in chapter 6, here consideration is restricted to linear theory. Since finite resistivity allows a much larger class of plasma motions, it may give rise to new types of instabilities. Assume a closed, in general toroidal, equilibrium with nested magnetic surfaces and consider a rational surface spanned by closed field lines. A narrow layer about this surface is topologically equivalent to the neutral sheet configuration shown in Fig. 4.3 with periodic boundary conditions in the y- and z-directions. On the rational surface $x = 0$ the magnetic field $\mathbf{B}$ is in the z-direction. Consider a plasma displacement $\boldsymbol{\xi} = \xi(x) \exp\{iky\}$.

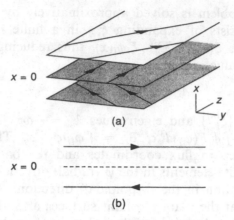

Fig. 4.3. Schematic drawing of a neutral sheet configuration (a), xy projection (b).

The ideal magnetic perturbation perpendicular to the magnetic surfaces vanishes at the resonant surface,

$$\mathbf{B}_{1\perp}^{id} = [\nabla \times (\boldsymbol{\xi} \times \mathbf{B})]_\perp = \mathbf{B} \cdot \nabla \boldsymbol{\xi} = 0 . \qquad (4.82)$$

However, taking the resistive term in eq. (4.2) into account, $\mathbf{B}_{1\perp}$ is in general finite. The difference is illustrated in Fig. 4.4. Instead of being only deformed the magnetic surfaces in a narrow layer around the resonant surface are disrupted and reconnected forming a chain of filaments, or magnetic islands in projection. The island size w is determined by the magnitude of field perturbation. Writing $\mathbf{B}_\perp = \mathbf{e}_z \times \nabla \psi$, the flux function in the vicinity of the resonant surface is given by

$$\psi(x, y) = \psi_0(x) + \widetilde{\psi}(x, y) \simeq \psi_0'' x^2/2 + \psi_1 \cos ky , \qquad (4.83)$$

where ψ_0'', ψ_1 are constant and chosen positive. The surface $\psi = \psi_1$ is the separatrix between disrupted and merely deformed surfaces, and there is a sequence of discrete points, where the magnetic field $B_\perp = |\nabla \psi|$ vanishes, alternately X-type and O-type neutral points or simply X- and O-points. The island size, defined as the distance between the two separatrix branches at the O-point, is easily calculated:

$$w = 4\sqrt{\psi_1/\psi_0''} . \qquad (4.84)$$

In order to analyse resistive eigenmodes, we have to consider both the equation of motion (4.1) and Faraday's law (4.2), since $\mathbf{B}_1$ can no longer be expressed in terms of $\boldsymbol{\xi}$ in explicit form. Since resistive motions are typically slower than ideal ones, incompressibility is usually assumed from the outset, together with a homogeneous density distribution $\rho = 1$. Hence the magnetic field has the dimension of a velocity, $B = v_A$. Neglecting $v_{\parallel}$,

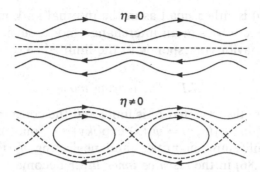

Fig. 4.4. Effect of a periodic perpendicular plasma displacement in a neutral sheet leading to field compression for $\eta = 0$ and field disruption for $\eta \neq 0$.

one writes $\mathbf{v}_\perp$ in terms of a stream function ϕ, $\mathbf{v}_\perp = \mathbf{b} \times \nabla\phi \simeq \mathbf{e}_z \times \nabla\phi$, assuming a strong field component in the z-direction. Since resistive instabilities are mainly driven by the parallel current $j_\parallel \simeq j_z$, one neglects $j_\perp$, hence $B_z = \text{const}$. This approximation corresponds to the reduced MHD equations (2.50), (2.51), which in linearized form become

$$\gamma \nabla_\perp^2 \phi_1 = \mathbf{B} \cdot \nabla j_1 + \mathbf{B}_1 \cdot \nabla j \tag{4.85}$$

$$\gamma \psi_1 = \mathbf{B} \cdot \nabla \phi_1 + \eta \nabla_\perp^2 \psi_1 - \eta_2 \nabla_\perp^{(4)} \psi_1 , \tag{4.86}$$

where we have written $\partial_t = \gamma$ instead of $-i\omega$ and also included a hyper-resistivity η_2. Dissipation is only important in a narrow region δ around the resonant surface where $\mathbf{B} \cdot \nabla \phi_1 = 0$, $\delta = O(\eta^\nu)$ or $O(\eta_2^{\nu'})$, $0 < \nu, \nu' < 1$. Hence the stability analysis leads to the following boundary layer problem: calculate separately the solution in the dissipative layer, using simplified geometry, and in the outside region, using simplified physics, and match both asymptotically to obtain a smooth overall solution, which yields the dispersion relation. It can be shown (Furth et al., 1963) that the resistive equations (4.85), (4.86) have only purely growing unstable solutions provided certain conditions on the resistivity are satisfied; see also section 4.7.3.

It is useful to distinguish between ideally stable modes and modes close to marginal ideal stability. In the first case slow growth allows the perturbed magnetic field ψ_1 to be finite at the resonant surface, varying weakly across the resistive layer, $\psi_1' \ll \psi_1/\delta$, but exhibiting a jump in the derivative, $\psi_1'' \simeq \psi_1 \Delta'/\delta$,

$$\Delta' = \lim_{\delta \to 0} \left[\psi_1'(\delta) - \psi_1'(-\delta) \right] / \psi_1 , \tag{4.87}$$

where the prime on ψ_1 denotes the derivative with respect to x. This allows us to simplify the analysis in the dissipative layer by assuming $\psi_1 = \text{const}$. Modes with this property are called tearing modes. In the second case $\psi_1(0)$ is either small as in the internal kink mode, or $\psi_1(x)$ is localized around $x \simeq 0$ as in an interchange mode, such that $\psi_1' \sim \psi_1/\delta$. Formally this behavior corresponds to the limit $\Delta' \to \infty$.

4.7.1   The tearing mode

Let us first discuss the case that Δ' is finite and the constant-ψ approximation is valid. Writing $\psi_1(x, y) = \psi_1(x)\, \exp(iky)$ etc., omitting the subscript zero on the equilibrium quantities, and neglecting η_2 for the moment, eqs (4.85) and (4.86) in the *resistive inner layer* become

$$\gamma\phi_1'' = ixkB'\psi_1'' - ikj'\psi_1 \tag{4.88}$$

$$\gamma\psi_1 = ixkB'\phi_1 + \eta\psi_1'' . \tag{4.89}$$

All terms in each of these equations are a priori of the same order. Nevertheless the j'-term in eq. (4.88) is usually omitted in the layer analysis, which formally speaking is only correct for $j' \ll B'\Delta'$, i.e. $\Delta'a \gg 1$, where a is the equilibrium scale height. A rigorous analysis, however, shows (Bertin, 1982) that the j'-term only leads to a small correction $\Delta' \to \Delta' + O(\eta^{2/5})$. Neglecting the j'-term in eq. (4.88), the functions ϕ_1, ψ_1 can be chosen to have definite parities in the resistive layer, $\phi_1(x)$ odd and $\psi_1(x)$ even. This fact makes the following order of magnitude analysis semi-quantitative. With the approximations $\phi_1'' \simeq -\phi_1/\delta^2$, $\psi_1'' \simeq \Delta'\psi_1/\delta$ eq. (4.88) and the two relations obtained from eq. (4.89) by equating the l.h.s. with either the first or the second term on the r.h.s. give γ and δ as functions of Δ',

$$\gamma \simeq \eta^{3/5} \left(\Delta'\right)^{4/5} \left(kB'\right)^{2/5} , \tag{4.90}$$

$$\delta \simeq \eta\Delta'/\gamma \simeq \eta^{2/5} \left(\Delta'\right)^{1/5} \left(kB'\right)^{-2/5} , \tag{4.91}$$

which are conveniently written in dimensionless form

$$\hat{\gamma} \simeq \hat{\eta}^{3/5}\hat{k}^{2/5}(\hat{\Delta}')^{4/5} ,$$

$$\hat{\delta} \simeq \hat{\eta}^{2/5}\hat{k}^{-2/5}(\hat{\Delta}')^{1/5} ,$$

with $\hat{\gamma} = \gamma\tau_A$, $\hat{\delta} = \delta/a$, $\hat{\eta} = \eta\tau_A/a^2 = \tau_A/\tau_\eta$, $\hat{k} = ka$, $\hat{\Delta}' = \Delta'a$, and $\tau_A^{-1} = B'$. Instead of $\hat{\eta}$ one often uses the Lundquist number $S = \hat{\eta}^{-1}$. A rigorous theory gives the numerical factor 0.55 in the expression of γ, eq. (4.90), as is derived in section 4.7.2, eq. (4.124).

If resistivity η is replaced by hyperresistivity η_2 then eq. (4.89) becomes

$$\gamma\psi_1 = ixkB'\phi_1 - \eta_2\psi_1^{(4)} . \tag{4.89a}$$

Using $\psi_1^{(4)} \simeq -\Delta'\psi_1/\delta^3$ and performing the same analysis as above, one obtains

$$\gamma \simeq \eta_2^{1/3} \left(kB'\Delta'\right)^{2/3} \tag{4.90a}$$

$$\delta \simeq \eta_2{}^{2/9} \left(\Delta'\right)^{1/9} \left(kB'\right)^{-2/9} . \tag{4.91a}$$

Hence the dependence of γ and δ on η_2 is weaker than on η. A rigorous theory gives the numerical factor 0.47 in eq. (4.90a); see eq. (4.128).

In the preceding analysis it was tacitly assumed that $\Delta' > 0$. In fact it can be shown by a rigorous solution of the inner-layer equations (4.88), (4.89) that the tearing mode is unstable if and only if $\Delta' > 0$ (Furth et al., 1963). As we will see in section 4.7.3, $-\Delta'$ is just the (normalized) energy δW of the tearing mode, hence $\Delta' > 0$ implies $\delta W < 0$, indicating instability.

Since $\hat{\gamma} \ll 1$, the inertia term in the ideal *outer region* is negligible, and hence the ideal equation becomes

$$\mathbf{B} \cdot \nabla j_1 + \mathbf{B}_1 \cdot \nabla j = 0 . \tag{4.92}$$

In the vicinity of the resonant surface the cross-field variation of ψ_1 dominates, such that eq. (4.92) assumes the form

$$\psi_1'' - \frac{K}{x} \psi_1 = 0 \tag{4.93}$$

with $\mathbf{B} \simeq xB'\mathbf{e}_y$, and $K = j'/B'$. The dominant terms in the solution of eq. (4.93) on both sides of the singularity are

$$\psi_1/\psi_1(0) = \begin{cases} f_1(x) + A_- f_2(x) , & x < 0 \\ f_1(x) + A_+ f_2(x) , & x > 0 , \end{cases} \tag{4.94}$$

$$f_1(x) = 1 + Kx \log|x| , \quad f_2(x) = x .$$

The coefficients $A_\pm$ are determined by integrating eq. (4.92) across the ideal region from the outer boundaries toward the singular surface. To match the ideal solution (4.94) to the resistive-layer solution one identifies the jump of the logarithmic derivatives of the two solutions across the layer, which gives Δ' in terms of the coefficients $A_\pm$,

$$\Delta' = \left(\psi_1'(0_+) - \psi_1'(0_-)\right)/\psi_1(0) = A_+ - A_- . \tag{4.95}$$

Note that the logarithmic singularity in ψ_1' cancels in Δ'.

The global geometry of the system enters only into the solution of the ideal equation (4.92). For a symmetric sheet pinch, the prototype of a tearing-unstable configuration, this equation takes the form

$$\psi_1'' - \left(k^2 + \frac{j'(x)}{B(x)}\right)\psi_1 = 0 . \tag{4.96}$$

The behavior is illustrated in Fig. 4.5a. Choosing the standard sheet pinch profile $B(x) = \tanh(x/a)$ and boundary conditions $\psi_1 = 0$ at $x = \pm\infty$

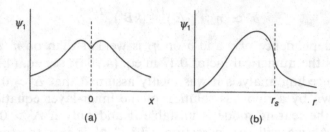

Fig. 4.5.   Tearing mode ψ_1 for (a) a symmetric sheet pinch, (b) a cylindrical pinch.

eq. (4.96) can be solved analytically :

$$\psi_1 = e^{-\hat{k}|\hat{x}|}(1 + \hat{k}^{-1}\tanh|\hat{x}|) \tag{4.97}$$

$$\hat{\Delta}' = 2(\hat{k}^{-1} - \hat{k}) \tag{4.98}$$

with $\hat{x} = x/a$, $\hat{k} = ka$, $\hat{\Delta}' = \Delta' a$. Hence the tearing mode is stable for large and unstable for small wavenumbers. It can be shown that this is valid for rather general profiles $B(x)$, in particular the long-wavelength behavior $\Delta' \propto k^{-1}$. Hence for sufficiently small k the assumption of constant ψ underlying the scaling results (4.90) and (4.91) must break down. We obtain an estimate of the value k_m, where a deviation is expected to occur, by setting $\hat{\Delta}' = \hat{\delta}^{-1}$ in eq. (4.91) and using $\hat{\Delta}' \simeq \hat{k}^{-1}$, which gives

$$\hat{k}_m \simeq \hat{\eta}^{1/4} . \tag{4.99}$$

The growth rate, which increases with decreasing $\hat{k}$ for $\hat{k} \lesssim 1$, $\hat{\gamma} \propto \hat{k}^{-2/5}$, reaches the maximum value $\hat{\gamma}_m$

$$\hat{\gamma}_m \simeq \hat{\gamma}(\hat{k}_m) \simeq \hat{\eta}^{1/2} . \tag{4.100}$$

It follows from the instability threshold $\hat{k} = 1$ that a sheet pinch of aspect ratio $L/a \gtrsim 10$ is unstable with respect to tearing modes. (Here $L = L_y$ is the (poloidal) extent of the sheet. In the theory of current sheets L is conventionally called "width" while the cross-sheet extent is called "thickness".) The result seems to contradict the appearance of stable current sheets of much larger aspect ratio often observed in nonlinear MHD processes. Such dynamic current sheets are, however, associated with an inhomogeneous mass flow along the sheet, which has a strong stabilizing effect (section 6.5).

In the cylindrical tokamak a resonant surface $r = r_s$ is determined by the equation

$$F \equiv \mathbf{k} \cdot \mathbf{B} = \frac{m}{r}B_\theta\left(1 - \frac{n}{m}q\right) = 0 . \tag{4.101}$$

While the relations (4.90), (4.91) derived in the inner layer remain valid with the replacement

$$kB' \rightarrow -F' = -m \, d \left(B_\theta / r \right) / dr \, ,^\dagger \tag{4.102}$$

the equation (4.96) in the ideal outer region now becomes

$$\frac{d^2\psi_1}{dr^2} + \frac{1}{r}\frac{d\psi_1}{dr} - \left(\frac{m^2}{r^2} + \frac{dj/dr}{B_\theta(r)\left(1 - nq(r)/m\right)} \right) \psi_1 = 0 \, . \tag{4.103}$$

This equation must in general be solved numerically for given current profile $j(r)$ and value $q(a)$ (or equivalently B_z) to determine Δ'. A typical solution is shown in Fig. 4.5b; most of the perturbed flux ψ_1 is concentrated inside the resonant surface which is reminiscent of a kink mode, while ψ_1 falls off as r^{-m} on the outside, where dj/dr becomes small. Note the logarithmic singularity of $d\psi_1/dr$ caused by $dj/dr|_{r_s} \neq 0$.

It is also interesting to note that in a cylindrical pinch, in contrast to the plane case, the wavenumber is $\widehat{k} \equiv (m/r_s)\, l_j \gtrsim 1$, where $l_j = j/|dj/dr|$ is the current density gradient scale. Hence in general only the lowest m modes can be unstable. Since the current density gradient at the singular surface r_s has a particularly strong weight in determining Δ', local steepening of $j(r)$ can destabilize the mode, while flattening leads to stabilization. Completely tearing-mode-stable j-profiles can thus be constructed (Glasser et al., 1977).

4.7.2 *The resistive internal kink mode*

As discussed in sections 4.5 and 4.6, the ideal internal kink mode is usually close to marginal stability, either weakly unstable or weakly stable. Finite resistivity is therefore expected to have a significant effect. Following Ara et al. (1978), we treat the mode in cylindrical geometry but account for toroidal geometry phenomenologically by introducing an ideal growth rate

$$\gamma_{id} = \lambda_H \tau_A^{-1} \, , \tag{4.104}$$

with $\tau_A^{-1} = (B_\theta/r)\, d\ln q/d\ln r|_{r_1} = |rF'|_{r_1}$, $F(r_1) = 0$. λ_H is of the order of W_1, where $W_1 = \delta W_c$, eq. (4.68), or $W_1 = \delta W_t$, eq. (4.79). In the ideal region far away from the inner layer around the resonant surface inertia and resistivity are negligible and the minimizing displacement ξ is obtained perturbatively from the Euler equation corresponding to the

---

† The minus sign results from our convention $m > 0$, $k_z = -n/R < 0$, which makes $F' < 0$ for the usual case of an increasing q-profile.

variation of δW, eq. (4.57),

$$\frac{d}{dr} r^3 F^2 \frac{d\xi_r}{dr} = g\xi_r ,$$ (4.105)

considering g as a small quantity. Writing

$$\xi_r = \xi_r^{(0)} + \xi_r^{(1)} ,$$

$$\xi_r^{(0)} = \begin{cases} \xi_0 & r < r_1 \\ 0 & r > r_1 , \end{cases}$$ (4.106)

integration of

$$\frac{d}{dr} r^3 F^2 \frac{d\xi_r^{(1)}}{dr} = g\xi_r^{(0)}$$

gives

$$\frac{d\xi_r^{(1)}}{dr} = \frac{W_1(\xi_r^{(0)}, \xi_r^{(0)})}{\xi_0 r^3 F^2}$$ (4.107)

with

$$W_1(\xi, \xi) = \int_0^a dr \, g \, |\xi|^2 .$$

$\xi_r^{(1)}$ is only important in the vicinity of the resonant surface. Integration of eq. (4.107) yields

$$\xi(x) = \begin{cases} \xi_0 \left(1 - c/x\right) & \text{for } x \to 0_- \\ -\xi_0 \, c/x & \text{for } x \to 0_+ , \end{cases}$$ (4.108)

with

$$x = (r - r_1)/r_1 , \ \xi = \xi_r/r_1 , \ c = W_1/(\xi_0 r_1^2 F')^2 , \ F = xF' .$$ (4.109)

Now consider eqs (4.88), (4.89), valid in the inner layer, which can be written in the following convenient dimensionless form:

$$\hat{\gamma}^2 \xi'' = x\psi'' ,$$ (4.110)

$$\psi = -x\xi + (\hat{\eta}/\hat{\gamma}) \psi'' ,$$ (4.111)

where $\xi = i\phi_1/\gamma r_1^2$, $\psi = \psi_1/r_1^3 |F'|$, $\hat{\gamma} = \gamma \tau_A$, $\hat{\eta} = \eta \tau_A/r_1^2 = S^{-1}$. When the resistivity is neglected, eqs (4.110), (4.111) can easily be integrated,

$$\xi_{id} = \frac{\xi_0}{2} \left(1 - \frac{2}{\pi} \arctan \frac{x}{\lambda_H}\right) ,$$ (4.112)

using the definition $\lambda_H = \gamma_{id} \tau_A$. Asymptotic matching of the logarithmic derivatives of the inner solution (4.112) and the outer solution (4.108) gives the ideal growth rate λ_H in terms of W_1:

$$\lambda_H = -\pi c = -\pi W_1/(\xi_0 r_1^2 F')^2 .$$ (4.113)

For $\widehat{\eta} \neq 0$ all terms in eqs (4.110), (4.111) are of the same order, which yields the scaling

$$\psi/\xi \sim x \sim \widehat{\gamma} \sim \widehat{\eta}^{1/3} \,. \tag{4.114}$$

Also in this general case the inner-layer equations (4.110), (4.111) can be solved analytically; for details see Ara et al. (1978). Matching to the ideal solution (4.108) gives the dispersion relation

$$\lambda_H = \frac{8\widehat{\gamma}}{\alpha^{9/4}} \frac{\Gamma((\alpha^{3/2} + 5)/4)}{\Gamma((\alpha^{3/2} - 1)/4)} \,, \qquad \alpha = \frac{\widehat{\gamma}}{\widehat{\eta}^{1/3}} \,. \tag{4.115}$$

This expression is easily evaluated in three special cases. For $\lambda_H \gg \widehat{\eta}^{1/3}$ application of the asymptotic properties of the Γ-function, $\Gamma(z + a)/\Gamma(z + b) \simeq z^{a-b}$, gives the growth rate of the ideal internal kink mode,

$$\widehat{\gamma} = \lambda_H \quad \text{for} \quad \lambda_H \gg \widehat{\eta}^{1/3} \,. \tag{4.116}$$

In the case of marginal ideal stability, $\lambda_H = 0$, as obtained for instance in the reduced MHD model, the argument of the Γ-function in the denominator of eq. (4.115) has to vanish, corresponding to the first pole $\alpha = 1$,

$$\widehat{\gamma} = \widehat{\eta}^{1/3} \quad \text{for} \quad \lambda_H = 0 \,. \tag{4.117}$$

In this case the inner layer solution has the form

$$\xi = \tfrac{1}{2}\xi_0(1 - \text{erf}(x/\widehat{\delta})) \tag{4.118}$$

$$\psi = \tfrac{1}{2}\xi_0 \left[x\,\text{erf}(x/\widehat{\delta}) - x + \tfrac{1}{2}\widehat{\delta}^2 \frac{d}{dx}\text{erf}(x/\widehat{\delta}) \right] \,, \tag{4.119}$$

where $\widehat{\delta} = \sqrt{2}\widehat{\eta}^{1/3} = \delta/r_1$ is the resistive layer width. Figure 4.6 compares the eigenfunctions in the resistive layer for the resistive kink mode with those of the $m \geq 2$ tearing mode. For large negative values of λ_H the dispersion equation (4.115) can also readily be solved, since in this case $\alpha \to 0$, which gives

$$\widehat{\gamma} = \left(\frac{\Gamma(1/4)}{2\Gamma(3/4)} \right)^{4/5} \widehat{\eta}^{3/5} |\lambda_H|^{-4/5} \quad \text{for} \quad \lambda_H < 0 \,, \quad |\lambda_H| \gg \widehat{\eta}^{1/3} \,. \tag{4.120}$$

Hence in the deeply MHD stable regime the resistive $m = 1$ mode assumes the properties of the tearing mode. In fact in this case λ_H is directly related to Δ'. From eqs (4.108), (4.113) one obtains

$$\xi = \begin{cases} (\xi_0/r_1)\,(1 + \lambda_H/\pi x) & x \to 0_- \\ (\xi_0/r_1)\,\lambda_H/\pi x & x \to 0_+ \,, \end{cases} \tag{4.121}$$

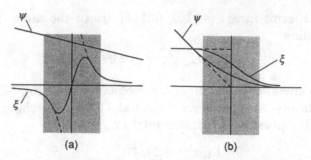

Fig. 4.6. Eigenfunctions ψ, ξ in the resistive layer: (a) tearing mode $m \geq 2$, (b) kink mode $m = 1$. The dashed lines indicate the ideal mode behavior, the shaded regions indicate the resistive layer.

and by use of the ideal relation $\psi = -x\xi$

$$\psi = \begin{cases} -(\xi_0/r_1)(x + \lambda_H/\pi) & x \to 0_- \\ -(\xi_0/r_1)\,\lambda_H/\pi & x \to 0_+ \, . \end{cases} \tag{4.122}$$

Therefore

$$\widehat{\Delta}' \equiv \psi'/\psi|_{x \to 0_+} - \psi'/\psi|_{x \to 0_-} = -\pi/\lambda_H \, . \tag{4.123}$$

Inserting the result into eq. (4.120) gives

$$\widehat{\gamma} = 0.55 \, \widehat{\eta}^{3/5} \, (\widehat{\Delta}')^{4/5} \, , \tag{4.124}$$

in agreement with eq. (4.90) with $\widehat{k} = 1$ and a specified numerical factor. Hence it appears that $\Delta' > 0$ always, implying tearing instability of the internal kink mode however stable the ideal mode. A sufficiently broad shoulder in the q-profile at $r \simeq r_1$ can, however, completely stabilize the resistive mode (section 8.1).

Choosing hyperresistivity instead of resistivity, eq. (4.111) becomes

$$\psi = -x\xi - (\widehat{\eta}_2/\widehat{\gamma}) \, \psi^{(4)} \, , \tag{4.125}$$

where $\widehat{\eta}_2 = \eta_2 \tau_A/r_1^4$. Also in this case the problem can be treated rigorously (Aydemir, 1990) using the low-m-number ballooning formalism developed by Pegoraro & Schep (1986). The dispersion relation is an algebraic expression, even simpler than the transcendental expression eq. (4.115) in the resistive case,

$$\lambda_H = \widehat{\gamma} \left(1 - \frac{1}{\alpha^{5/2}}\right) \, , \quad \alpha = \frac{\widehat{\gamma}}{\widehat{\eta}_2^{1/5}} \, . \tag{4.126}$$

For $\lambda_H \gg \widehat{\eta}_2^{1/5}$ one obtains again the ideal growth rate $\widehat{\gamma} = \lambda_H$, in the ideally marginal case one has $\alpha = 1$,

$$\widehat{\gamma} = \widehat{\eta}_2^{1/5} \quad \text{for } \lambda_H = 0 \, , \tag{4.127}$$

while for large negative values of λ_H, implying $\alpha \ll 1$, the growth rate becomes

$$\hat{\gamma} = 0.47\,\hat{\eta}_2^{1/3}(\hat{\Delta}')^{2/3} \quad \text{for } \lambda_H < 0, \quad |\lambda_H| \gg \hat{\eta}_2^{1/5}\,, \tag{4.128}$$

using relation (4.123), which agrees with the scaling result, eq. (4.90a). The η_2-dependence of the growth rate is weaker than the η-dependence in the corresponding regimes. The nonlinear behavior and its implications for the time scale of dynamic processes such as the sawtooth collapse will be discussed in sections 6.6.2 and 8.1, respectively.

4.7.3 Resistive energy principle

The primary goal of stability theory is to decide whether a configuration is unstable and to obtain a qualitative picture of the unstable mode structure and an estimate of the growth rate, rather than the precise properties of the eigenmodes. Therefore the energy principle of ideal MHD has been of considerable practical value. The argument is even more valid in discussing resistive instabilities, since growth rates for resistive modes are often influenced strongly by additional nonideal processes such as diamagnetic effects and may also be modified nonlinearly at very low amplitude. By contrast, simple estimates of the linear *mode energy* are often found to provide valuable information about the final saturated state. Therefore a generalization of the ideal energy principle to the resistive case would be very useful.

Unfortunately this is in general not possible. The linear eigenmode equation of a large class of dissipative systems can be written in the form

$$N\ddot{\Psi} + M\dot{\Psi} + Q\Psi = 0\,, \tag{4.129}$$

where Ψ is the state vector and N, M are positive hermitian operators. If Q is hermitian too, the stability problem reduces to considering the properties of Q alone. In particular the system is stable if and only if Q is positive, i.e.

$$\delta W \equiv (\Psi, Q\Psi) > 0$$

for all Ψ (Barston, 1969). In general, however, Q contains both a hermitian and an antihermitian part, which implies in particular that the system may have a real frequency $\omega \neq 0$ at marginal stability. Hence instead of considering only Q, corresponding to an energy principle, the full normal mode equation (4.129) has to be solved. However, an interesting special case allowing an exact resistive energy principle has been discussed by Tasso & Virtamo (1980). It was shown that in the cylindrical tokamak approximation corresponding to the reduced MHD model, and only in this limit, Q becomes hermitian if the resistivity profile corresponds to resisitive equilibrium $\eta j = $ const. and either the resistivity perturbation η_1 is convected by the fluid

$$\eta_1 = -\xi \cdot \nabla \eta \tag{4.130}$$

or η remains constant on the perturbed magnetic surfaces,

$$\mathbf{B} \cdot \nabla \eta_1 = -\mathbf{B}_1 \cdot \nabla \eta . \tag{4.131}$$

For the usual case of a collisional resistivity where η is a function of temperature, the first case corresponds to vanishing heat conductivity, the second to infinite parallel one (Tasso, 1975). If the state vector is chosen as

$$\Psi = \begin{pmatrix} u_1 \\ \psi_1 \end{pmatrix} , \tag{4.132}$$

where u_1 is the stream function of the displacement vector, $\xi = \mathbf{e}_z \times \nabla u_1$, and ψ_1 the flux function, $\mathbf{B}_1 = \mathbf{e}_z \times \nabla \psi_1$, the operator Q has the form

$$Q^{(f)} = \begin{pmatrix} j'(\psi)(\mathbf{B} \cdot \nabla)(\mathbf{B}_\perp \cdot \nabla) & -j'(\psi)(\mathbf{B}_\perp \cdot \nabla) \\ j'(\psi)(\mathbf{B}_\perp \cdot \nabla) & -\nabla_\perp^2 \end{pmatrix} \tag{4.133}$$

in the first case, eq. (4.130) (superscript f for "fluid"), and

$$Q^{(s)} = \begin{pmatrix} 0 & 0 \\ 0 & j'(\psi)(\mathbf{B} \cdot \nabla)^{-1}(\mathbf{B}_\perp \cdot \nabla) - \nabla_\perp^2 \end{pmatrix} \tag{4.134}$$

in the second case, eq. (4.131) (superscript s for "surface"). The corresponding energy integrals are

$$\delta W^{(f)} = \int d^3x \{(\nabla_\perp \psi_1)^2 \tag{4.135}$$
$$-j'(\psi)\left[(\mathbf{B} \cdot \nabla u_1)(\mathbf{B}_\perp \cdot \nabla u_1) - 2\psi_1 \mathbf{B}_\perp \cdot \nabla u_1\right]\}$$

$$\delta W^{(s)} = \int d^3x [(\nabla_\perp \psi_1)^2 + j'(\psi)\psi_1 (\mathbf{B} \cdot \nabla)^{-1}(\mathbf{B}_\perp \cdot \nabla \psi_1)] . \tag{4.136}$$

The last expression (4.136) is formally identical with the energy integral for ideal reduced MHD, eq. (4.47), written in terms of ψ_1, the difference being that in contrast to the ideal case ψ_1 may be finite at the resonant surface.

We now show that $\delta W^{(f)}$ can always be made negative corresponding to the rippling instability, a quasi-electrostatic ($\psi_1 \simeq 0$) localized mode driven by the resistivity gradient (Furth et al., 1963). We introduce Hamada-like coordinates ψ, θ, ζ (Hamada, 1962), where $\zeta = z/R$ is the "toroidal" angular coordinate, and θ defined by

$$d\theta = 2\pi \, (dl/B_\perp) \Big/ \oint dl/B_\perp$$

on a surface $\psi =$ const. is the poloidal angular coordinate, such that field lines are straight in θ, ζ:

$$\mathbf{B} \cdot \nabla = \frac{B_z}{qR}(\partial_\theta + q\,\partial_\zeta) \tag{4.137}$$

$$\mathbf{B}_\perp \cdot \nabla = \frac{B_z}{qR}\partial_\theta . \tag{4.138}$$

Choosing a trial function $\psi_1 = 0$, $u_1 = u(\psi)e^{i(n\zeta - m\theta)}$, $\delta W^{(f)}$ in eq. (4.135) becomes

$$\delta W^{(f)} = \int j'(\psi)\left(\frac{B_z}{qR}\right)^2 m\,(nq - m)\,|u|^2 d^3x . \tag{4.139}$$

Since $nq - m$ changes sign at the resonant surface, concentration of u on the negative side of the integrand gives $\delta W^{(f)} < 0$. Hence a cold plasma, where eq. (4.130) holds, is always resistively unstable. In most laboratory plasmas, however, the rippling mode is confined to the cool edge zone of the plasma column. The plasma interior is usually sufficiently hot such that eq. (4.131) and hence $\delta W^{(s)}$ are the more appropriate approximations.

The singularity which can arise in the second term in the integral (4.136) is treated as the principal value to ensure the hermiticity of Q. It implies that ψ is symmetric in the vicinity of the resonant surface. $\delta W^{(s)}$ can be regarded as the energy of a tearing-mode-like perturbation. Consider a mode (m, n) in a circular plasma cylinder. Using eqs (4.137), (4.138), it can be seen by integration by parts that $\delta W^{(s)}$ is proportional to Δ',

$$\delta W^{(s)} = 4\pi^2 R \int r\,dr\left(\left|\frac{d\psi_1}{dr}\right|^2 + \frac{m^2}{r^2}|\psi_1|^2 + \frac{j'(\psi)}{1 - nq/m}|\psi_1|^2\right)$$

$$= -4\pi^2 Rr_1\,|\psi_1(r_1)|^2\,\Delta' , \tag{4.140}$$

where $\int = \lim_{\varepsilon \to 0}\int_0^{r_1 - \varepsilon} + \int_{r_1 + \varepsilon}^a$, and ψ_1 following eq. (4.103) is the tearing-mode solution in the ideal region. Using a Ritz-Galerkin technique similar to those in ideal MHD as described in section 4.6, $\delta W^{(s)}$ in eq. (4.136) has been evaluated numerically for different current profiles $j(r)$ in a circular cylinder (Kerner & Tasso, 1982), and also for noncircular cross-section cylinders, where coupling of different m-values occurs (Kerner & Tasso, 1983).

4.7.4 *The toroidal tearing mode*

For cylindrical configurations of finite aspect ratio or toroidal configurations the approximations inherent in the reduced MHD equations, in particular the inner-layer equations (4.88), (4.89), are not valid. When effects of curvature and plasma pressure are included, the analysis in the resistive layer becomes much more involved; see Coppi et al. (1966) for the cylindrical case and Glasser et al. (1975) for the toroidal case. The main new effect is plasma compressibility, which introduces a stabilizing mechanism associated with parallel plasma motion. Here we mention only the result for a large-aspect-ratio, circular-cross-section toroidal plasma column. The dispersion relation becomes

$$\Delta' = \gamma^{5/4}\eta^{-3/4}f(q)\left(1 - \frac{\pi}{4}\frac{D_R(q,\beta)}{\gamma^{3/2}}\eta^{1/2}\right), \qquad (4.141)$$

where f and D_R are functionals of q and β, $D_R \propto dp/dr$. For $D_R > 0$ there is always an unstable root (for finite Δ'), the resistive interchange mode $\gamma \propto \eta^{1/3}$. Since the factor $\gamma^{5/4}\eta^{-3/4} \propto \eta^{-1/3}$ becomes large for $\eta \to 0$, the growth rate becomes

$$\gamma = \eta^{1/3}\left(\pi D_R/4\right)^{2/3}. \qquad (4.142)$$

The case $D_R = 0$ corresponds to the usual tearing mode, eq. (4.90), where instability requires $\Delta' > 0$. For $D_R < 0$ we find a modified tearing mode with a complex eigenvalue γ and a finite threshold $\Delta_0 > 0$,

$$\Delta_0 \propto \eta^{-1/3}\,|D_R|^{5/6}. \qquad (4.143)$$

The stabilization for $0 < \Delta' < \Delta_0$ in the case of negative D_R is due to favorable average curvature. The regime $D_R < 0$ is typical for hot toka-mak plasmas. The analytic theory of Glasser et al. has been confirmed and extended to general aspect ratio and plasma cross-section by computational studies solving the full MHD equations in toroidal geometry without ordering assumption (Hender et al., 1987).

Nonlinearly the curvature effect, such as other stabilizing effects arising in the linear theory when including more sophisticated physical processes in the layer, tends to become weak at finite but still small amplitudes, as is discussed in section 5.4.2.

5

Nonlinear evolution of
MHD instabilities

The study of linear stability of plasmas had for a long period been carried by the conception that only stable configurations can exist in nature, since instability would lead to destruction of the equilibrium and loss of plasma confinement, which would be the faster the larger the growth rate. Statements like: "all plasmas (meaning real inhomogeneous plasma configurations) are unstable", sometimes pronounced by plasma theoreticians in the heyday of instability theory, seemed to imply that magnetic fusion research is basically a futile endeavor. The development in experimental plasma physics during the past two decades proved this conception thoroughly wrong. Tokamak discharges may exist, well confined, in spite of the presence of instabilities, which often lead only to a slight change of the plasma profiles and a certain increase of plasma and energy transport (and which may even have beneficial effects such as the removal of impurities by the sawtooth process). Thus in order to judge the effect of an instability it is evidently necessary to calculate or at least estimate its nonlinear behavior, in particular the saturation level. It will turn out that linear mode properties, in particular growth rates, often have little to say about the nonlinear behavior.

As a general rule an instability is found to be the more "dangerous", i.e. its effect on the plasma configuration is the more detrimental, the longer the wavelength (global modes). Linear growth rates, however, often have the opposite tendency, being larger at shorter wavelengths, e.g. ballooning modes. In addition, the evolution to large amplitudes usually involves magnetic reconnection even in the case of ideal instabilities requiring finite resistivity, since the ideal constraint of flux conservation would lead to artificial saturation at low level.

In the conventional nonlinear theory one starts from an unstable equilibrium and calculates the ensuing nonlinear evolution and saturation of a small initial pertubation. However, this approach is often rather unrealistic. Since typical times for nonlinear saturation, even in the case of a resistive instability, are much shorter than the time scales of equilibrium

85

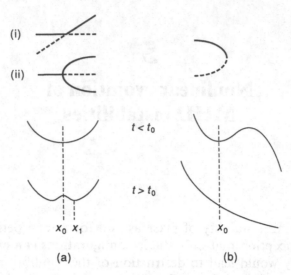

Fig. 5.1. Critical behavior of the equilibrium evolution. (a) bifurcation: (i) trans-critical or (ii) pitchfork bifurcation, mechanical analog of (ii); (b) catastrophe: loss of equilibrium.

evolution, the unstable mode will grow, as soon as the marginal point is passed. Two different situations may occur, as illustrated schematically in Fig. 5.1. Either the equilibrium bifurcates, in the simplest cases by a transcritical bifurcation (i) or a pitchfork bifurcation (ii), Fig. 5.1a, or a catastrophe occurs, corresponding to a (local) loss of equilibrium, Fig. 5.1b.

The mechanical analogs in Fig. 5.1 can, however, give only a very crude picture of the actual development, since in the unstable case the instability itself modifies the equilibrium, i.e. the quasi-potential. Hence in case (b) the resulting dynamics of the system may "dig a local well", thus keeping the system close to x_0, since the actual equilibrium evolution is only controlled by some global parameters, leaving much freedom for local profile modifications. Therefore bifurcation studies indicate only the possible onset of dynamics at $t = t_0$ but not their final extent. To obtain a reliable picture dynamical studies are required including the slow phase of equilibrium evolution.

The conventional approach of following the nonlinear development from an initial unstable state has nevertheless some justification. In the case of equilibrium bifurcation it implies starting at some point on the unstable dashed curve, from where the instability takes the system to the corresponding point on the stable branch. In this scenario only the final state is relevant. However, it may also occur that owing to additional processes not included in the resistive MHD model the instability is temporarily suppressed, such that the system continues to move on the dashed curve up to some point, where instability finally sets in. Since such stabilizing effects are often switched off at very small amplitude (e.g. diamagnetic effects; see Biskamp, 1979), the instability subsequently grows at the finite MHD growth rate which would give rise to a truly disruptive process. Such behavior can also be due to a very low initial perturbation level ("supercooled system"). If the instability threshold is exceeded, the unstable mode amplitude requires many exponential growth times until it reaches an observable level, during which period the system has evolved well into the unstable regime.

In this chapter we therefore follow the conventional approach, restricting consideration primarily to instability saturation levels. In addition, we remain within the geometry determined by the most unstable mode, i.e. a one-dimensional equilibrium develops a two-dimensional dynamic structure, ignoring the presence of modes with different orientations of **k** (different helicities in a cylindrical configuration). The chapter also excludes dynamic processes involving fast magnetic reconnection. Discussion of such processes, for instance the resistive kink mode, is postponed to chapter 6.

The lowest-order nonlinear approximation, the quasi-linear theory, where only the reaction of the unstable modes on the equilibrium profiles is taken into account, is introduced in general terms in section 5.1. The next two sections deal with the nonlinear behavior of ideal instabilities. External kink modes, section 5.2, are usually regarded as the most violent and dangerous processes in tokamak-like plasmas. We first derive a nonlinear energy integral, then discuss the special case of shearless configurations which may allow the development of magnetic bubbles, and finally consider the behavior in more general sheared systems. In section 5.3 the theory of the nonlinear ideal internal kink mode is presented, which reveals the limits of ideal MHD and the necessity for allowing magnetic reconnection. The final two sections are devoted to the nonlinear tearing mode. Section 5.4 gives the theory in the small-amplitude regime for low β as well as the generalization to finite β and magnetic curvature. In section 5.5 we discuss the saturation of the tearing mode, comparing predictions of a simple quasi-linear theory with numerical simulation results for different equilibrium current profiles and considering the influence of self-consistent resistivity profile evolution.

5.1 The quasi-linear approximation

The quasi-linear approximation is the simplest model to describe the nonlinear evolution of an instability. Consider a dynamical system, given by the quantity $f(\mathbf{x}, t)$ following an equation of the form

$$\partial_t f + Lf = Nf^2 \,, \tag{5.1}$$

where L, N are certain differential operators. In the vicinity of an equilibrium state f_0 one writes

$$f = f_0 + \varepsilon f_1 \,, \quad \varepsilon \ll 1 \,. \tag{5.2}$$

In general the perturbation f_1 has a more complicated spatial structure than the equilibrium f_0, e.g. $f_0(x)$ and $f_1(x, y)$, such that

$$\langle f \rangle = f_0 \,, \quad \langle f_1 \rangle = 0 \,, \tag{5.3}$$

where $\langle \; \rangle$ indicates the spatial average over equilibrium surfaces $x = $ const. Splitting eq. (5.1) into an average and a fluctuating part gives

$$\partial_t f_0 = \varepsilon^2 \langle Nf_1^2 \rangle \,, \tag{5.4}$$

$$\partial_t f_1 + Lf_1 = 2Nf_0 f_1 + \varepsilon(Nf_1^2 - \langle Nf_1^2 \rangle) \,. \tag{5.5}$$

Here the equilibrium equation $Lf_0 = Nf_0^2$ and the condition $\langle Nf_0 f_1 \rangle = Nf_0 \langle f_1 \rangle = 0$ are used, assuming for formal simplicity that N does not act on the coordinate y. The first step involved in the quasi-linear approximation consists of neglecting the ε-term in eq. (5.5), which thus becomes a linear equation for the perturbation f_1. The second step is the assumption that the time variation of the equilibrium f_0 due to the perturbation is slow compared with typical time scales of f_1

$$\partial_t f_0 / f_0 \ll \partial_t f_1 / f_1 \,, \tag{5.6}$$

which allows us to solve eq. (5.5) in the WKB-approximation. Then $f_1(t)$ can be written as a superposition of eigenmodes

$$f_1(t) = \sum_j a_j(t) g_j(\varepsilon^2 t) \,, \tag{5.7}$$

where the complex amplitudes a_j are given by

$$a_j(t) = a_j(0) \exp \left\{ -i \int_0^t \omega_j d\tau \right\} \,,$$

and the frequencies $\omega_j(\varepsilon^2 t)$ and the normalized eigenfunctions $g_j(\mathbf{x}, \varepsilon^2 t)$ are determined by the instantaneous equilibrium, varying on the slow time scale $f_0(\varepsilon^2 t)$. Equation (5.4) can be written in a more familiar way by formal integration of eq. (5.5),

$$f_1 = 2 \sum_j (L - i\omega_j)^{-1} a_j g_j N f_0 \,,$$

and insertion into eq. (5.4):

$$\partial_t f_0 = N (DNf_0) \,, \tag{5.8}$$

which is the well-known quasi-linear diffusion equation with the diffusion coefficient

$$D = 2 \sum_j \langle |a_j g_j|^2 \rangle Re\{(L - i\omega_j)^{-1}\} \,. \tag{5.9}$$

The quasi-linear model is well known from microscopic plasma theory, where the conditions for the validity of this approximation have been discussed repeatedly without, however, reaching a final conclusion. One particular problem is associated with the δ-function arising from the resonance $Re\{(L - i\omega_k)^{-1}\} \propto \delta(\omega_k - kv)$, the dominant contribution in the usual limit of weakly growing modes $\gamma_k \ll \omega_k$. To avoid a singular diffusion coefficient requires the presence of unstable modes with phase velocities ω_k/k covering a finite range Δv in velocity space, which would also make the perturbation behave in a random way and allow interpretation of the average $\langle \, \rangle$ in a statistical sense. But even if this condition is satisfied, the problem remains that the most unstable mode will finally dominate the spectrum, if the initial fluctuation or noise level is low compared with the saturation amplitude, thus again generating an effective single-mode situation corresponding to a singular quasi-linear diffusion coefficient. In spite of such conceptual difficulties, however, the quasi-linear model has proved to give reasonable estimates in a variety of wave–particle interaction problems.

In ideal and resistive MHD unstable modes usually have $Re\{\omega\} = 0$, if equilibrium flows are negligible. As a consequence the time scale ordering assumption (5.6) is not satisfied because in the phase directly preceding saturation γ is of the order of the equilibrium diffusion rate. In addition there is often only one unstable mode, such that the nonlinear process is not random but strictly coherent. The formal divergence of the diffusion coefficient (5.9) in the limit $\gamma \to 0$ is removed by expressing D in terms of the time-integrated fluctuation, practically speaking the displacement ξ instead of the velocity v_1. Since D is computed with the actual eigenfunctions $g_j(x, t)$ taking the equilibrium modification into account, an initial singularity in g_j is in general smoothed due to a local modification of the equilibrium profile and D will be a smooth function.

Often interest is not in the dynamic evolution of the instability but only in the saturated state. Quasi-linear stabilization occurs if $\gamma = 0$ for the most unstable mode. Since diffusion flattens the profile f_0, a simple approximation of the saturated state is to modify the original equilibrium by inserting a "plateau",

$$N f_0(x) \propto \frac{d}{dx} f_0(x) = 0 \,,$$

the extent of which is determined by the condition $\gamma = 0$.

It is of no practical value to write down the general quasi-linear equations for MHD. In the few cases, where quasi-linear theory has actually been applied simplified equations tailored to the particular problem are used. Even if a nearby equilibrium exists, the relaxation cannot always be properly described by quasi-linear theory. An example is the ideal internal kink mode (section 5.3) where an exact calculation of the neighboring helical equilibrium gives a much smaller amplitude than the quasi-linear estimate.

The tearing instability is the best-known example of problems where quasi-linear theory, in different variants, gives reasonably reliable estimates. It explains the modification of the instability properties from exponential to algebraic growth which already occurs at very low amplitude, as well as the final saturation. Because of the coherence of the system in the nonlinear evolution the averaging procedure in eq. (5.3) can be performed in a more refined way; see e.g. Rutherford's theory (section 5.4).

5.2   Nonlinear external kink modes

External kink modes are generally regarded as the most virulent MHD instabilities arising in a plasma column. On the one hand, the linear growth rate is large and the instability does not, or not strongly, depend on the distributions of pressure and field within the plasma. On the other hand, free surface modes are not expected to saturate at low amplitude but to lead to strong, possibly disruptive plasma deformations. Such behavior has been evident and troublesome in various kinds of pinch experiments. Also, the most sophisticated and tamed pinch configuration, the reversed-field pinch (for details see chapter 9), would be very sensitive to external kinks if these were not stabilized by a conducting wall close to the plasma. In a tokamak, external kink modes are usually stable even without a nearby conducting wall, choosing the safety factor q_a appropriately to avoid possible windows of instability, as discussed in section 4.5, but plasma conditions may accidently slip into one of these unstable regions giving rise to a disruption (section 8.2).

We restrict the discussion to tokamak-like configurations. Since external kink modes are driven by the lowest-order potential energy $O(\varepsilon^2)$, eq. (4.67), they are usually considered in the framework of reduced MHD, the formal simplicity of which is particularly convenient for a nonlinear treatment.

5.2.1 Nonlinear energy integral for free boundary modes

The energy conservation relation of reduced MHD, eq. (2.52), was derived with the assumption that plasma of homogeneous density fills the entire system. In the case of external modes, however, one usually assumes that the plasma is separated from a conducting wall by a vacuum region of finite extent. Equation (2.52) should therefore be generalized (Kadomtsev & Pogutse, 1974). Restriction to helical symmetry is very convenient. For any physical quantity f one has

$$f(r,\theta,z) = f(r,\theta - \alpha z) , \quad \alpha = n/mR , \tag{5.10}$$

corresponding to the helicity of the fundamental linearly unstable mode (m,n) under consideration. The reduced MHD equations can then be written in terms of the helical flux function introduced in eq. (3.15). Since $B_z = B_0 = $ const. in the framework of reduced MHD, we have $A_\phi = rB_0/2$ in eq. (3.15). Since in tokamak theory ψ usually denotes the poloidal flux function $\psi = -A_z$, $\mathbf{B}_p = \mathbf{e}_z \times \nabla\psi$, we denote the helical flux function by ψ_*,

$$\psi_* = \psi - (r^2/2)\alpha B_0 , \tag{5.11}$$

$$\partial_r\psi_* = B_\theta - r\alpha B_0 = B_\theta(1 - nq/m) \equiv B_* , \tag{5.12}$$

$$\nabla^2\psi_* = j - 2\alpha B_0 . \tag{5.13}$$

Hence eqs (2.49), (2.51) become

$$(\partial_t + \mathbf{v} \cdot \nabla)\,\psi_* = 0 , \tag{5.14}$$

$$(\partial_t + \mathbf{v} \cdot \nabla)\,\nabla^2\phi = \mathbf{e}_z \times \nabla\psi_* \cdot \nabla\nabla^2\psi_* , \tag{5.15}$$

where $\nabla \equiv \left(\partial_r, r^{-1}\partial_\theta\right)$ and $\mathbf{v} = \mathbf{e}_z \times \nabla\phi$. Multiplying eq. (5.14) by $\nabla^2\psi_*$ and eq. (5.15) by ϕ, adding and integrating over the plasma volume, one obtains after integration by parts

$$\frac{1}{2}\int d^3x\partial_t[(\nabla\phi)^2 + (\nabla\psi_*)^2]$$
$$- \oint d\mathbf{F} \cdot (\partial_t\psi_*\nabla\psi_* + \phi\partial_t\nabla\phi + \phi\mathbf{v}\nabla^2\phi) = 0 , \tag{5.16}$$

using $d\mathbf{F} \cdot \mathbf{B} = d\mathbf{F} \times \nabla\psi_* = 0$, since the plasma boundary S is a flux surface. The first term in the surface integral can be written in the form

$$\oint \partial_t\psi_*\nabla\psi_* \cdot d\mathbf{F} = -\oint \mathbf{v} \cdot \nabla\psi_*\nabla\psi_* \cdot d\mathbf{F}$$
$$= -\oint (\nabla\psi_*)^2 \mathbf{v} \cdot d\mathbf{F} . \tag{5.17}$$

Since for any scalar quantity K we have

$$\frac{d}{dt}\int Kd^3x = \int d^3x\partial_tK + \oint K\mathbf{v} \cdot d\mathbf{F} ,$$

the magnetic terms in eq. (5.16) become

$$\frac{1}{2}\int d^3x \partial_t \, (\nabla\psi_*)^2 - \oint \partial_t\psi_*\nabla\psi_* \cdot d\mathbf{F}$$

$$= \frac{d}{dt}\int d^3x \frac{(\nabla\psi_*)^2}{2} + \oint \frac{(\nabla\psi_*)^2}{2}\mathbf{v}\cdot d\mathbf{F} \, , \qquad (5.18)$$

using eq. (5.14) and $\nabla\psi_* \parallel d\mathbf{F}$. To evaluate the kinetic terms in eq. (5.16) the change of the tangential velocity is required. Integration of eq. (5.15) gives, using $\nabla^2\phi = \mathbf{e}_z \cdot \nabla \times \mathbf{v}$,

$$\partial_t\mathbf{v} + \mathbf{v}\cdot\nabla\mathbf{v} = -\nabla\psi_*\nabla^2\psi_* - \nabla P \, . \qquad (5.19)$$

The generalized pressure P can be obtained by comparison with the equation of motion written in the usual form

$$\partial_t\mathbf{v} + \mathbf{v}\cdot\nabla\mathbf{v} = -\nabla\left(p + \frac{B^2}{2}\right) + \mathbf{B}\cdot\nabla\mathbf{B} \, . \qquad (5.20)$$

Since $\mathbf{B}\cdot\nabla = \mathbf{e}_z \times \nabla\psi_* \cdot \nabla$, $\mathbf{B} = \mathbf{e}_z \times \nabla\psi + B_0\mathbf{e}_z$, one finds

$$\mathbf{B}\cdot\nabla\mathbf{B} = \nabla\frac{(\nabla\psi_*)^2}{2} - \nabla\psi_*\nabla^2\psi_* - \alpha B_0\nabla\psi_* \, , \qquad (5.21)$$

and therefore

$$\nabla P = \nabla\left(p + \frac{B^2}{2} - \frac{(\nabla\psi_*)^2}{2} + \alpha B_0\psi_*\right) \, . \qquad (5.22)$$

Using this result in eq. (5.19) we can rewrite the second term in the surface integral in eq. (5.16):

$$\oint \phi\partial_t\nabla\phi \cdot d\mathbf{F} = \oint \phi\partial_t\mathbf{v} \cdot (\mathbf{e}_z \times d\mathbf{F}) \qquad (5.23)$$

$$= -\oint \phi\mathbf{v}\cdot\nabla\nabla\phi \cdot d\mathbf{F} - \oint \phi(\mathbf{e}_z \times d\mathbf{F})\cdot\nabla\left(p + \frac{B^2}{2} - \frac{(\nabla\psi_*)^2}{2}\right) \, .$$

Considering the kinetic terms, one can show directly by decomposition into tangential and normal components in the plane perpendicular to $\mathbf{e}_z$ and integration by parts that

$$\oint \phi(\mathbf{v}\cdot\nabla\nabla\phi - \mathbf{v}\nabla^2\phi) \cdot d\mathbf{F} = \oint d\mathbf{F}\cdot\mathbf{v}\,(\nabla\phi)^2 \, /2 \, . \qquad (5.24)$$

The somewhat lengthy calculation is not given here, since the term is only of minor importance. The pressure p in the second term on the r.h.s. of eq. (5.23) is eliminated by using the pressure balance on the plasma boundary

$$p + \frac{B^2}{2} = \frac{B_V^2}{2} \, . \qquad (5.25)$$

The simplest way of expressing B_V^2 by ψ_*^V is by applying the relation for the vacuum field:

$$\nabla B_V^2/2 = \mathbf{B}_V \cdot \nabla \mathbf{B}_V$$
$$= \nabla \frac{(\nabla \psi_*^V)^2}{2} + \alpha B_0 \nabla \psi_*^V , \qquad (5.26)$$

where eq. (5.21) and $\nabla^2 \psi_*^V = -2\alpha B_0$ have been used. Since only the tangential derivative enters in the last term in eq. (5.23), we can apply the pressure balance (5.25) to introduce the vacuum field contribution (5.26) into this term and obtain after integration by parts

$$\oint \phi \nabla \left(p + \frac{B^2}{2} - \frac{(\nabla \psi_*)^2}{2} \right) \cdot (\mathbf{e}_z \times d\mathbf{F})$$
$$= \oint d\mathbf{F} \cdot \mathbf{v} \left(\frac{(\nabla \psi_*^V)^2}{2} - \frac{(\nabla \psi_*)^2}{2} \right) , \qquad (5.27)$$

which differs from zero only in the presence of a surface current. Combination of expressions (5.18), (5.23), (5.24) and (5.27) in the energy balance (5.16) yields

$$\frac{d}{dt} \frac{1}{2} \int_{\Omega_P} d^3x [(\nabla \phi)^2 + (\nabla \psi_*)^2] = -\frac{1}{2} \oint_S d\mathbf{F} \cdot \mathbf{v}(\nabla \psi_*^V)^2 . \qquad (5.28)$$

This result is easily understood. The change of the plasma energy is due to the work performed against the vacuum field pressure. The last term in eq. (5.28) can be written in several different forms using the relation

$$\frac{d}{dt} \int_{\Omega_V} d^3x (\nabla \psi_*^V)^2 = \int_{\Omega_V} d^3x \partial_t (\nabla \psi_*^V)^2 - \oint_S d\mathbf{F} \cdot \mathbf{v}(\nabla \psi_*^V)^2 , \qquad (5.29)$$

where the minus sign results from the choice of the surface normal pointing into the vacuum region V. Integration by parts of the first term on the r.h.s. gives

$$\int_{\Omega_V} d^3x \nabla \psi_*^V \cdot \partial_t \nabla \psi_*^V = \oint d\mathbf{F} \cdot \psi_*^V \partial_t \nabla \psi_*^V - \int_{\Omega_V} d^3x \psi_*^V \partial_t \nabla^2 \psi_*^V , \qquad (5.30)$$

where the last term vanishes since $\nabla^2 \psi_*^V = \text{const.}$ The surface term in eq. (5.30) contains the contributions from both the plasma boundary S and the outer conducting wall S_w, which are both flux surfaces. Normalizing ψ_* such that its value on the plasma boundary vanishes, $\psi_{*a} = 0$, and noting that

$$\oint_{S_w} d\mathbf{F} \cdot \nabla \psi_*^V = I + \text{const.} ,$$

where I is the total plasma current (considering a plasma cylinder of unit length), the energy balance (5.28) can be written in the following form:

$$\frac{d}{dt}(W_P - W_V) + \psi_w \frac{dI}{dt} = 0 \tag{5.31}$$

$$W_P = \frac{1}{2}\int_{\Omega_P} d^3x[(\nabla\phi)^2 + (\nabla\psi_*)^2]\,, \quad W_V = \frac{1}{2}\int_{\Omega_V} d^3x(\nabla\psi_*^V)^2\,.$$

Two special cases of plasma discharge operation are approximately realized in experiments. Either the plasma current is maintained constant by injecting poloidal flux from the external coil system, i.e. ψ_w varies in time. In this case eq. (5.31) gives

$$W_P - W_V = \text{const.} \tag{5.32}$$

Or the flux ψ_w is constant, which is the case when plasma motions occur on time scales which are short compared with the response time of the external circuit. In this case eq. (5.31) can also be integrated to give

$$W_P - W_V + \psi_w I = \text{const.} \tag{5.33}$$

Note that here I varies in time.

The negative sign of the vacuum field contribution W_V in eq. (5.31), which is somewhat unexpected, is connected with the fact that ψ_*^V is not an independent dynamic variable but results from the surface term in eq. (5.28). One can write the energy integral also in a form where W_V appears with a positive sign. Performing the integration by parts in eq. (5.30) in the opposite way, eq. (5.33) for the case $\psi_w = \text{const.}$ assumes the alternative form

$$W_P + W_V - 2\alpha B_0 \int_{\Omega_V} \psi_* d^3x = \text{const.} \tag{5.34}$$

In our derivation of the energy relation it might appear somewhat unnatural to reintroduce the pressure, which is not an independent quantity in incompressible theory, instead of remaining within the ϕ, ψ_* formalism of the reduced equations, in particular, since the result (5.31) is again written in terms of ϕ, ψ_* only. However, one has to realize that the reduced equations assuming homogeneous plasma density $\rho = 1$ are only valid within the plasma. Therefore a generalization is necessary to treat the plasma–vacuum system and the procedure given above appears to be the most convenient and intuitive one.

As an alternative to the assumption of a strict vacuum region surrounding the plasma, the model of a highly resistive plasma with the same density as the weakly resistive bulk plasma is sometimes used, which is particularly convenient in numerical studies, since formally no distinction between plasma and vacuum regions is required. If the reduced equations

(2.49), (2.51) are used, one adds only a resistive term $\eta(\psi_*)j$ on the r.h.s. of eq. (2.49), where $\eta(\psi_*)$ is a quasi-step function increasing within a narrow layer from a small to a very large value. Under these conditions resistive energy dissipation is negligible except in the case of a surface current, which would be affected by the "vacuum" resistivity. But even in the absence of a surface current this pseudo-vacuum model is not identical with a genuine plasma–vacuum system. Since the pseudo-vacuum carries a mass density, there will be a "vacuum" contribution to the kinetic energy, whenever the plasma boundary is moving. Hence in this case the integration domain of the kinetic energy in the energy balance relation (5.31) has to be extended up to the wall.

5.2.2 Vacuum bubbles

It has been suggested by Kadomtsev & Pogutse (1974) that the disruptive instability, or major disruption, in a tokamak plasma (section 8.2) is caused by an external kink mode. The physical picture is that the plasma column becomes strongly convoluted as the result of the kink instability, eventually capturing one or several tubular cavities called vacuum bubbles, which leads to an inflation of the outer plasma boundary and possible termination of the discharge due to intense limiter contact. Though this process has subsequently been shown by numerical simulation to require rather special nontypical conditions, namely a nearly constant resonant q-profile in the plasma column, it constitutes an example of a strongly nonlinear MHD process described by a simple and elegant theory and hence is interesting also in its own right. In addition, qualitatively similar processes may actually occur whenever q is nearly constant over an extended radial region in the plasma, for instance the flat central profile $q \simeq 1$, which has been suggested as a model of the sawtooth disruption (section 8.1).

The theory is based on the energy equation (5.31), the magnetic part of which serves as an effective potential energy and will be denoted by W. The general tendency is that a system evolves toward a state of minimum W. This state is in general not cylindrically symmetric, or if cylindrical, cannot be reached dynamically by a sequence of cylindrical states but requires helical intermediate states. The initial cylindrical plasma is either linearly unstable or linearly stable. In the latter case a finite amplitude perturbation is required, since there is in general a potential barrier separating the initial state from the absolute minimum state. For a sheared magnetic field in the plasma, the nonlinear motion will be strongly restricted by flux conservation, such that in the framework of ideal MHD the minimum energy state cannot in general be reached. Though it is shown in the following chapters that in real systems fast reconnection processes can often effectively eliminate the ideal flux constraint, the present discussion is confined to the particularly simple and theoretically convenient case of unsheared systems.

Consider a cylindrical plasma column with constant q-profile, $q = q_0 = m/n$, such that the helical field B_*, defined in eq. (5.12), vanishes throughout the plasma. Hence ψ_* is constant and we choose the constant such that in the plasma

$$\psi_* = 0 \; .$$

Because of $\nabla\psi_* = 0$ in the plasma the potential energy W consists only of the vacuum part W_V. Since only ψ_*^V enters the following discussion we omit for convenience the superscript V. In the plasma the current density j is constant, but there may be a surface current, which implies a discontinuity of the azimuthal field at the plasma boundary $r = a$,

$$B_{\theta a} = \xi B_{\theta 0} = \xi a\alpha B_0$$
$$q_a/q_0 = nq_a/m = \xi^{-1} \; , \tag{5.35}$$

where the subscript a indicates the value immediately outside the plasma boundary. If $\xi > 1$, i.e. $q_a < m/n$, a resonant surface exists in the vacuum as illustrated in Fig. 5.2, which is a necessary condition for external kink instability. In fact one can easily see that for $\xi > 1$ small perturbations grow, while for $\xi < 1$ the system is stable. Consider the vacuum "pressure" $(\nabla\psi_*)^2/2 = B_*^2/2$. Since

$$\psi_* = \alpha B_0 \left(\xi a^2 \ln \frac{r}{a} + \frac{a^2 - r^2}{2} \right) , \tag{5.36}$$

one obtains

$$\frac{d}{dr} \frac{B_*^2}{2} = -(\alpha B_0)^2 \left(\frac{\xi a^2}{r} - r \right) \left(\frac{\xi a^2}{r^2} + 1 \right)$$
$$= -(\alpha B_0)^2 a(\xi^2 - 1) \quad \text{for } r = a \; . \tag{5.37}$$

Hence the pressure B_*^2 decreases away from the boundary for $\xi > 1$ (instability) and increases for $\xi < 1$ (stability). $\xi = 1$ corresponds to marginal stability. A more detailed stability analysis gives the instability range

$$\frac{(m+1)(a/b)^{2m} + m - 1}{(m-1)(a/b)^{2m} + m + 1} < \frac{nq_a}{m} = \xi^{-1} < 1 \; , \tag{5.38}$$

where $r = b$ is the position of the conducting wall.

It follows from eq. (5.38) that the lower instability boundary increases with m, while near the upper boundary all m are unstable, as illustrated in Fig. 5.3. Hence for nq_a/m close to the lower boundary of the fundamental

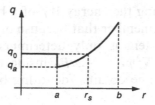

Fig. 5.2. *q*-profile as considered in the theory of Kadomtsev and Pogutse.

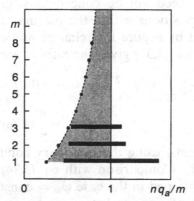

Fig. 5.3. Instability range (gray area) for mode number m, according to eq. (5.38). The bars indicate the ranges of energetically favorable bubble states ($\psi_w = $ const.) for $N = 1, 2, 3$.

mode (m_0, n_0) low-amplitude saturation (neighboring equilibrium) is in principle possible owing to coupling to stable harmonics (lm_0, ln_0), $l > 1$, while a strongly nonlinear behavior is expected near the upper stability boundary $q_a = m/n$, where all harmonics are unstable.

Instability will initially generate a flute-like corrugation of the plasma boundary. As the motion continues, vacuum parts penetrate deeply into the plasma region until the outer plasma boundary becomes circular again with a radius $a_* > a$, such that the plasma contains a number of helical cavities, "bubbles" in cross-section; see Fig. 5.4. Whether this process actually occurs can only be decided by a full dynamic computation which will be discussed later.

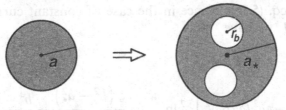

Fig. 5.4. Increase of plasma radius $a \to a^* > a$ by capture of vacuum bubbles.

Let us start by calculating the energy W_b of a bubble of circular cross-section with radius r_b. (Remember that because of $\nabla\psi_* = 0$ in the plasma, the energy W of the system is only determined by the vacuum field contribution W_V.) Since $\nabla^2\psi_* = -2\alpha B_0$ inside the bubble and $\psi_* = 0$ on the bubble boundary, one obtains (denoting bubble quantities by the subscript b)

$$\psi_{*b}(r) = \alpha B_0(r_b^2 - r^2)/2 . \tag{5.39}$$

Note that the bubble need not be concentric with the external plasma boundary. In fact all positions within the plasma are energetically equivalent, being connected by a pure interchange motion. In the case of the constraint $I = \text{const.}$, eq. (5.32) gives the bubble energy contribution

$$W_b = -\tfrac{1}{2}\int_{V_b}(\nabla\psi_*)^2\,d^3x \;=\; -\alpha B_0\int_{V_b}\psi_{*b}d^3x$$
$$= -\frac{\pi}{4}(\alpha B_0)^2\,r_b^4 , \tag{5.40}$$

after integration by parts, using the boundary condition $\psi_* = 0$ and the relation $\nabla^2\psi_* = -2\alpha B_0$. Comparison with eq. (5.34) shows that the same expression of W_b is obtained in the case $\psi_w = \text{const.}$

Note that a bubble of noncircular cross-section but identical volume has a higher energy, since the integral $\int\psi_*d^3x$ is extremal for circular cross-section if ψ_* obeys the equation $\nabla^2\psi_* = \text{const.}$ In addition, for a given available area $\pi(a_*^2 - a^2)$ the configuration with the lowest possible number of bubbles has the lowest energy, since with $N\pi r_b^2 = \pi(a_*^2 - a^2)$,

$$W_{bN} = NW_{b1} = -\frac{1}{N}\frac{\pi}{4}(\alpha B_0)^2\,(a_*^2 - a^2)^2 . \tag{5.41}$$

Hence coalescence of two bubbles lowers the energy. Without symmetry restrictions the lowest energy state has just one bubble.

The energy contribution W_e from the external vacuum region depends on the external circuit. First consider the case of constant current I, where the magnetic field B_* outside the plasma column conserves its initial value,

$$B_* = \alpha B_0\left(\frac{\xi a^2}{r} - r\right) ,$$

obtained from eq. (5.36). Hence in the case of constant current, denoted by superscript I, we have

$$W_e^I = -\pi\int_{a_*}^b B_*^2 r dr \tag{5.42}$$

$$= -\pi a^2(a\alpha B_0)^2\left[\xi^2\ln\frac{b}{a_*} - \frac{\xi(b^2 - a_*^2)}{a^2} + \frac{b^4 - a_*^4}{4a^4}\right] .$$

The total energy of a plasma of radius a_* enclosing N bubbles of radius $r_b = [(a_*^2 - a^2)/N]^{1/2}$ is

$$W^I = W_e^I + W_{bN} \, ,$$

which in order to simplify notation can be written in normalized form with $\rho = a_*/a$ and $\widehat{W}^I(\rho = 1) = 0$:

$$\widehat{W}^I(\rho) = \frac{\xi^2}{2} \ln \rho^2 - \xi(\rho^2 - 1) + \frac{\rho^4 - 1}{4} - \frac{(\rho^2 - 1)^2}{4N} \, . \tag{5.43}$$

Note that $\widehat{W}^I$ does not depend on the wall radius b.

In the case of constant flux ψ_w the external field changes, B_* increasing with increasing plasma radius a_*, since the same flux is enclosed in a smaller volume. Evaluation of eq. (5.33) or (5.34) gives the total energy denoted by the superscript ψ,

$$\begin{aligned} W^\psi &= W_e^\psi + W_{bN} \\ &= \pi a^2 (a\alpha B_0)^2 \left[\xi^2 \ln \frac{b}{a_*} - \frac{b^4 - a_*^4}{4a^4} - \frac{(a_*^2 - a^2)^2}{4a^4 N} \right] \, , \end{aligned} \tag{5.44}$$

where ξ now depends on a_* determined by the condition $\psi_*(b) = \psi_w$,

$$\xi = \left(\frac{\psi_w}{a^2 \alpha B_0} + \frac{b^2 - a_*^2}{2a^2} \right) \Big/ \ln \frac{b}{a_*} \, . \tag{5.45}$$

Writing W^ψ in normalized form analogous to $\widehat{W}^I$ in eq. (5.43) with $\hat{b} = b/a$ yields

$$\widehat{W}^\psi = -\frac{\xi^2(\rho)}{2} \ln \rho^2 + \frac{\rho^4 - 1}{4} - \frac{(\rho^2 - 1)^2}{4N} + \left(\xi^2(\rho) - \xi^2(1) \right) \ln \hat{b} \, , \tag{5.46}$$

where $\xi(\rho)$, defined in eq. (5.45), can be expressed by $\xi(1) = \xi_0$, the value in the original cylindrical configuration:

$$\xi(\rho) = \left(\xi_0 \ln \hat{b} - \frac{\rho^2 - 1}{2} \right) \Big/ \ln \frac{\hat{b}}{\rho} \, . \tag{5.47}$$

The behavior of $\widehat{W}^I$ and $\widehat{W}^\psi$ is illustrated in Fig. 5.5 for $N = 2$ and three values of ξ and ξ_0, respectively. Within a certain range of ξ, the potential energy W has a local minimum at some value ρ_m. If $W(\rho_m) < W(1) = 0$ (and $\rho_m < \hat{b}$ in the case of constant current) then the bubble state ρ_m is energetically favorable. This state is in general separated from the cylindrical state $\rho = 1$ by a potential barrier, $d\widehat{W}^{I,\psi}/d\rho = (\xi - 1)^2 \geq 0$ at $\rho = 1$.

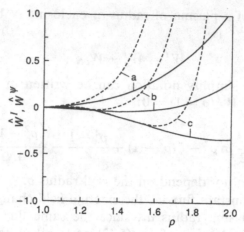

Fig. 5.5. Quasi-potential $\widehat{W}^I(\rho)$ (solid lines) and $\widehat{W}^\psi(\rho)$ (dashed lines) for $N = 2$ and three values of ζ and ζ_0, respectively: (a) ζ, $\zeta_0 = 0.5$ (no bubble state); (b) ζ, $\zeta_0 = 1$, (c) ζ, $\zeta_0 = 1.5$ (bubble states possible).

We note that the range of $\zeta = m/nq_a$, where energetically favorable bubble states exist, does not coincide with the range of linear instability given in eq. (5.38) and plotted in Fig. 5.3. Identifying the number of bubbles N with the mode number m_0 of the fundamental mode $(m, n) = (m_0, 1)$, the range of bubble states can be inserted into the stability diagram as done in Fig. 5.3 for $N = 1, 2, 3$. In particular, bubble states are also possible for $\zeta < 1$, i.e. $nq_a/m > 1$, where the corresponding linear modes are stable. In this range "hard" onset of bubble formation may occur, triggered by an accidental finite external perturbation. If the system slowly evolves with q_a decreasing toward m_0, such a process becomes more and more probable. This is a simple model of the sudden precursorless onset of a major disruption observed in tokamak plasmas, particularly if q_a is gradually lowered towards 2 (section 8.2).

As seen in Fig. 5.5, the system is somewhat more stable to bubble formation for $\psi_w = $ const. (dashed lines) than for $I = $ const. (full lines). This is to be expected, since in the former case flux compression produced by the expansion of the plasma has a stabilizing effect. The difference is particularly strong in the case of a single bubble $N = 1$, where for $I = $ const. the plasma expands up to the wall, since the ρ^4-terms cancel in $W^I(\rho)$.

The potential barrier separating cylindrical from bubble states can in principle be overcome by the helical instability breaking the quasi-cylindrical symmetry inherent in the energy integral W. The linear instability does, however, not guarantee that the minimum-energy state is actually reached, since the instability may saturate in a helical state if the barrier is too high.

The theory of vacuum bubbles has been confirmed in essential parts by numerical simulations (Rosenbluth et al., 1976). Starting with the reduced equations (5.14), (5.15), one notes that $\psi_* = 0$ in the plasma results in $\nabla^2\phi = 0$, i.e. the plasma motion is irrotational. Hence $\mathbf{v}$ is most conveniently described by a scalar potential χ, $\mathbf{v} = \nabla\chi$, obeying Laplace's equation

$$\nabla \cdot \mathbf{v} = \nabla^2\chi = 0 . \tag{5.48}$$

Since conditions in the plasma interior are uniform, the problem is reduced to computing the motion of the plasma boundary $\mathbf{r}_a$,

$$\dot{\mathbf{r}}_a = \nabla\chi .$$

Here χ is determined by eq. (5.48) with $\chi = \chi(\mathbf{r}_a)$ on the plasma boundary. The change of $\chi(\mathbf{r}_a)$ follows from the equation (one introduces a damping term with suitably chosen ν to optimize relaxation to the minimum-energy state)

$$\left(\frac{d\chi}{dt} - \nu\chi\right)_{\mathbf{r}=\mathbf{r}_a} = -\tfrac{1}{2}\left(\partial_n\psi_*\right)^2 , \tag{5.49}$$

which is easily derived from eq. (5.20) since $\mathbf{B} \cdot \nabla B = 0$ and $p + B^2/2 =$ const. in the plasma. ψ_* in the vacuum region is obtained from the equation $j = 0$,

$$\nabla^2\psi_* = -2\alpha B_0 \tag{5.50}$$

with the condition $\psi_* = 0$ on the plasma boundary $\mathbf{r}_a$ and $\psi_* = \psi_w$ on the conducting wall, where ψ_w is either fixed or determined by the constraint of constant current $\oint \partial_r\psi_* d\theta = $ const.

Both equations (5.48) and (5.50) are solved numerically by using Green's theorem expressing $\chi(\mathbf{r}_a)$, $\psi(\mathbf{r}_a)$ by surface integrals. The calculations show that bubble states are formed if q_a is sufficiently close to the resonance $q_0 = m_0/n_0$ of the mode considered, even for q_a slightly larger than q_0, where the mode is linearly stable and the process needs a finite initial perturbation. For lower q_a the instability is not able to penetrate the potential barrier and saturates by mode-coupling effects before the bubble state is reached, which can be related to the linear stability of higher harmonics; see Fig. 5.3.

5.2.3 *Effect of magnetic shear*

The shearless resonant equilibria carrying a surface current, which have been assumed in the theory of vacuum bubbles, are highly idealized and not typical for real plasma discharges. A finite value of the resistivity would rapidly smooth out such a singular current distribution. Only the special case $\xi = 0$, where the surface current vanishes and the resonant surface coincides with the plasma boundary, seems to have some practical significance, for instance in configurations with a central flat $q \simeq 1$ profile, which are discussed in section 8.1 as one possible model of the sawtooth collapse. In general, however, magnetic shear is finite, $dq/dr \neq 0$, and it is therefore important to consider the nonlinear behavior of external kink modes for sheared configurations. In this case the problem becomes much more complex, such that the analytical approach is no longer feasible and one has to resort to a numerical treatment.

The problem was first investigated by Rosenbluth et al. (1976) using a generalization of the method described above for the shearless case, eqs (5.48)–(5.50). They considered the $(m, n) = (2, 1)$ mode for a parabolic current profile $j(r)$ with $j(a) = 0$, no skin current and a ratio $q_a/q_0 = 2$ with several values of $q_a < 2$. The nonlinear evolution was found to be much less dramatic than in the shearless case, the plasma column developing a quasi-elliptical shape of moderate elongation, a result which led to the impression that for realistic plasma configurations external kinks are relatively harmless. This is, however, true only for sufficiently strong shear, as has been shown by Dnestrovskij et al. (1985, 1987). These authors studied the same problem both by considering equilibrium bifurcations and by dynamical simulation. In the equilibrium approach the equation

$$\nabla^2 \psi_* = j(\psi_*) - 2\alpha B_0$$

is solved with a specified parametrized current profile,

$$j(\psi_*) = \begin{cases} j_0 (\psi_{*a} - \psi_*)^\kappa & \psi_* < \psi_{*a} \\ 0 & \psi_* > \psi_{*a}, \end{cases}$$

where $\psi_* = \psi_{*a}$ at the plasma boundary, and j_0 and κ are determined by prescribing values of q_a and q_a/q_0. Figure 5.6 illustrates the computed equilibrium properties in the e–v plane, where e is the ellipticity of the plasma cross-section in the helical state and $v = 2/q_a$. For $v > v_c$, where $v_c = v_c (q_a/q_0)$ is the bifurcation point, cylindrical ($e = 1$) equilibria are stable with respect to the $(2, 1)$ mode. If v falls below v_c, the cylindrical equilibrium becomes unstable. For $q_a/q_0 = 1$, corresponding to a uniform current profile, no stable helical states exist, since the bifurcation curve in Fig. 5.6 is inclined to higher v. In the entire range $1 \leq q_a/q_0 \lesssim 2$, with the upper value depending somewhat on the position of the conducting wall, bifurcation curves rise steeply with decreasing v, giving rise to values of the ellipticity which in a real plasma device would result in limiter contact. Only for $q_a/q_0 > 2$, corresponding to rather strongly peaked current profiles, bifurcated helical states are only slightly elongated, $e \lesssim 1.1$.

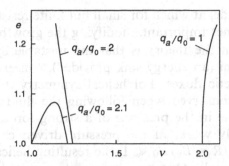

Fig. 5.6. Ellipticity e of the helical minimum energy state as a function of $v = 2/q_a$ for three values of q_a/q_0 (from Dnestrovskij et al., 1987).

Hence the $(2, 1)$ external kink mode in fact appears to be harmless for sufficiently strong shear $q_a/q_0 \gtrsim 2$, in agreement with the cases considered by Rosenbluth et al., whereas for weaker shear the resulting large plasma elongation would probably lead to discharge termination. (Note that the stability with respect to helicities other than $m/n = 2$ is not considered in these investigations.)

In several tokamak experiments stable operation at very low safety factor $q_a \sim 1.5$ has been achieved (Maeda et al., 1979; Barsukov et al., 1982; Burrell et al., 1983). To reach such a state the discharge must cross a range $(1.5 \lesssim q_a \lesssim 2)$ of external kink instability. The results on bifurcated helical equilibria just discussed, requiring a sufficiently peaked current profile $q_a/q_0 \gtrsim 2$ for mild elongation, suggest that $q_0 < 1$ in the experiments. This point is discussed in the context of sawtooth oscillations (section 8.1).

5.3   Nonlinear theory of the ideal internal kink mode

Ideal internal kink modes are stable, see eq. (4.67), except for $m = 1$, which is usually called *the* internal kink mode and which may be weakly unstable owing to pressure effects or noncircular cross-section. Nonlinearly the ideal kink mode is found to saturate at small amplitude corresponding to a helical neighboring equilibrium. Since this relaxed state is singular containing a current sheet, the treatment cannot strictly speaking be confined to the framework of ideal MHD, as an arbitrarily small amount of resistivity is expected to change this state. But since the nonlinear theory developed by Rosenbluth et al. (1973) constitutes one of the rare examples of a quasi-exact analytic theory of a nonlinear MHD process, which has been verified quantitatively by numerical simulations, it is outlined here. Seen from a more practical point of view it gives an upper limit of the amplitude, at which for small but finite resistivity current sheet reconnection becomes important, modifying the growth rate of the mode.

The basic idea in the theory is that the unstable system relaxes into a helical equilibrium (an energy sink provided), conserving both poloidal and toroidal magnetic fluxes. For helical symmetry this implies that the helical flux ψ_* is conserved when following the fluid motion, the latter being incompressible in the presence of a strong toroidal field. Since the instability is usually weak, in the pressure driven case (see eq. (4.68)) $\gamma \tau_A = O(\varepsilon^2)$, $\varepsilon = r_1/R \sim B_\theta/B_0 \ll 1$, the resulting helical displacement ξ_0 is expected to be small. One now constructs a neighboring equilibrium $\psi_*(r, \theta)$, which is related to the initial state $\psi_0(r)$ by the relation

$$\psi_*(r, \theta) = \psi_0 (r - \xi(r, \theta)) \ . \tag{5.51}$$

To lowest order in ε the equilibrium equation is obtained from the reduced equation of motion (5.15), which implies that $\nabla^2 \psi_*$ is a flux function,

$$\nabla^2 \psi_* = I (\psi_*) \ . \tag{5.52}$$

Incompressibility requires the cross-sectional area enclosed by a flux surface to be invariant:

$$\int_{\psi_*} r d\theta dr = \int_{\psi_0} r d\theta dr \ . \tag{5.53}$$

Equations (5.52), (5.53) are solved in the vicinity of the resonant surface $r = r_1$, where $\mathbf{k} \cdot \mathbf{B} = r^{-1} \psi_0' = 0$. This boundary layer solution is then matched to the linear solution away from the resonant surface, which determines the helical amplitude ξ_0.

Since in the narrow boundary layer one has $\nabla^2 \psi_* \simeq \partial_r^2 \psi_*$, eq. (5.52) can be integrated to give

$$(\partial_r \psi_*)^2 = F(\psi_*) + G(\theta) \ , \tag{5.54}$$

where the unknown functions F, G have to be determined. It is convenient to replace the coordinate ψ_* by a geometric quantity, the distance x of the corresponding flux surface ψ_0 in the unperturbed state from the resonant radius r_1. $x = 0$ denotes the resonant surface, such that the displacement $\xi(x, \theta)$ shifting the surface labeled x to $r(x, \theta)$ is given by the relation

$$\xi(x, \theta) = r(x, \theta) - r_1 - x \ . \tag{5.55}$$

Expanding $\psi_0(r)$ around the resonant radius r_1 yields

$$\psi_0(r) = \psi_0(r_1) + \tfrac{1}{2} x^2 \psi_0''(r_1) \ , \tag{5.56}$$

since $\psi_0'(r_1) = 0$. Hence

$$d\psi = x dx \psi_0''(r_1) \ . \tag{5.57}$$

Further integration of eq. (5.54) and use of eqs (5.55), (5.57) gives

$$\xi(x,\theta) = \int_0^x dx' \left(\frac{x'}{\sqrt{f(x')+g(\theta)}} - 1 \right) + h(\theta) , \qquad (5.58)$$

introducing the normalized functions $f = F/\psi_0''(r_1)^2$ etc. From the conservation of area, eq. (5.53), in particular the differential area between surfaces numbered x and $x+dx$, one obtains to lowest-order in ε

$$\oint r\partial_x r d\theta/2\pi \simeq r_1 \oint \partial_x r d\theta/2\pi = r_1 , \qquad (5.59)$$

where we use the notation $\oint d\theta \equiv \int_0^{2\pi} d\theta$. Differentiation with respect to x and averaging over θ of eq. (5.58) and use of eqs (5.55), (5.59) gives

$$\oint \frac{x}{\sqrt{f(x)+g(\theta)}} d\theta/2\pi = 1 . \qquad (5.60)$$

Since x changes sign across the resonant surface, being negative inside and positive outside, the sign of the square root must be chosen accordingly to satisfy eq. (5.60) for all x. From eq. (5.54) one sees that a change of sign of $\sqrt{f+g}$ in general implies a jump of the helical field $\partial_r \psi_*$ and hence of B_θ, i.e. a sheet current at the resonant surface.

In order to match the inner-layer solution (5.58) to the ideal linear solution given in eq. (4.121)[*]

$$\xi(x,\theta) = \begin{cases} \xi_0 \left(1 + \lambda_H r_1/\pi x\right) \cos\theta & x < 0 \\ \xi_0 \left(\lambda_H r_1/\pi x\right) \cos\theta & x > 0 , \end{cases} \qquad (5.61)$$

with $\lambda_H \tau_A^{-1} = \gamma$, the linear growth rate, and $\tau_A^{-1} = \psi_0''(r_1)$, we write ξ in the following asymptotic form:

$$\xi(x,\theta) = \begin{cases} h(\theta) + \int_0^\infty dx' \left(\frac{x'}{\sqrt{f+g}} - 1 \right) + \dfrac{1}{2}\dfrac{g(\theta)}{x} & x \to \infty \\ h(\theta) - \int_0^\infty dx' \left(\frac{x'}{\sqrt{f+g}} - 1 \right) + \dfrac{1}{2}\dfrac{g(\theta)}{x} & x \to -\infty . \end{cases}$$

$$(5.62)$$

This can easily be verified using the asymptotic behavior of $f(x) \to x^2$ for $|x| \to \infty$ evident from eq. (5.60). Considering the limits $\lim_{x\to-\infty} \xi(x,\theta) = \xi_0 \cos\theta$, $\lim_{x\to+\infty} \xi(x,\theta) = 0$, one eliminates $h(\theta)$, which is not needed in the following, and obtains an integral equation for $g(\theta)$

---

[*] In eq. (4.121) the quantities ξ_0, x are normalized to r_1.

$$\xi_0 \cos \theta = -2 \int_0^\infty dx' \left(\frac{x'}{\sqrt{f(x') + g(\theta)}} - 1 \right)$$

$$= - \int_0^\infty \frac{\oint (f + g)^{-3/2} d\theta/2\pi}{\left(\oint (f + g)^{-1/2} d\theta/2\pi \right)^3} \left(\frac{1}{\sqrt{f + g}} - \oint \frac{d\theta/2\pi}{\sqrt{f + g}} \right) df ,$$

(5.63)

where the integration variable x has been substituted by f using relation (5.60). Equation (5.63) is in fact an integral equation for $\hat{g}(\theta) = g(\theta)/\xi_0^2$, since the substitution $\hat{f} = f/\xi_0^2$, $\hat{g} = g/\xi_0^2$ gives

$$\int_0^\infty \frac{\oint \left(\hat{f} + \hat{g} \right)^{-3/2} d\theta/2\pi}{\left(\oint \left(\hat{f} + \hat{g} \right)^{-1/2} d\theta/2\pi \right)^3} \left(\frac{1}{\sqrt{\hat{f} + \hat{g}}} - \oint \frac{d\theta/2\pi}{\sqrt{\hat{f} + \hat{g}}} \right) d\hat{f} + \cos \theta = 0 .$$

(5.64)

While eq. (5.58) and its asymptotic forms are exact solutions of the nonlinear equilibrium equation (5.52) such that $g(\theta)$ contains all harmonics, connection to the ideal linear solution (5.61) is valid only for the first harmonic $m = 1$. This does *not* imply $g \propto \cos \theta$; instead $g(\theta)$ is determined by the integral equation (5.63). Asymptotic matching of the ideal solution, eq. (5.61), with the first harmonic of the inner-layer solution, eq. (5.62), determines the amplitude ξ_0:

$$\oint \frac{g(\theta) \cos \theta d\theta}{\xi_0} = \xi_0 \oint \hat{g}(\theta) \cos \theta d\theta = \frac{\lambda_H r_1}{\pi} .$$

(5.65)

It is obviously difficult to solve eq. (5.64) exactly, which could only be achieved numerically. However, using a variational principle Rosenbluth et al. obtained a quite accurate estimate of the first harmonic $\oint \hat{g}(\theta) \cos \theta d\theta \simeq 0.075$, such that eq. (5.65) becomes

$$\xi_0/r_1 \simeq \frac{13.3}{\pi} \lambda_H = O(\varepsilon^2) .$$

(5.66)

It should be noted that quasi-linear theory applied in a naive way would give $\xi_0/r_1 = O(\varepsilon)$. The singularity at $r = r_1$, however, makes this procedure meaningless, since in the perturbation expansion higher-order terms contain successively worse divergencies.

The result has been verified by numerical simulations of the nonlinear saturated state of the internal kink mode (Park et al., 1980), where the ideal incompressible but non-reduced MHD equations are solved in a straight system of finite length, i.e. finite effective aspect ratio $\varepsilon^{-1} = R/a$. The singular current sheet predicted by the theory is numerically controlled by superimposing a small external helical magnetic field B_a, such that the current sheet density j_s increases with decreasing B_a , $j_s \propto B_a^{-1}$. Varying ε, the extrapolated value of $\lim_{\varepsilon \to 0}(\xi_0/\varepsilon^2)$ is found to agree well with the theoretical prediction calculated from eqs (5.66), (4.68), (4.113).

The behavior of $\psi_*(r,\theta)$, $j(r,\theta)$ is very similar to the initial phase in the nonlinear evolution of the resistive kink mode (section 6.6.2). As mentioned above, the relaxed state of the ideal internal kink mode discussed here has no proper physical meaning, since the singular sheet current gives rise to magnetic reconnection, however small the value of the resistivity, the latter determining only the time scale of the subsequent resistive relaxation.

5.4 The small-amplitude nonlinear behavior of the tearing mode

The tearing instability presented in section 4.7.1 is one of the most important instabilities in plasma physics, frequently invoked to explain dynamic processes in both laboratory and astrophysical systems. It is, however, not an instability in the usual sense. In the classical picture of instability evolution a perturbation grows exponentially at essentially the linear growth rate up to some finite amplitude, where nonlinear effects lead to saturation which is either a new equilibrium or some quasi-stationary dynamic, possibly turbulent state. The practical advantage of such instability behavior is that it allows us to estimate observable dynamic time scales, which necessarily relate to nonlinear processes, in terms of the linear growth rate, which can be computed rather easily and reliably. The picture no longer applies when the growth rate is modified long before saturation levels are reached, as is the case for the tearing instability. In fact, nonlinear effects start to reduce the growth rate at practically unobservably small amplitudes in such a way that the further growth is no longer exponential but algebraic. In this section we derive the small-amplitude nonlinear growth law for the tearing mode for both the usual low-β case and the case of finite β and curved field lines. The discussion of the nonlinear behavior of the resistive kink mode where a different algebraic law is found is postponed to section 6.6.2.

5.4.1 Standard low-β case

The basic nonlinear theory of the tearing mode has been developed by Rutherford (1973). Let us first derive an estimate of the amplitude, at which the exponential growth is modified, by considering the quasi-linear equations for the change of the equilibrium profile $\delta\psi_0 = \psi_0(x,t) - \psi_0(x,0)$ and the kinetic perturbation $\tilde{\phi}$ in terms of the magnetic perturbation $\tilde{\psi}$. From the incompressible 2-D MHD equations

$$(\partial_t + \mathbf{v} \cdot \nabla)\psi = \eta j - E , \qquad (5.67)$$

$$\rho(\partial_t + \mathbf{v} \cdot \nabla)\nabla^2\phi = \mathbf{B} \cdot \nabla j , \qquad (5.68)$$

one obtains

$$\partial_t \delta \psi_0 - \eta \delta j_0 = -\langle \widetilde{\mathbf{v}} \cdot \nabla \widetilde{\psi} \rangle \, , \tag{5.69}$$

$$\rho \partial_t \nabla^2 \widetilde{\phi} - \widetilde{\mathbf{B}} \cdot \nabla \delta j_0 = \widetilde{\mathbf{B}} \cdot \nabla j_0 + \mathbf{B}_0 \cdot \nabla \widetilde{j} \, , \tag{5.70}$$

which is the quasi-linear extension of the linearized equations (4.85), (4.86). E is the constant applied electric field introduced in eq. (5.67) to ensure resistive equilibrium, $E = \eta j_0(0)$. To point out the inertia effect the mass density ρ, usually assumed unity, has been written explicitly. With $\widetilde{\psi} = \psi_1(x, t) \cos ky$, $\widetilde{\phi} = \phi_1(x, t) \sin ky$, $\nabla^2 \widetilde{\phi} \simeq \partial_x^2 \phi_1 \sin ky$, and the constant-$\psi_1$ approximation $\partial_x \ln \psi_1 \ll \partial_x \ln \phi_1$ one obtains in the vicinity of the resonant surface

$$\partial_t \delta \psi_0 - \eta \delta j_0 = k \psi_1 \partial_x \phi_1 / 2 \tag{5.71}$$

$$\rho \partial_t \partial_x^2 \phi_1 - k \psi_1 \partial_x \delta j_0 = k j_0' \psi_1 - x k B_0' j_1 \, . \tag{5.72}$$

Since the diffusion time across the resistive layer δ, δ^2/η, is faster than the growth time γ^{-1}, $\gamma \delta^2/\eta = O(\eta^{2/5})$, the first term on the l.h.s. in eq. (5.71) is negligible and δj_0 can be obtained explicitly. Inserting the result into eq. (5.72) gives

$$\left(\rho \partial_t + \frac{k^2 \psi_1^2}{2\eta} \right) \partial_x^2 \phi_1 = k j_0' \psi_1 - x k B_0' j_1 \, . \tag{5.73}$$

The amplitude ψ_{1c}, above which the nonlinear term on the l.h.s. dominates over the inertia term, is easily estimated. Using expressions (4.90) and (4.91) for γ and δ one obtains $\psi_{1c}^2 \simeq \delta^4 B_0'^2$, or with the definition of the island width $w = 4\sqrt{\psi_1/B_0'}$, eq. (4.84),

$$w_c = 4\sqrt{\psi_{1c}/B_0'} \sim \delta \, . \tag{5.74}$$

Hence for amplitudes such that the magnetic island width exceeds the resistive layer width inertia becomes negligible, and the equation of motion degenerates to the equilibrium equation $\mathbf{B} \cdot \nabla j = 0$, i.e. $j = j(\psi)$, following Ampère's law:

$$\nabla^2 \psi = j(\psi, t) \, . \tag{5.75}$$

However, this does not imply that the velocity can be completely neglected. Instead it is now determined by the diffusion process and the incompressibility condition in such a way that j remains a flux function, $j = j(\psi)$. The time evolution of the magnetic perturbation for $w > w_c$ is calculated from eq. (5.67) averaged over flux surfaces ψ,

$$\langle \partial_t \psi \rangle_\psi = \eta j(\psi) - E \, , \tag{5.76}$$

where

$$\langle f \rangle_\psi = \oint f \frac{dl}{|\nabla \psi|} \Big/ \oint \frac{dl}{|\nabla \psi|} . \tag{5.77}$$

In general, eqs (5.75), (5.76) must be solved numerically, an elegant numerical method having been developed by Grad et al. (1975). An analytic solution can, however, be obtained in the small-amplitude limit, where ψ assumes the approximate form

$$\psi(x, y, t) = \psi_0(x) + \psi_1(t) \cos y , \tag{5.78}$$

$$\psi_0(x) = \psi_0'' x^2/2 , \quad \psi_0'' > 0 .$$

Neglecting the x-dependence of ψ_1 in the vicinity of the island reflects the constant-ψ property of the tearing mode valid for small island width, which should, however, exceed the threshold given by eq. (5.74). Equation (5.76) is solved by matching the solution asymptotically to the linear solution in the outer region,

$$\frac{1}{\pi} \int_{-\infty}^{+\infty} dx \int_{-\pi}^{\pi} dy \cos y \, j(\psi) = d\psi_1/dx|_{-\infty}^{+\infty} = \Delta' \psi_1 . \tag{5.79}$$

The fact that higher harmonics of ψ are neglected in eq. (5.78) does *not* imply that these are also negligible in j, i.e. j cannot be computed from the approximate form of ψ. But in the asymptotic matching (5.79) the first harmonic is sufficient to provide the small-amplitude solution $\psi_1(t)$. To evaluate integrals such as

$$\int_{-\infty}^{+\infty} dx \int_{-\pi}^{\pi} dy f = \int_{\psi_{min}}^{\infty} d\psi \oint f \frac{dl}{|\nabla \psi|} , \tag{5.80}$$

use is made of the small-amplitude approximation eq. (5.78)

$$|\nabla \psi| \simeq |x| \psi_0'' = \sqrt{2\psi_0''} \sqrt{\psi - \psi_1 \cos y} , \tag{5.81}$$

so that for instance

$$\oint \frac{\cos y \, dl}{\sqrt{\psi - \psi_1 \cos y}} \simeq 4 \int_0^{y_{max}} \frac{\cos y \, dy}{\sqrt{\psi - \psi_1 \cos y}} , \tag{5.82}$$

where

$$y_{max} = \begin{cases} \pi & u > 1 \\ \arccos u & 1 > u > -1 , \end{cases} \tag{5.83}$$

$$u = \psi/\psi_1 .$$

The way of integration in eq. (5.80) is illustrated in Fig. 5.7. Equation (5.76), integrated over space, now assumes the following form, using eq. (5.79),

$$\frac{4}{\pi} A \sqrt{\frac{\psi_1}{2\psi_0''}} \dot{\psi}_1 = \eta \Delta' \psi_1 \tag{5.84}$$

where

$$A = \int_{-1}^{\infty} du \left(\int_0^{y_{max}} \frac{\cos y \, dy}{\sqrt{u - \cos y}} \right)^2 \bigg/ \int_0^{y_{max}} \frac{dy}{\sqrt{u - \cos y}} \tag{5.85}$$

$$= 2 \int_1^{\infty} du \frac{\left(uK\left(\frac{2}{u+1}\right) - (u+1)E\left(\frac{2}{u+1}\right) \right)^2}{\sqrt{u+1}\, K\left(\frac{2}{u+1}\right)}$$

$$+ \sqrt{2} \int_{-1}^1 du \frac{\left(K\left(\frac{u+1}{2}\right) - 2E\left(\frac{u+1}{2}\right) \right)^2}{K\left(\frac{u+1}{2}\right)} ,$$

and K, E are the complete elliptic integrals (see e.g. Abramowitz & Stegun, 1965). Numerical evaluation gives

$$A \simeq 1.827 .$$

Introducing the island size $w = 4\sqrt{\psi_1/\psi_0''}$, one finally obtains

$$\dot{w} = \frac{\pi}{2} \frac{\sqrt{2}}{A} \eta \Delta' = 1.22 \eta \Delta' . \tag{5.86}$$

Hence for $w_c < w \ll w_s$, where w_s is the saturation island size, the island grows linearly in time $w \propto t$. The physical process governing the growth in this phase is the diffusive broadening of the current perturbation j_1, which can be seen from the following simple argument. We have approximately $j_1 \simeq \psi_1 \Delta'/d$, where $d \simeq \delta$, the resistive layer width, for ψ_1 in the linear instability regime $w < \delta$, and $d \simeq w$ for $w > \delta$, since the current perturbation is concentrated within the island, falling off rapidly outside the separatrix. Hence ψ_1 follows the simplified diffusion equation

$$\dot{\psi}_1 = \eta j_1 \simeq \eta \frac{\Delta'}{w} \psi_1 , \tag{5.87}$$

which is essentially identical with eq. (5.86) using the definition of w.

The numerical factor in eq. (5.86) varies somewhat in the literature. The value in the original paper by Rutherford (1973) is close to unity, but cannot be verified or falsified owing to the imprecise definition of the respective integrals as given there. In a later investigation extending the small-amplitude result (5.86) to the island saturation regime (White et al., 1977), the factor obtained was 1.66, and it was claimed that a more careful evaluation of the integrals in Rutherford's theory gives $\pi/2$

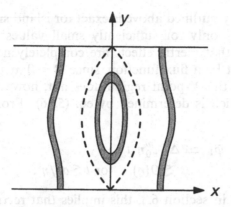

Fig. 5.7. Illustration of the space integration in eq. (5.80).

instead of unity. However, in view of the relative complexity of these integrals such a simple exact result is not very probable. The evaluation of Rutherford's theory has therefore been repeated as given above, resulting in the intermediate value 1.22, which also agrees with the number given by Somon (1984).

Within the framework of the approximation of $\psi(x, y, t)$, eq. (5.78), we can also calculate from eq. (5.76) the current distribution $\delta j(\psi)$,

$$\delta j(\psi) = j(\psi) - E/\eta = \dot{\psi}_1 \langle \cos y \rangle_\psi / \eta \qquad (5.88)$$

where

$$\langle \cos y \rangle_\psi = \begin{cases} u - (u+1) \dfrac{E\left(\frac{2}{u+1}\right)}{K\left(\frac{2}{u+1}\right)} & u > 1 \\[4mm] 1 - 2\dfrac{E\left(\frac{u+1}{2}\right)}{K\left(\frac{u+1}{2}\right)} & -1 < u < 1, \end{cases} \qquad (5.89)$$

$$u = \psi/\psi_1 .$$

$\delta j(\psi)$ has a cusp-like singularity at the separatrix $\psi = \psi_s = \psi_1$, i.e. $u = 1$. Considering the elliptic integrals E, K in the vicinity of this point we obtain the behavior

$$\langle \cos y \rangle_\psi \simeq 1 + \frac{4}{\ln|1-u|} , \quad |1-u| \ll 1 , \qquad (5.90)$$

which shows that $\delta j(\psi_s)$ is finite, but that the derivative diverges from both sides. (We should, however, note the logarithmic character of the singularity, hence $|1 - u|$ has to be very small for eq. (5.90) to be valid.)

Rutherford's theory outlined above is exact for island size $w_c < w \ll w_s$, a range which exists only for sufficiently small values of η. However, this does not imply that inertia effects are completely negligible. On the separatrix j need not be a flux function, since $\mathbf{B} = 0$ at the X-point. The detailed behavior in the X-point region does not, however, influence the global evolution, which is determined by eq. (5.76). From eq. (5.86) one finds

$$\dot{\psi}_1 \simeq \Delta'^2 \psi_0'' \eta^2 t \tag{5.91}$$
$$\lesssim O(\eta) \quad \text{for } t \lesssim a^2/\eta \, .$$

As will be discussed in section 6.3, this implies that reconnection is slow and no macroscopic current sheet is formed. The flow in the vicinity of the X-point can be calculated approximately; see eqs (6.33), (6.34).

The theory should be compared with numerical solutions of the full dynamic equations, (5.67), (5.68). Figures 5.8–5.10 illustrate simulations of the tearing mode in a plane sheet pinch with $B_{y0}(x) = \tanh(x/a)$, $k_0 a = 0.5$ (corresponding to $\Delta' a = 3$, see eq. (4.98)), $\eta = \eta_0/j_0(x)$, $\eta_0 = E = 10^{-5}$. The numerical scheme uses finite differences in the x-direction with a sufficiently fine grid and Fourier analysis in the y-direction. Figure 5.8 shows the evolution of the island size. Plotted is the normalized quantity $\dot{w}/\eta\Delta'$ for different values of the maximum mode number M used in the computations. While in the phase of exponential growth, $t \lesssim 1500\,\tau_A$, all curves coincide as expected since only the first harmonic is important, the subsequent behavior depends on M with convergence reached for $M \gtrsim 12$. Hence many Fourier components are required for proper spatial resolution of the current density and the vorticity. After a transition phase the Rutherford regime $\dot{w} \simeq$ const. is clearly established for $t \gtrsim 12\,000\,\tau_A$. The subsequent slow decay of the $\dot{w}$-curves in Fig. 5.8, particularly visible for $M = 1$, reflects the fact that the island size is a finite though still small fraction of the equilibrium scale size, and can be ascribed to the beginning nonlinear decrease of Δ', $\Delta'(w) < \Delta'$, as discussed in section 5.5. Reducing η_0 from 10^{-5} to 10^{-6}, where the Rutherford regime starts at smaller island size (since the resistive layer width is smaller, $\delta \propto \eta^{2/5}$), the corresponding $M = 1$ curve is nearly horizontal. Note that $\dot{w}$ in the simple quasi-linear approximation $M = 1$ agrees qualitatively with the exact solution, differing only by a factor of order one. This behavior is due to the relative smoothness of the configuration, owing to the slow dynamics, and does not hold in more rapid reconnection processes such as the nonlinear resistive kink mode discussed in section 6.6.2.

Figure 5.9a gives a scatter plot of j vs. ψ taken from a simulation with $\eta_0 = 10^{-6}$ and $M = 24$, which indicates that $j = j(\psi)$ in good approximation. The left-hand branch, $\psi_{min} \leq \psi \leq \psi_s$, corresponds to

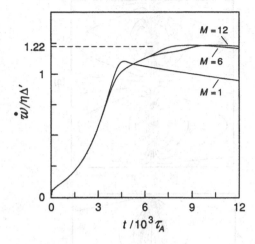

Fig. 5.8. Evolution of the magnetic island size w in a plane sheet pinch with $k_0 a = 0.5$, $\eta_0 = 10^{-5}$. $t \lesssim 1500\,\tau_A$ linear regime, $t \gtrsim 3000\,\tau_A$ Rutherford regime, $\dot{w} = $ const. M indicates the maximum mode numbers used in the computations.

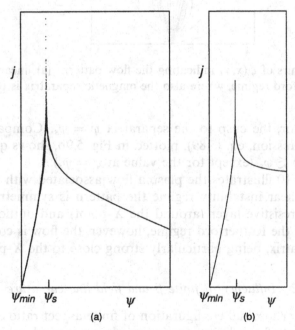

Fig. 5.9. (a) Scatter plot of j vs. ψ for an island state in the Rutherford regime, (b) analytic result $j(\psi)$, eq. (5.89).

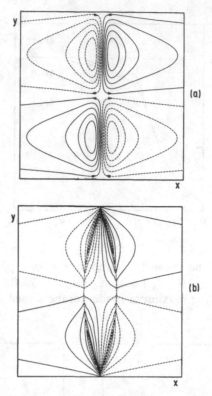

Fig. 5.10.   Contours of $\phi(x, y)$ indicating the flow pattern: (a) linear regime, (b) nonlinear Rutherford regime, where also the magnetic separatrix is plotted.

the island interior, the cusp to the separatrix $\psi = \psi_s$. Comparison with the analytic expression, eq. (5.89), plotted in Fig. 5.9b, shows quantitative agreement for $\psi \lesssim \psi_s$ (except for the value at $\psi = \psi_s$).

Finally Fig. 5.10 illustrates the plasma flow associated with the tearing mode. In the linear instability regime the pattern is symmetric between inflow into the resistive layer (around the X-point) and outflow (around the O-point). In the Rutherford regime, however, the flow is concentrated along the separatrix, being particularly strong close to the X-point.

5.4.2   *Influence of finite β and field line curvature*

In a cylindrical or toroidal configuration of finite aspect ratio and plasma β, the properties of the linear tearing mode are considerably modified compared with the case of large aspect ratio and low β to which the reduced MHD equations apply. In particular, field line curvature and plasma pressure have a stabilizing influence, which gives rise to a finite instability threshold $\Delta'_0 > 0$ (eq. (4.143)). With increasing island size, however, the stabilizing effects become weaker, as has been shown by Somon (1984) in the cylindrical case and by Kotschenreuther et al. (1985) in the toroidal case. The basic mechanism can be demonstrated in a simple plane system where curvature is modelled by adding an effective gravitational force $\mathbf{F}_g = -\kappa p$, $\kappa = \kappa \mathbf{e}_x$ = curvature vector (Fig. 5.11). Since for islands exceeding the resistive layer width inertia is negligible, the tearing mode evolves as a sequence of equilibria satisfying the model equilibrium equation

$$j e_z \times \mathbf{B} - \kappa p \, e_x = 0 . \tag{5.92}$$

On applying the operator $\mathbf{e}_z \cdot \nabla \times$, one obtains

$$\mathbf{B} \cdot \nabla j + \partial_y (\kappa p) = 0 , \tag{5.93}$$

$j \equiv j_z \simeq j_{\parallel}$, generalizing the reduced equilibrium equation $\mathbf{B} \cdot \nabla j = 0$. Assuming $p = p(\psi)$, the solution of eq. (5.93) can be written in the form

$$j = F(\psi) + x \kappa dp/d\psi , \tag{5.94}$$

where $F(\psi)$ is an arbitrary function. Note that j is no longer a flux function. The time evolution is determined by eq. (5.76), which now becomes

$$\langle \partial_t \psi \rangle_\psi = \eta \langle j \rangle_\psi - E .$$

Using this equation, one can eliminate the unknown function $F(\psi)$,

$$j = \frac{1}{\eta} \langle \partial_t \psi \rangle_\psi + (x - \langle x \rangle_\psi) \, \kappa \frac{dp}{d\psi} - \frac{E}{\eta} . \tag{5.95}$$

Now proceed as in the previous section. With the small-amplitude form of $\psi(x, y)$, eq. (5.78), and the asymptotic matching condition, eq. (5.79), one obtains

$$\psi_1 \Delta' = \frac{1}{\pi} \int_{\psi_{min}}^{\infty} d\psi \oint \frac{dl}{|\nabla \psi|} \cos y \left[\frac{1}{\eta} \psi_1 \langle \cos y \rangle_\psi + (x - \langle x \rangle_\psi) \, \kappa \frac{dp}{d\psi} \right] . \tag{5.96}$$

The first term on the r.h.s. is essentially identical with the l.h.s. of eq. (5.84). To evaluate the second term one notes that the contribution from the island region vanishes, since $\langle x \rangle_\psi = \langle x \cos y \rangle_\psi = 0$ for $\psi < \psi_1$ because of the symmetry of the flux surfaces. Outside the separatrix one has

$$\frac{dp}{d\psi} = \frac{dp}{dx_0} \Big/ \frac{d\psi}{dx_0} \begin{cases} < 0 & x < 0 \\ > 0 & x > 0 , \end{cases}$$

where $x_0 = x_0(\psi)$ is defined by $\psi = \frac{1}{2}\psi_0'' x_0^2$. Hence for $\psi > \psi_1$ the contributions of the $x_0 < 0$ and $x_0 > 0$ branches of the flux surface ψ do not cancel but add to each other. Note that

$$\oint \frac{dl}{|\nabla \psi|} x \cos y \simeq \oint dy \cos y = 0 . \tag{5.97}$$

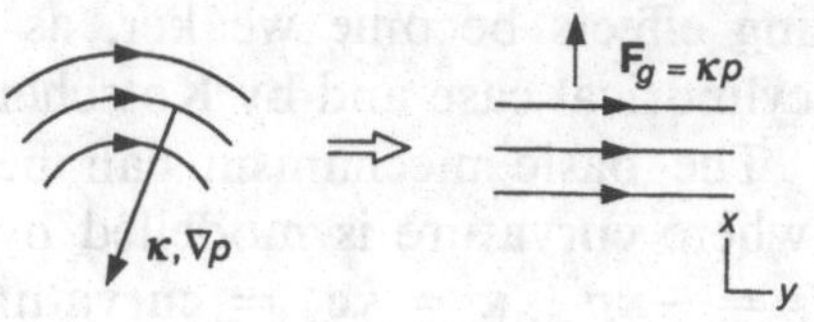

Fig. 5.11. Modelling of the magnetic curvature effect by a gravitational force in a plane system.

With the change of variable $\psi \to u = \psi/\psi_1$ the second term on the r.h.s. of eq. (5.96) becomes

$$-\frac{4}{\pi}\sqrt{\frac{\psi_1}{2\psi_0''}} \int_1^\infty du \frac{\kappa}{\psi_0''} \frac{dp}{dx_0} \frac{\langle x \rangle_\psi}{x_0} \int_0^\pi \frac{\cos y \, dy}{\sqrt{u - \cos y}}, \qquad (5.98)$$

which depends on the p-profile in the vicinity of the separatrix. If one assumes that the pressure gradient $dp/dx_0 = p'$ is constant outside the separatrix, the final equation for the evolution of the island width becomes

$$\dot{w} = \eta \left(1.22\Delta' + 7.53 \frac{\kappa p'}{\psi_0''^2} \frac{1}{w} \right), \qquad (5.99)$$

where the numerical factor in the second term results from the evaluation of the integral in eq. (5.98):

$$\int_1^\infty du \frac{\langle x \rangle_\psi}{x_0} \int_0^\pi \frac{\cos y \, dy}{\sqrt{u - \cos y}}$$

$$= \pi \int_1^\infty \left[\frac{du}{\sqrt{u}} \int_0^\pi \frac{\cos y \, dy}{\sqrt{u - \cos y}} \Big/ \int_0^\pi \frac{dy}{\sqrt{u - \cos y}} \right] \simeq 1.72 \,.$$

For $\kappa p' > 0$ the second term on the r.h.s. of eq. (5.99) is destabilizing, corresponding to a "tearing-interchange" mode, while for $\kappa p' < 0$ it is stabilizing. It gives rise to a finite threshold of Δ' for the linear tearing mode, which can be estimated by setting $w \sim \delta$ in eq. (5.99), $\Delta' \sim \kappa p'/(\psi_0''^2\delta)$. The stabilizing effect, however, decreases with increasing w. Hence if the initial perturbation produces an island exceeding the width $w_* \sim \kappa p'/\psi_0''^2\Delta'$, the stabilizing effect is switched off and the perturbation will grow in time.

Let us relate the result to cylindrical and toroidal geometry, where $\psi_0''^2 = [rd(B_\theta/r)/dr]^2 = (B_\theta/r)^2 s^2$, evaluated at the resonant surface r_s, and s = shear parameter, eq. (3.10). In the cylindrical case $\kappa_c = -(B_\theta^2/B_0^2)\mathbf{e}_r/r$ and hence $p' > 0$, a pressure profile inverted locally at the resonant surface, is required in order to increase tearing mode stability. (A rigorous evaluation obtained by Somon (1984) gives an additional factor of 2, $\kappa p'/\psi_0''^2 \to -2rp'/B_0^2s^2$.) In the toroidal case the field-line-averaged curvature is positive for $q > 1$ with an approximate value $\kappa_n \simeq \kappa_c(1 - q^2)$ as included in Mercier's criterion eq. (4.73). Hence for $q > 1$ there is increased tearing mode stability for a normal pressure profile $p' < 0$. The critical island width w_*, above which the stabilizing effect is negligible, is usually rather small,

$$w_*/a \sim \beta \, .$$

Hence only for tearing modes with small saturation island width $w_s < w_*$, where w_s is computed ignoring the curvature effect, would this effect lead to stability, while for more robust modes with larger islands, $w_s \gtrsim 0.1$, the curvature effect becomes weak.

5.5 Saturation of the tearing mode

Although the nonlinear growth of the tearing mode as given by eq. (5.86) or (5.99) occurs formally on the global resistive time scale $\tau_\eta = \eta/a^2$, the actual saturation time $\sim \tau_\eta w^2/a^2$ is considerably shorter, since the typical island size is $w/a \sim 0.1$. Hence in general the saturated islands adjust quasi-instantaneously to the slow changes of equilibrium profiles. In fact the appearance of the tearing mode during equilibrium evolution should be regarded as an equilibrium bifurcation with the helical state having a lower energy than the cylindrical one. From a practical point of view it is therefore more important to determine the saturation amplitude of the tearing mode or the corresponding island width w_s as a function of the equilibrium quantities than to describe the actual dynamic relaxation process.

There is, however, a basic ambiguity in how the two corresponding equilibrium states, the cylindrical and the helical one, should be related, because of the lack of a strict local constraint such as flux conservation. The only tie between these states is the resistivity profile (apart from possible global constraints such as the total current). Hence, starting from the cylindrical state, the relaxation process and the final helical state depend on the behavior of the resistivity during relaxation, i.e. on the thermal transport, assuming a classical resistivity behavior, where $\eta = \eta(T) \propto T^{-3/2}$ is a function of the temperature T (see Braginskii, 1965). Only for sufficiently small islands $w_s/a < 0.1$ is this transport effect weak, while for large islands it is expected to play an important role.

Saturation levels w_s must in general be computed numerically. Numerical simulation by solving the dynamic equations (5.67), (5.68) can in principle provide a quasi-exact result for some finite S-value ($S < 10^6$ to give a number), since in two-dimensional fluid dynamic problems of this kind, viz. nonturbulent quasi-stationary flows, spatial resolution can usually be chosen adequately fine. Several numerical codes have been developed (e.g. White et al., 1977; Biskamp & Welter, 1977; Waddell et al., 1978) to study the nonlinear evolution of tearing modes, considering also the effect of a dynamic resistivity $\eta(\mathbf{x}, t)$.

In practice, however, dynamic simulations are used only to give certain reference points in order to check the accuracy of simpler semi-quantitative methods of calculating w_s, which are often sufficient because of the uncertainties in the corresponding experimental observations. The approximations are centered around the quantity Δ', which measures the energy of the linear tearing mode, eq. (4.140).

5.5.1   *Quasi-linear theory of tearing-mode saturation*

A theory of the saturation of the tearing instability has been developed by White et al. (1977) for sufficiently small saturation island width w_s, which extends Rutherford's theory, eq. (5.86), into the saturation regime,

$$\dot{w} = C\eta \left(\Delta'(w) - \alpha'w \right) .^{\dagger} \qquad (5.100)$$

(Owing to the particular approximations used the coefficient given by White, $C = 1.66$, differs somewhat from the exact value $C = 1.22$ in eq. (5.86).) $\Delta'(w)$ is the finite-amplitude generalization of the linear quantity $\Delta' = \Delta'(0)$,

$$\Delta'(w) = (d\psi_1/dr|_{r_+} - d\psi_1/dr|_{r_-})/\psi_1(r_s) , \qquad (5.101)$$

where $r_\pm = r_s \pm w/2$, and ψ_1 is the solution of the linear outer region equation (4.103) using the initial current profile. The coefficient α' in eq. (5.100) is a rather complicated expression depending primarily on the gradients $d j_0/dr$ and $d\eta/dr$. A simple approximate interpretation of this effect will be given in section 5.5.3. Saturation occurs for

$$\Delta'(w_s) - \alpha'w_s = 0 . \qquad (5.102)$$

Let us first consider the case of a symmetric tearing mode in a plane sheet pinch $B_y(x) = \tanh(x/a)$. From the solution $\psi_1(x)$ given in eq. (4.97) one obtains the expression

$$\Delta'(w) = \frac{2}{a}e^{-kw/2} \left[\frac{1}{ka \cosh^2(w/2a)} - ka - \tanh \frac{w}{2a} \right] , \qquad (5.103)$$

while the coefficient α' is given by White et al. as $\alpha' = (ka)^2 - 0.78$. Figure 5.12 shows the solution of eq. (5.102) (full line) and simulation results (full circles). The agreement is reasonably good for $w/a \lesssim 2$, w

---

† Here the notation α' is used instead of α in the paper by White to avoid confusion with the helical pitch, eq. (5.10).

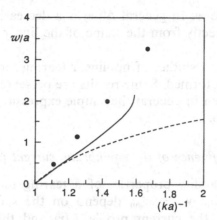

Fig. 5.12. Quasilinear solutions w of a sheet pinch, $\Delta'(w) - \alpha'w = 0$ (full curve), $\Delta'(w) = 0$ (dashed curve), and simulation results (full circles).

increasing roughly linearly with $(ka)^{-1} - 1$, which is somewhat surprising, since the theory makes use of the small island approximation $w/a < 1$. For comparison we have also plotted the result $\Delta'(w_s) = 0$ (dashed line), neglecting the α'-term, which agrees with the full quasi-linear solution of eq. (5.102) for $w/a \lesssim 0.5$, but gives an obviously poorer approximation for higher values of w/a. Incidentally, replacing the island size w in eq. (5.102) by its azimuthal average $\bar{w} = w \int_{-\pi/2}^{\pi/2} \cos y \, dy/\pi = 2w/\pi$, which in a quasi-linear approach appears to be the more appropriate quantity and amounts to multiplying the full curve in Fig. 5.12 by a factor of $\pi/2$, would lead to a perfect match with the simulation points, for $w/a \lesssim 1.5$.

We now turn to the geometrically more complicated case of helical tearing modes in a cylindrical pinch. We note that here the equilibrium scale height is of the order of the resonant radius r_s and geometry enforces $w/r_s < 1$. Hence one expects that the α'-term in eq. (5.102) is small such that the quasi-linear prediction of the saturation island size is given by

$$\Delta'(w_s) = 0 \,, \tag{5.104}$$

which in fact is used in most practical applications (e.g. Carreras et al., 1979a).

The physical meaning of eq. (5.104) can easily be understood. The dominant nonlinear effect is the modification δj_0 of the equilibrium current density, which corresponds to a flattening of the current profile j_0 across the island width, while j_0 is essentially unchanged outside the island region and the total current remains constant. Considering the linear outer region equation (4.103), the flattening of j_0 implies that the solution ψ_1 is determined by integrating the equation only up to the separatrix. $\Delta'(w)$ may be interpreted as the remaining free energy of the tearing mode at the finite amplitude w. In general $\Delta'(w)$ is a decreasing function of w, which can be seen directly from the shape of the linear eigenfunction ψ_1, Fig. 4.5.

Numerous simulation studies of nonlinear tearing modes in cylindrical plasmas have been performed. Some results are presented in the following section. As will be seen in general the simple expression $\Delta'(w_s) = 0$ gives a good approximation.

5.5.2   *Influence of the equilibrium current profile*

In a tokamak the stability properties of a tearing mode (m, n) and its saturation island width $w_s = w_{mn}$ depend on the safety factor profile $q(r)$ or, equivalently, the current profile $j_0(r)$ and the position of the resonant radius r_s, $q(r_s) = m/n$. We assume a fixed resistivity profile $\eta(r) = \eta_0/j_0(r, t = 0)$, ignoring for the moment the effect of a dynamic resistivity response, which will be treated in the subsequent section 5.5.3.

Let us first discuss the general behavior for smooth, bell-shaped current profiles. Consider a class of profiles introduced by Furth et al. (1973):

$$q(r) = q_0 \left(1 + \left(\frac{r}{r_0} \right)^{2\nu} \right)^{1/\nu} \tag{5.105}$$

or equivalently

$$j_0(r) = \frac{j_0}{\left(1 + \left(\frac{r}{r_0} \right)^{2\nu} \right)^{1+1/\nu}} . \tag{5.106}$$

By solving eq. (4.103) numerically, $\Delta'(w)$ and, in particular, w_{mn} can be determined for any position of r_s. Results are summarized in Fig. 5.13, showing $j_0(r)$ with $r_0 = 0.5$ and the saturation island width of the $(2, 1)$ mode, normalized to its resonant radius r_s, $w_{21}(r_s)/r_s$, as a function of r_s, for (a) $\nu = 1$; (b) $\nu = 2$; (c) $\nu = 4$. The main feature is that w_{21}/r_s is roughly constant or $w_{21} \propto r_s$ for r_s inside the current channel and then falls off the more abruptly the sharper the decay of the current profile. The maximum value of w_{21} is reached for r_s just outside the maximum gradient of the current density. Since this point is located at larger radii for more square-shaped profiles and since $w \propto r_s$, the maximum island size is larger for a square-shaped than for a peaked profile.

As seen in Fig. 5.13, the simple quasi-linear prediction $\Delta'(w) = 0$ agrees surprisingly well with the exact simulation results, indicated by full circles, even for what appear to be large islands. A discrepancy is found, however, for small r_s in the central flat part of the current profiles, where quasi-linear theory predicts constant normalized island size $w/r_s \simeq 0.35$, while

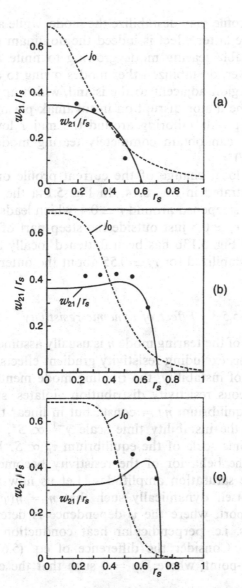

Fig. 5.13. Normalized island width $w_{21}(r_s)/r_s$ for (a) $v = 1$ ("peaked profile"); (b) $v = 2$ ("rounded profile"); (c) $v = 4$ ("flat profile"). Quasi-linear result $\Delta'(w) = 0$ (full line), simulation results (full circles), current profiles (dashed lines).

simulations show that islands grow toward the inside down to the center $r = 0$, resulting in $w/r_s \to 1$. Such behavior is related to the vacuum bubbles in a shearless plasma column (section 5.2.2).

The stability of the tearing mode depends rather sensitively on the current density gradient in the vicinity of the resonant surface. A local steepening of the profile may destabilize the mode, while a flattening leads to stabilization. The latter effect is indeed the dominant self-stabilization process of an unstable tearing mode growing to finite amplitude. This process may, however, destabilize other modes owing to a current profile steepening in the region adjacent to the island, which appears to play an important role in the major disruption in tokamak plasmas (section 8.2). By careful current profile tailoring around the major low-m-number rational surfaces one can obtain completely tearing mode stable profiles (Glasser et al., 1977).

The effects of a local change of the current profile on the saturation island size are illustrated in Fig. 5.14. In Fig. 5.14a the $v = 1$ profile of Fig. 5.13a is locally steepened around $r \simeq 0.4$, which leads to a substantial increase of w_{21} for $r_s \simeq 0.5$ just outside the steep part of j. In Fig. 5.14b the $v = 2$ profile of Fig. 5.13b has been flattened locally for $r \simeq 0.5$. As a result the mode is stabilized for $r_s \simeq 0.55$ about the outer edge of the flat part.

5.5.3 Effect of dynamic resistivity

In the linear theory of the tearing mode η is usually assumed homogeneous and constant in time, excluding resistivity gradient effects, which give rise to a different type of instability, the rippling mode mentioned in section 4.7.3. A homogeneous resistivity distribution violates, strictly speaking, the condition for equilibrium $\eta j = $ const., but in linear theory this effect is negligible, since the instability time scale $\gamma^{-1} \propto S^v$, $v < 1$, is faster than the resistive time scale of the equilibrium $\tau_\eta \propto S$. In the nonlinear regime, however, the behavior of the resistivity becomes important, in particular for large saturation amplitudes. Let us now assume that the resistivity adjusts itself dynamically such that $\eta = \eta(\psi)$ owing to fast parallel heat transport, where the ψ-dependence is determined by slow transport processes, i.e. perpendicular heat conduction and local heat sources and sinks. Consider the difference of eq. (5.67) taken at the X-point and the O-point, where $\nabla\psi = 0$ such that the convective terms vanish:

$$\frac{d}{dt}(\psi_X - \psi_O) = \eta_X j_X - \eta_O j_O . \tag{5.107}$$

Since $\psi_X - \psi_O = 2\psi_1$, $j_X - j_O \simeq 2\psi_1 \Delta'/w$ and $\eta_X - \eta_O \simeq 2\psi_1 d\eta/d\psi$, one obtains approximately

$$\dot{w} \simeq \eta\Delta' + j\frac{d\eta}{d\psi}w . \tag{5.108}$$

The $d\eta/d\psi$-term, which is related to the α'-term in eq. (5.100), represents the effect of a dynamic resistivity behavior.

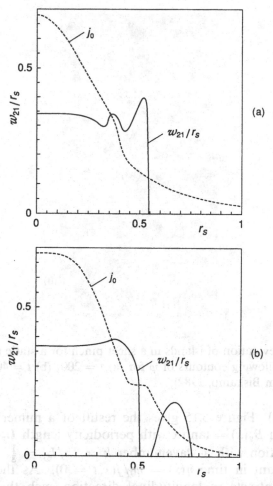

Fig. 5.14. Effect of local profile modification. Quasi-linear island size $w_{21}(r_s)/r_s$ (full line), current profile (dashed line): (a) peaked profile, steepened at $r \simeq 0.4$; (b) rounded profile, flattened at $r \simeq 0.5$.

Let us first discuss the case of tearing modes in a *plane sheet pinch*. The physical meaning of eq. (5.107) can easily be understood. The change of the magnetic flux in the island $\psi_X - \psi_O$ is determined by the difference between the flux injection rate $\eta_X j_X$ due to reconnection of outside flux at the X-point and the dissipation rate $\eta_O j_O$ due to resistive diffusion in the island. Lowering η_O with respect to η_X, $d\eta/d\psi > 0$ (island heating), leads to an enhanced growth and larger saturation island size, while increasing η_O (island cooling) slows down the growth and reduces the saturation width.

Numerical simulations have been performed comparing the behavior for a static resistivity profile $\eta(x)$ with that of a dynamic resistivity model

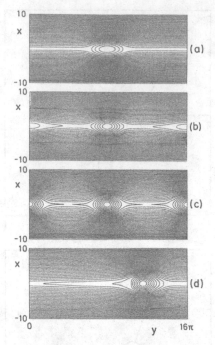

Fig. 5.15.   Time evolution of islands in a sheet pinch for a static resistivity profile $\eta(x) = \eta_0/j_0(x)$, showing contours of ψ for (a) $t = 200$; (b) $t = 400$; (c) $t = 1000$; (d) $t = 8000$ (from Biskamp, 1982).

(Biskamp, 1982). Figure 5.15 gives the result of a numerical simulation of a sheet pinch $B_y(x) = \tanh x$ with periodicity length $L_y = 16\pi$ and an initial perturbation with a wavenumber $k = 2\pi/L_y = \frac{1}{8}$. The resistivity profile is constant in time $\eta(x) = \eta_0/j(x, t = 0)$. As the island grows in width, it contracts in longitudinal direction such that the effective wavenumber becomes $k_{eff} \simeq 0.5$, Fig. 5.15a. The remaining extended sheet is, however, still tearing unstable leading to growth of a second similar island, Fig. 5.15b, c. Both finally coalesce into a single non-symmetric one which moves with constant speed; Fig. 5.15d. Strong parallel flows prevent further tearing. (Note that the resistivity $\eta(x)$ does not allow static two-dimensional equilibria since $\eta(x)j(\psi) \neq$ const.) The stabilizing effect of parallel flows and the process of island coalescence are discussed in more detail in sections 6.5 and 6.6.1. The islands tend to have an approximately circular internal shape corresponding to the natural shape of a plasma pinch.

In the case of a dynamic resistivity η was computed using a convection diffusion equation,

$$\partial_t \eta + \mathbf{v} \cdot \nabla \eta = \kappa_\parallel \nabla_\parallel^2 \eta + \kappa_\perp \nabla_\perp^2 \eta \; ,$$

with $\kappa_\parallel \gg \kappa_\perp$, which gives rise to $\eta \simeq \eta(\psi)$. In this case a completely different asymptotic behavior is found. Since plasma of high resistivity is convected from outside into the X-point region, $\eta_X > \eta_O$ and the current density is concentrating more and more strongly in the island center. The final state consists of a single island of maximum possible wavelength $\lambda = L_y$ and self-similar shape $w_s/\lambda \simeq 0.4$. Obviously islands are much wider than in the case of a static resistivity with $\eta_X = \eta_O$. The normalized island width w_s/L_y corresponds roughly to the asymptotic value extrapolated for small k from simulation points in Fig. 5.12. The final configurations are practically in static resistive equilibrium $\eta(\psi)j(\psi) \simeq$ const. Owing to the self-consistent change of the resistivity profile the initial sheet pinch is completely disrupted. For large L_y the final state approximately consists of a line current flowing at the island center. In this case the solution can be given analytically:

$$\psi(x, y) = \frac{L_y}{2\pi} \ln \left(\cosh \frac{2\pi x}{L_y} + \cos \frac{2\pi y}{L_y} \right), \qquad (5.109)$$

corresponding to the special case $c \to \infty$ of the class of solutions (3.57) of the equilibrium equation (3.53).

We now consider the effect of a dynamic resistivity in the case of *helical tearing modes* in a cylindrical plasma which is of considerable interest in tokamak physics. The corresponding reduced MHD equations (5.67), (5.68) are written in terms of the helical flux function ψ_*, as in eqs (5.14), (5.15). Equation (5.107) is still valid in ψ_*, but since $\nabla^2 \psi_* = j - 2\alpha B_0$, one has $\psi_{*0}''(r_s) < 0$ for an increasing q-profile, since $\psi_{*0}'(r) = B_{\theta 0}(1 - nq/m)$, assuming j_0 to be positive. Hence ψ_* has a maximum at the O-point, in particular $\psi_{*X} - \psi_{*O} < 0$, such that $(d/dt)(\psi_{*X} - \psi_{*O}) > 0$ implies *shrinking* of the island, and

$$\frac{d\eta}{d\psi_*} < 0 \qquad \text{for } \eta_X > \eta_O .$$

Hence eq. (5.108) leads to the opposite dependence on $\eta_X - \eta_O$ as in the plane case.

Simulations demonstrating this effect are illustrated in Figs 5.16, 5.17. The initial current profile is the rounded profile ($\nu = 2$) with $r_0 = 0.5$, $r_s = 0.65$, and the resistivity distribution $\eta(\psi_*)$ is defined in such a way that the profile along a ray across the X-point remains constant, $\eta(r, \theta_X) = \eta_0/j_0(r, t = 0)$. The resistivity in the island η_w is still free. For $t < 10^3 \tau_A$ we assume a flat distribution $\eta_w(\psi_*) = \eta_X$, the value at the X-point. For $t > 10^3 \tau_A$ the island resistivity is shaped,

$$\eta_w (\psi_*) = \eta_X (1 + \epsilon (\psi_* - \psi_{*X})) .$$

Choosing $\epsilon > 0$, i.e. $\eta_O > \eta_X$ corresponding to island cooling, gives rise

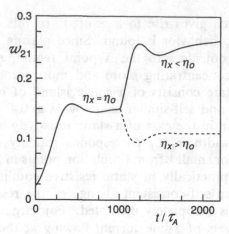

Fig. 5.16.   Island size w_{21} for the rounded ($v = 2$) initial current profile, $r_s = 0.6$, $\eta = \eta(\psi)$. $0 \leq t \leq 10^3 \tau_A$: $\eta = \eta_X$ in the island interior. $t > 10^3 \tau_A$: $\eta_O > \eta_X$ (full curve), $\eta_O < \eta_X$ (dashed curve).

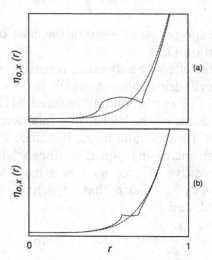

Fig. 5.17.   Radial η-profiles across O-point (full line) and X-point (dashed line) for the final states in Fig. 5.16. (a) $\eta_O > \eta_X$, (b) $\eta_O < \eta_X$.

to larger islands (the upper branch in Fig. 5.16), whereas $\epsilon < 0$, i.e. $\eta_O < \eta_X$, corresponding to island heating, leads to smaller islands (the lower, dashed branch). This behavior was first observed by Biskamp & Welter (1977) and White et al. (1977). The corresponding η-profiles are plotted in Fig. 5.17. One can see that a relatively small change of η_w (~ 10 per cent) has a rather dramatic effect on the island size. The effect of the island resistivity profile is associated with the growth of the $m = 2$ precursor oscillation in a disruption (section 8.2.3).

6

Magnetic reconnection

There is hardly a term in plasma physics exhibiting more scents, facets and also ambiguities than does magnetic reconnection or, simply, reconnection. It is even sometimes used with a touch of magic. The basic picture underlying the idea of reconnection is that of two field lines (thin flux tubes, properly speaking) being carried along with the fluid owing to the property of flux conservation until they come close together at some point, where by the effect of finite resistivity they are cut and reconnected in a different way. Though this is a localized process, it may fundamentally change the global field line connection as indicated in Fig. 6.1, permitting fluid motions which would be inhibited in the absence of such local decoupling of fluid and magnetic field. Almost all nonlinear processes in magnetized conducting fluids involve reconnection, which may be called the essence of nonlinear MHD.

Because of the omnipresence of finite resistivity in real systems resistive diffusion takes place everywhere in the plasma, though usually at a slow rate. Reconnection theory is concerned with the problem of *fast* reconnection in order to explain how in certain dynamic processes very small values of the resistivity allow the rapid release of a large amount of free magnetic energy, as observed for instance in tokamak disruptions or solar flares. By fast we mean faster than the resistive time τ_η associated with the average gradient scale L, $\tau_\eta/\tau_A = S \equiv \eta^{-1}$. Hence a fast process is defined to occur with a time rate $\sim O(\eta^\nu)$ $0 \leq \nu < 1$. (In the literature this attribute is sometimes reserved to the special case $\nu = 0$.)

Fast reconnection is basically a localized process. The fundamental structure is a current sheet, the basic dynamics of which is introduced in section 6.1.

Conventional reconnection theory has been focused on 2-D stationary magnetic configurations. It is mainly associated with either one of two different lines of thought, Petschek's slow shock model and Syrovatskii's current sheet model, which are discussed in section 6.2. Since these theories deal effectively only with the ideal outer region, ignoring the so-called diffusion region, the narrow layer where resistivity is important, we call them quasi-ideal models. While Petschek's theory is now known to

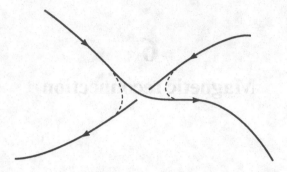

Fig. 6.1. Change of field line topology due to reconnection.

be incorrect in the limit of small η, providing at most a phenomenological reconnection model in the presence of anomalous, i.e. locally enhanced, resistivity, current sheets by contrast represent a fundamental feature of weakly resistive MHD, and Syrovatskii's theory accounts for many properties observed in dynamic resistive MHD systems.

In section 6.3 we summarize the scaling properties of stationary current sheet configurations as observed in numerical simulations.

Section 6.4 gives a refined theory of dynamical current sheets. We first investigate the properties in the central part, the vicinity of the "X-point", and then discuss the rather complicated behavior in the edge region.

The question of the tearing stability of such current sheets naturally arises. As numerical simulations indicate, dynamic current sheets are considerably more stable than static sheets, allowing a much larger aspect ratio A = sheet width/thickness. A qualitative theory of the stability threshold is given in section 6.5.

Section 6.6 presents examples of two-dimensional reconnecting systems, the coalescence of magnetic islands, the nonlinear evolution of the resistive kink mode, and the process of plasmoid formation.

In section 6.7 we consider some generalizations of the reconnection concept arising in fully three-dimensional systems.

Finally in section 6.8 the asymptotic behavior in the large Reynolds number limit is discussed, where time-dependent, presumably turbulent reconnection processes prevail.

6.1   Current sheets: basic properties

6.1.1   *Sweet-Parker current sheet model*

If fluid volumes carrying oppositely directed magnetic fields are pushed together the fluid is squeezed sideways along the field and the fields ap-

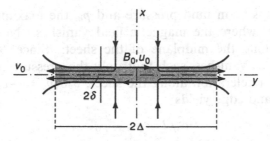

Fig. 6.2. Dynamic (Sweet-Parker) current sheet.

proach each other, until resistive diffusion becomes important. The result is a quasi-stationary dynamic current sheet, called a Sweet-Parker current sheet (Sweet, 1958; Parker, 1963), which is the simplest reconnection configuration (see Fig. 6.2). As usual we assume incompressible motions with homogeneous mass density, $\rho = 1$. The quasi-one-dimensional configuration is characterized by six parameters. These consist of three dynamic quantities: the magnetic field B_0 immediately outside the sheet called the upstream field (the downstream field at the sheet edges is small in this quasi-one-dimensional configuration), the upstream flow u_0 perpendicular to the field and the downstream flow v_0 along the field; two geometric quantities: the width* Δ and the thickness δ; and finally the resistivity η.

These quantities are connected by three relations derived from the continuity equation, Ohm's law, and the equation of motion, assuming stationarity. Integrating the continuity equation $\nabla \cdot \rho \mathbf{v} = 0$ over the inflow and outflow surfaces gives

$$u_0 \Delta = v_0 \delta . \tag{6.1}$$

Consider Ohm's law along the x-axis. Stationarity requires $E_z = \text{const}$. In the upstream region outside the sheet, where the current density is small, the resistive term in Ohm's law is negligible, while in the current sheet center, where $j = j_m$ is large and the velocity vanishes, the resistive term dominates, which gives the relation

$$u_0 B_0 = \eta j_m \simeq \eta \frac{B_0}{\delta} . \tag{6.2}$$

Since usually $u_0 \ll B_0$, as will be seen a posteriori, the inertia term is negligible in the force balance across the sheet. Hence $\partial_x \left(p + B^2/2 \right) = 0$, which gives

$$B_0^2/2 = p_m - p_0 . \tag{6.3}$$

---

* In this chapter we use the terminology common in reconnection theory calling Δ the current sheet width instead of length (the length should be measured in the third direction z).

Here p_0 is the upstream fluid pressure and p_m the maximum pressure in the sheet center, where the magnetic field vanishes. Now consider the force balance along the midplane of the sheet. Since B_x is negligible, the magnetic force vanishes, such that only the pressure force is present, leading to fluid acceleration along the sheet, $v_y \partial_y v_y = -\partial_y p$. Integration between center and edge yields

$$v_0^2/2 = p_m - p_0 . \qquad (6.4)$$

Here the current sheet edge $y = \Delta$ is *defined* by the vanishing of the pressure difference across the sheet. In reality the edge region has a complicated structure, as is shown in section 6.3, where a more quantitative description of a current sheet is given. Replacing $p_m - p_0$ by $B_0^2/2$ we obtain the important result that the downstream flow velocity equals the upstream Alfvén speed,

$$v_0 = B_0 = v_A . \qquad (6.5)$$

Relations (6.1) and (6.2) can be used to express two of the remaining variables by the other three. Choosing the latter as B_0, Δ, η one obtains

$$\frac{u_0}{v_A} \equiv M_0 = \left(\frac{\eta}{B_0 \Delta} \right)^{1/2} \equiv S_0^{-1/2} , \qquad (6.6)$$

$$\frac{\delta}{\Delta} \equiv A^{-1} = S_0^{-1/2} . \qquad (6.7)$$

Here we have introduced the Mach-number M_0, which is conventionally used as a dimensionless measure of the reconnection rate, implying steady state conditions at least locally in the sheet such that $M_0 \propto \eta j_m$[†]. A is called the aspect ratio and S_0 is the Lundquist number of the current sheet. The Sweet-Parker reconnection rate $M_0 = S_0^{-1/2}$ is a characteristic quantity for a current sheet. If B_0 and Δ are of the order of the global field intensity and spatial scale of the magnetic configuration, S_0 is the global Lundquist number, which in most cases of practical interest is very large, typically 10^{10} in the solar corona, such that the Sweet-Parker rate would lead to reconnection times in a solar flare many orders of magnitude longer than observed. Obviously a single quasi-stationary current sheet of an aspect ratio $A \sim 10^5$ is a very implausible configuration.

It was in fact in flare theory where the need for a faster reconnection process first became evident. Here Petschek's slow shock mechanism (Petschek, 1964) seemed to provide a simple and elegant solution. This

---

[†] Calling $M = u/B$ the reconnection rate has led to some confusion, since properly speaking the reconnection rate is the rate of flux change at the X-point, $\partial_t \psi_x = \eta j_m$, which equals the product uB for stationary conditions.

configuration, too, contains a current sheet, where reconnection occurs; however, its width Δ is not given by the overall system size but forced by the configuration to be very small $\Delta \sim O(\eta)$. Hence the reconnection rate becomes essentially independent of the resistivity, which is what observations seem to demand. Unfortunately, however, Petschek's model incorporates a fundamental inconsistency invalidating the theory in the most interesting regime of small resistivity, as discussed in section 6.2.2. Instead it is found that in an intermediate resistivity regime reconnection occurs in fact in macroscopic current sheets, in the original sense of the Sweet-Parker reconnection model, while for smaller values of η the reconnection process becomes nonstationary, leading to fully developed turbulence in the limit of $\eta \to 0$.

Current sheets are formed under quite general conditions. In particular the field must not be strictly antiparallel, vanishing at the neutral line, but only a particular component, called the poloidal field, has to change sign. Hence there is in general a finite axial field B_z in the sheet region. Where in a sheared magnetic configuration current sheets form depends on the plasma flow as excited for instance by an MHD instability. Several examples are presented in section 6.6.

A strong axial field $B_z \gg B_y \sim B_0$ provides also a solid justification of the incompressibility assumption, which might otherwise appear doubtful within the sheet, where the fluid velocity becomes large. The criterion for incompressible motion is $v_{\parallel} \ll c_s$ for parallel flow and $v_{\perp} \ll v_A$ for perpendicular flow. In the absence of an axial field, incompressibility is valid only for the case of high β, where $p_0 \gg B_0^2/2$, such that $p_m - p_0 \ll p_0$. High-β plasmas, which can be confined only by gravitation, are not very interesting, however, since the typical plasma properties result from the interaction with magnetic fields. By contrast, a strong axial component, corresponding to the case of magnetically confined plasma with relatively low β, automatically guarantees incompressibility. Even if the flow reaches the Alfvén speed, meaning the poloidal Alfvén speed, eq. (6.5), it is still slow compared with the axial one, and hence constitutes essentially an incompressible $\mathbf{E} \times \mathbf{B}$ convection.

6.1.2   *Effects of hyperresistivity and viscosity*

Besides resistivity further dissipation processes may be important, changing the current sheet scaling laws. Let us first consider hyperresistivity η_2, introduced in section 2.6, which is related to the (anomalous) electron viscosity and may become an important dissipative effect in Ohm's law in a high-temperature plasma. Equation (6.2) is replaced by

$$u_0 B_0 = \eta_2\, \partial_x^2 j_m \simeq \eta_2 \frac{B_0}{\delta^3} \tag{6.8}$$

while the force balance and hence the relation $v_0 = B_0$ remain unchanged. Insertion into the mass conservation relation (6.1) gives the Mach number

$$M_0 = \left(\frac{\eta_2}{B_0 \Delta^3} \right)^{1/4}. \tag{6.9}$$

Much as in the case of linear tearing modes, eqs (4.90a), (4.91a), the η_2-dependence of M_0 is weaker than that on $\eta = \eta_1$, eq. (6.6), making the reconnection time scale less sensitive to the actual value of the transport coefficient.

Fluid viscosity v, corresponding to ion viscosity in a plasma, could also be an important effect, modelling for instance dissipation due to gyro-resonance in a nearly collisionless plasma. The case of finite viscosity has been considered by Park et al. (1984). The force balance along the sheet now becomes $v_y \partial_y v_y - v \partial_x^2 v_y = -\partial_y p$, since $\nabla^2 v_y \simeq \partial_x^2 v_y$. Integrating between the center and the edge of the sheet, using $\int_0^\Delta \partial_x^2 v_y dy \simeq -v_0 \Delta/2\delta^2$ (the minus sign arises from $\partial_x^2 v_y < 0$ for $v_y > 0$) and eq. (6.3), one obtains

$$v_0^2 + (v\Delta/\delta^2)v_0 = B_0^2$$

or, with $v_0 = \eta \Delta/\delta^2$ from eqs (6.1), (6.2),

$$v_0 = B_0 \left(1 + \frac{v}{\eta} \right)^{-1/2} \tag{6.10}$$

and hence

$$M_0 = S_0^{-1/2} \left(1 + \frac{v}{\eta} \right)^{-1/4}, \tag{6.11}$$

$$A = S_0^{1/2} \left(1 + \frac{v}{\eta} \right)^{-1/4}, \tag{6.12}$$

generalizing eqs (6.6), (6.7). Thus the modifications introduced by viscosity are relatively mild. While for $v \lesssim \eta$ its effect is negligible, $v \gg \eta$ results in a broader sheet with reduced inflow and outflow velocities.

6.2   Quasi-ideal models of stationary reconnection

6.2.1   *Driven reconnection*

The concept of driven reconnection plays an important role in conventional reconnection theory. Originally the term refers to externally forced systems in contrast to closed systems where internal reconnection processes occur spontaneously, for instance as a result of some instability. The concept can, however, be applied much more generally. If we assume that the reconnection is localized in space, we may restrict consideration to a small region of linear dimensions L around this location instead of the entire system of size Λ, $L \ll \Lambda$. On the other hand L should be large compared with the scales of the dissipative structures, $L \gg \Delta$ in the case of a single current sheet, so that these are not affected by the artificial boundaries of the subsystem. The main advantages of restricting attention to the subsystem L are that it allows us to simplify the geometry and also to assume stationarity, even for a nonstationary global system. Since the coupling to the latter occurs by the boundary conditions imposed on the subsystem, and these boundary conditions change on the global time scale $\sim \Lambda/v_A$, while the subsystem adjusts to these changes on the much faster time scale $\sim L/v_A$, we may consider the latter in steady state (if such a state exists).

In this sense the subsystem constitutes a stationary reconnection configuration. As shown for instance in Figs 6.3 and 6.5, fluid and magnetic field are injected from above and below, while the fluid leaves the system laterally carrying along the reconnected field. The small region, where dissipation processes, in particular resistivity, are important, is called the *diffusion region*, which is surrounded by the *ideal external region* where dissipation effects are negligible. The interpretation of such a model system has, however, given rise to some confusion. Since the fundamental issue in reconnection theory is to account for rapid processes, a particular figure of merit of a theoretical configuration is a weak dependence of the reconnection rate on the resistivity, or no dependence at all. In stationary driven reconnection this point needs further specification, since the reconnection rate is determined by the boundary conditions for the inflow velocity and magnetic field intensity and hence is independent of η by definition. For the reconnection process to be independent of η one has to require instead, that at fixed boundary conditions the *configuration* remains unchanged if η is varied. A consequence of such behavior would be that the ratio of outflow energy flux to input energy flux should be independent of η (essentially unity) and so should the ratio of energy dissipation rate to input energy flux (essentially zero).

6.2.2 *Petschek's slow shock model*

The model proposed by Petschek (1964) at a symposium on solar flares was almost immediately accepted as a major breakthrough in the theory of reconnection, serving as the basic concept for the following two decades. In fact most papers (at least in the western hemisphere) on the subject of reconnection deal with one or the other variant of Petschek's model, notably the review article by Vasyliunas (1975), or a subsequent review by Forbes & Priest (1987). Only in recent years has the basic inconsistency in the theory become apparent. Because of its historical importance

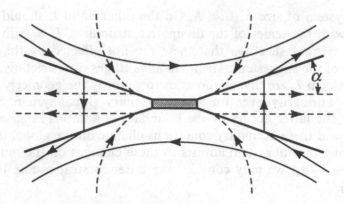

Fig. 6.3.   Schematic drawing of Petschek's reconnection configuration.

Petschek's model is described briefly, before we point out where the theory is in error, both conceptually and formally.

The configuration is illustrated in Fig. 6.3. It is characterized by two pairs of slow mode shocks standing back to back against the upstream flow, deviating it by roughly 90^0 into the downstream cone, where the magnetic field is weak. The current and vorticity are concentrated in the shock fronts and the central diffusion layer, while in the external region velocity and magnetic field are irrotational, $\nabla \times \mathbf{v} = \nabla \times \mathbf{B} = 0$, which together with $\nabla \cdot \mathbf{v} = \nabla \cdot \mathbf{B} = 0$ results in $\nabla^2 \psi = \nabla^2 \phi = 0$. The shocks derive their properties from the slow magnetosonic mode, discussed in section 4.2. This is primarily a longitudinal compressible mode, which at finite amplitude steepens to form a shock. Its pecularity is to survive with finite phase velocity in the incompressible limit, $\omega^2 = k_\parallel^2 v_A^2 = k^2 B_n^2$, where B_n is the component normal to the wave front. Hence for a given flow speed a plasma flow can always become supersonic with respect to this mode if the angle between wavefront and magnetic field is made sufficiently small.

The jump conditions across a slow mode shock in the incompressible limit can easily be derived. Equations $\nabla \cdot \mathbf{v} = \nabla \cdot \mathbf{B} = 0$ give

$$[v_n] = [B_n] = 0, \tag{6.13}$$

where $[f]$ indicates the change across the shock. From the normal component of the equation of motion follows the continuity of the total pressure

$$[p + B^2/2] = 0, \tag{6.14}$$

while from the tangential component follows

$$v_n [v_t] = B_n [B_t] . \tag{6.15}$$

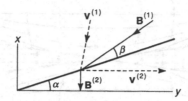

Fig. 6.4. Illustration of **B** and **v** at the Petschek slow shock (heavy line) and definition of angles α (downstream cone) and β, $B_n = B \sin \beta$.

Using these results in Ohm's law for steady-state $\mathbf{v} \times \mathbf{B} = \text{const.}$ yields

$$v_n^2 = B_n^2 . \tag{6.16}$$

If we require that for a given upstream magnetic field $\mathbf{B}^{(1)}$ the velocity and the magnetic field in the downstream cone are homogeneous as indicated in Fig. 6.4, $v_x^{(2)} = B_y^{(2)} = 0$, the angles α, β are determined as well as the tangential component $v_t^{(1)}$ of the upstream velocity. For small Mach number $M \simeq v^{(1)}/B^{(1)} \ll 1$ the downstream cone is narrow $\alpha \ll 1$, $\beta = \alpha$, and $v^{(2)} = B^{(1)}$, the upstream Alfvén speed.

The crucial (and basically wrong, as we shall see) assumption in Petschek's model is that the diffusion region is small, a tiny current sheet of dimensions $\Delta \sim \delta \sim \eta$, adjusting smoothly to the external configuration, where the latter is completely determined by the outer boundary conditions. (If this were true, then by the definition given in section 6.2.1 the reconnection rate would in fact be independent of η.) To determine the ideal external configuration the equations $\nabla^2 \psi = 0$, $\nabla^2 \phi = 0$ are solved in the limit $\Delta \to 0$ using the jump conditions at the shock. Petschek gives an analytical solution in the limit $M \ll 1$. Since the downstream flow equals the upstream Alfvén speed, it follows from mass conservation that $M \simeq \alpha$, the angle of the downstream cone. Petschek obtains the maximum reconnection rate achievable in his model

$$M_{max} \simeq \left(\ln \frac{L}{\Delta} \right)^{-1} \simeq (\ln S)^{-1} . \tag{6.17}$$

This weak S- or η-dependence is simply due to the fact that the magnetic field is weaker in front of the diffusion layer than the asymptotic field, assumed to be homogeneous, which is used in the definition of M in eq. (6.17).

More general solutions are obtained by relaxing the condition of irrotational $\mathbf{v}$ and $\mathbf{B}$, which allows for instance configurations with constant upstream magnetic field, as shown in Fig. 6.5b, with $M_{max} = O(1)$ independent of η. Special similarity solutions of this type, given by Sonnerup (1970) and Yeh & Axford (1970), consist of three regions (per quadrant) of homogeneous velocity and magnetic field, separated by two

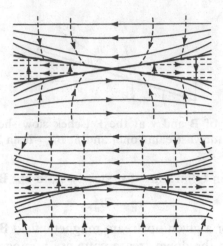

Fig. 6.5. Petschek-type reconnection configurations: magnetic field lines (full), stream lines(dashed), slow mode shocks (heavy). (a) Petschek's original configuration ("fast mode expansion"). (b) Sonnerup-type configuration ("slow mode expansion").

singular lines, the Petschek shock and a second line in the upstream region, in which the current and vorticity of the upstream region are concentrated. A wider class of analytical solutions, containing Petschek's and Sonnerup's as special cases, has been given by Priest & Forbes (1986).

By calling all these configurations, which are usually strictly distinguished in the literature, Petschek-like, it is indicated that their differences are in fact minor compared with the common basic assumption concerning the effect of the diffusion region. In fact all are solutions of the ideal external region which essentially ignore the dynamics in the diffusion region, whence the term quasi-ideal reconnection models. It is true that Petschek includes a treatment of the diffusion layer, which however can be regarded only as an interpolation between the origin and the external solution, assuming, for instance, a linear increase of the perpendicular field component $B_x \propto y$, while a rigorous treatment shows $B_x \propto y^3$, as is discussed in section 6.3.1.

Petschek's model is based on the analogy with a system of two supersonic gas streams hitting head-on and being deflected sidewise by shocks with the same geometry as the Petschek shocks. The physics of the central stagnation point, where the shocks join, is however quite different in both cases. While it is an ordinary flow stagnation point in the case of a nonmagnetic neutral fluid, it is the location of intense dissipation and magnetic diffusion, owing to high current density in the case of a magnetized conducting fluid. In the latter case the flow is supersonic only with respect to the slow mode, while it is subsonic with respect to the magnetosonic mode, the phase velocity of which is in fact infinite in the incompressible approximation. If resistive diffusion is reduced by decreasing η, the field is locally compressed in front of the diffusion layer, which is communicated upstream modifying the entire upstream configuration, in contrast to a supersonic gas stream where no signal can propagate upstream.

In fact, from a plasma physics point of view Petschek's concept contradicts intuition, which tells us that pushing two volumes of highly conducting plasma with opposite magnetic fields against each other produces a flat configuration with a current sheet and that pushing faster makes the configuration more flattish instead of further opening up the cone for sidewise ejection, $M = \alpha$, as predicted in a Petschek-like model. Intuition is corroborated by numerous numerical simulations, all exhibiting formation of current sheets for sufficiently small η, which become longer instead of shorter if η is reduced. Scaling laws of current sheet configurations are discussed in section 6.3.

The crucial deficiency of reconnection models of the Petschek-type is the ignorance or inappropriate treatment of the diffusion region. A correct theory requires the solution of the boundary layer problem, matching the inner resistive solution computed in simplified geometry to the external ideal solution. Petschek's external solution is correct and even stable, but it does not match to the diffusion layer for small η. Eliminating the problem of the diffusion layer by using a resistivity which is locally strongly enhanced in the vicinity of the X-point, $\eta_X = O(1)$ ("anomalous resistivity"), a Petschek-like configuration is in fact set up as seen in the simulations by Sato & Hayashi (1979). Hence Petschek's model is not a self-consistent reconnection model in the limit of small η. Because of the complicated structure of the diffusion layer, which is discussed in section 6.4, it appears quite hopeless to solve the matching problem analytically. Quasi-exact stationary solutions for relevantly small values of η have, however, been obtained by numerical simulation (e.g. Biskamp, 1986a), and are presented in sections 6.3 and 6.4. All these configurations are strongly dependent on η, and become nonstationary if η falls below some threshold.

The fact that numerical simulations of driven reconnection do not produce a Petschek-like configuration for small η has sometimes been attributed to an inappropriate choice of the boundary conditions. While the discussion of boundary conditions, in particular the actual freedom in their choice, is deferred to section 6.3, we here point out only that the simulations themselves effectively invalidate this argument: (a) Various kinds of boundary conditions have been used in driven reconnection simulations, none of which lead to a Petschek-type configuration for small η. However, switching on an anomalous resistivity to eliminate the diffusion layer problem, Petschek-type configurations are set up quite independently of the particular choice of the boundary conditions. (b) Various simulations of self-consistent reconnecting systems have been performed, such as the process of island coalescence (section 6.6.1) or the nonlinear resistive kink mode (section 6.6.2), where no internal boundary conditions that could possibly affect the reconnection process have to be imposed. All develop extended current sheets for small η.

6.2.3 Syrovatskii's current sheet solution

An alternative school of thought, with adherents mainly in the eastern hemisphere, originated from Syrovatskii's theory of current sheet formation (Syrovatskii, 1971). Like Petschek's model this is also a quasi-ideal, quasi-stationary approach, dealing only with the ideal solution, which may however exhibit sheet-like singularities. Though Syrovatskii's theory does not describe real configurations with high reconnection rates in the limit of small η, it provides a qualitatively correct picture for not-too-strong external driving.

The basic equations are somewhat different from those of two-dimensional incompressible MHD, to which the major part of this chapter is confined, using vanishing plasma pressure $p = 0$ instead. The main assumption is that all currents in the system are localized in isolated points and sheets. Hence ψ satisfies Laplace's equation

$$\nabla^2 \psi = 0, \tag{6.18}$$

such that ψ is a harmonic function and one can use complex analysis. The solution is determined by the boundary conditions. If these change in time then ψ obtains a parametric time dependence $\psi(x, y, t)$, which then determines the perpendicular component $\mathbf{v}_\perp$ of the velocity from the frozen-in condition,

$$\frac{d\psi}{dt} \equiv \partial_t \psi + \mathbf{v} \cdot \nabla \psi = 0 \,,$$

$$\mathbf{v}_\perp = -\partial_t \psi \nabla \psi / |\nabla \psi|^2 \,, \quad \mathbf{B} = \mathbf{e}_z \times \nabla \psi \,, \tag{6.19}$$

while the parallel component $\mathbf{v}_\parallel$ is calculated from the equation

$$\frac{d\mathbf{v}}{dt} \times \nabla \psi = 0 \,, \tag{6.20}$$

which follows from the equation of motion using $p = 0$. (The latter equation, however, implies that the current density and hence the Lorentz force does not vanish identically. Hence eq. (6.18) has to be regarded as an approximation in the sense that the effect of the distributed currents is small compared with that of the sheet currents.) The flow is in general not incompressible. We should note, however, that **v** is determined a posteriori, which reduces the theory to eq. (6.18), i.e. an equilibrium boundary value problem. The important point is that smooth solutions do in general not exist, since any change of the boundary conditions leading to a change of ψ, $\partial_t \psi \neq 0$, at a neutral point $\nabla \psi = 0$ gives rise to a singular current.

To discuss this singularity in more detail it is convenient to introduce the potential $F(z)$ in the complex plane $z = x + iy$,

$$F(z, t) = \psi(x, y, t) + i\chi(x, y, t), \qquad (6.21)$$

which is analytic in the region considered except for isolated singular points and branch cuts. The conjugate harmonic function χ can be determined by using the Cauchy-Riemann relations

$$\partial_x \chi = -\partial_y \psi, \quad \partial_y \chi = \partial_x \psi. \qquad (6.22)$$

From eq. (6.21) we obtain the magnetic field in the form

$$dF/dz = B_y + iB_x, \qquad (6.23)$$

which can be seen by choosing a special direction of the derivative, e.g. dF/dx, since owing to eqs (6.22) the complex derivative is independent of this choice.

Let $z = 0$ be the position of a neutral point of the magnetic configuration at time $t = 0$, $dF/dz|_{z=0} = 0$. In the vicinity of this point the complex potential is

$$F(z, t) = \frac{\alpha(t)}{2} z^2 + \beta(t), \qquad (6.24)$$

restricting consideration to the only practically relevant case of second-order neutral points. (Note that owing to eq. (6.18) there are only X-type nonsingular neutral points.) If the change of the boundary conditions for ψ is such that $d\beta/dt \neq 0$, this implies a nonvanishing electric field $E_z = \partial_t \psi$ at the neutral point indicating that the potential $F(z, t)$ cannot remain analytical at this point.

The most natural assumption, namely the induction of a line current at the neutral point,

$$F(z, t) = \frac{\alpha(t)}{2} z^2 + \beta(t) + \frac{I(t)}{2\pi} \ln z, \qquad (6.25)$$

is not allowed, since this would give rise to the appearance of an O-type singular neutral point (Fig. 6.6), implying a change of magnetic topology in a finite region, which is prohibited by condition (6.19). Here $I(t)$ is the total current generated in the plasma, which is determined by the boundary conditions with the initial condition $I(0) = 0$.

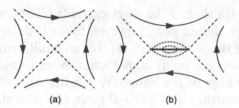

Fig. 6.6. Generation of a singular current at an X-point. (a) Initial nonsingular configuration; (b) effect of an induced singular line current in the original X-point, leading to a fictitious O-point and two adjacent X-points. The heavy line indicates the actually arising sheet current.

The only admissible alternative is a solution with a branch cut corresponding to a current sheet. The location of a branch cut is determined by that of the fictitious neutral points arising by the addition of a line current in the original X-point. As indicated in Fig. 6.6b, the cut passes through this point and the two adjacent X-points.

We can now discuss the structure of the field in the presence of a current sheet. At distances large compared with the width of the cut the potential $F(z, t)$ has approximately the form (6.25). The conditions at the cut are that **B** does not intersect the cut, since field lines remain continuous if they are so initially. Hence the cut is a line $\psi = $ const. Assume a straight cut extending along the y-axis between the points $y = \pm b$. The solution for the complex potential which has the asymptotic form (6.25) for $|z|/b \gg 1$ and $\psi = 0$ at the cut is given by

$$F(z) = \frac{\alpha}{2} z \sqrt{z^2 + b^2} + \frac{I}{2\pi} \ln \frac{z + \sqrt{z^2 + b^2}}{b} \qquad (6.26)$$

with the derivative

$$\frac{dF}{dz} = B_y + iB_x = \frac{\frac{I}{2\pi} + \alpha \frac{b^2}{2} + \alpha z^2}{\sqrt{z^2 + b^2}} . \qquad (6.27)$$

While the magnetic potential $\psi(x, y) = Re\{F(z)\}$ is continuous, the magnetic field B_y has a jump across the cut, the line density of the current carried by the sheet,

$$J(y) \equiv B_y(0_+, y) - B_y(0_-, y) = 2 \left(\frac{I}{2\pi} + \alpha \frac{b^2}{2} - \alpha y^2 \right) \Big/ \sqrt{b^2 - y^2} \qquad (6.28)$$

with

$$\int_{-b}^{b} J(y)\, dy = I .$$

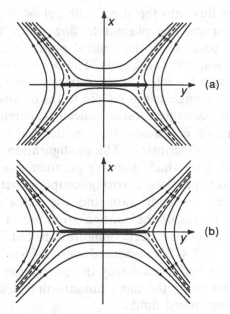

Fig. 6.7. Contours of the magnetic potential $\psi = Re\{F\}$, where F is given by eq. (6.26); heavy line = current sheet, dashed line = separatrix. (a) General case $y_0 < b$ exhibiting singularities at the current sheet endpoints; (b) limiting regular case $y_0 = b$ (from Syrovatskii, 1971).

The current distribution (6.28) shows an interesting feature. It is positive, i.e. in the direction of the total current I, in the center part $|y| < y_0$ and negative, i.e. in the opposite direction, in the outer parts $|y| > y_0$, where $y_0^2 = (I/2\pi\alpha) + b^2/2$. The points $z_{\pm} = (0, \pm y_0)$ are neutral points of the magnetic field, $dF/dz = 0$, where the separatrix branches off the y-axis, while the current sheet continues along the y-axis (Fig. 6.7a). At the end points $|y| = b$ the current density $J(y)$ becomes singular, giving rise to infinitely large magnetic fields. Only in the special case $I = \pi\alpha b^2$, where the neutral points coincide with the current sheet endpoints, does the singularity vanish (Fig. 6.7b). This value is in fact an upper limit of the current I (or a lower limit of the sheet width b for given I), since for larger values the points $z_{\pm}$ would form isolated neutral points – the separatrix would be similar to the dashed line drawn in Fig. 6.6b – which because of the frozen-in property is topologically not possible.

The velocity field $\mathbf{v}$ corresponding to the magnetic configuration (6.27), which is determined by eqs (6.19), (6.20), cannot be given, as it seems, in simple analytical form, but must be computed numerically, even in the stationary case $\partial_t\psi = $ const. The qualitative behavior close to the current sheet can however be easily understood. It follows from eq. (6.19) that there is a net plasma flux into the sheet $v_x(0_+, y) = -v_x(0_-, y) \neq 0$. Mass conservation then requires the plasma to flow along the sheet, leaving the sheet at high speed close to the neutral points $z_\pm$, while the plasma flow vanishes at the singular sheet ends. The downstream velocity in the cone formed by the two branches of the separatrix is of the order of the upstream flow; the cone angle is not related to the downstream flow speed. Evidently Syrovatskii's configuration is basically different from Petschek's.

Syrovatskii's approach is a quasi-static model and does not provide a self-consistent dynamic description. The configuration is independent of the reconnection rate $\partial_t \psi$, which is a free parameter in the theory, while a fully resistive theory predicts a strong coupling between the current sheet width b, the reconnection rate and the value of the resistivity. But the qualitative features of the configuration are in surprisingly good agreement with typical current sheet configurations obtained from resistive simulations; see section 6.4. The essential merit of Syrovatskii's theory is to describe in a simple and elegant way the generation of current sheets, a process which seems to be the most fundamental dynamical feature in highly conducting magnetized fluids.

6.3 Scaling laws in stationary current sheet reconnection

In order to determine the η-dependence of a stationary driven reconnection configuration the stationary resistive MHD equations must be solved for given inflow and outflow boundary conditions. Unfortunately it appears that the problem is too complicated to permit analytical solutions without severe approximations. In particular the matching of the solution in the diffusion layer to that in the ideal external region is virtually impossible, since the complicated shape of the diffusion region (section 6.4) seems to make the problem nonseparable, i.e. truly two-dimensional.

Hence we have to resort to numerical methods. In this case there is no particular advantage in restricting consideration to the stationary problem. In fact the simplest and most reliable way is to follow the system dynamically from an initial state for fixed boundary conditions until a stationary state is reached, which automatically eliminates unstable solutions. In the past the objections against such a purely numerical treatment have been that the limited spatial resolution would not allow us to obtain sufficiently accurate solutions in the most interesting range of small η, where small-scale structures are expected to be important, and that even if we had such a solution it would correspond only to one point in parameter space and we would not know how it changes if parameters are varied. For the problem of two-dimensional stationary driven reconnection these objections are no longer valid. Accurate numerical solutions in the relevant parameter regime can now be obtained and since there are only two essential parameters their scaling laws can be obtained from a rather small number of computer runs.

Since primary interest is in understanding the qualitative behavior, one can choose the simplest possible geometry having up–down and right–left symmetry, such that only a quadrant must actually be computed. The computational system is indicated in Fig. 6.8. The basic equations to be solved are the 2-D incompressible MHD equations, given here once more for convenience:

$$\partial_t \psi + \mathbf{v} \cdot \nabla \psi = \eta \nabla^2 \psi , \tag{6.29}$$

$$\partial_t \omega + \mathbf{v} \cdot \nabla \omega = \mathbf{B} \cdot \nabla j + \nu \nabla^2 \omega , \tag{6.30}$$

$$\mathbf{B} = \mathbf{e}_z \times \nabla \psi , \quad \mathbf{v} = \mathbf{e}_z \times \nabla \phi ,$$
$$j = \nabla^2 \psi , \quad \omega = \nabla^2 \phi .$$

Boundary conditions have to be assigned to ψ, ϕ, j, ω. While at the internal boundaries, the x-axis and the y-axis, boundary conditions follow from the imposed symmetry, ψ, j being symmetric, ϕ, ω antisymmetric, conditions at the upper ($x = L_x$) and the left-hand ($y = L_y$) boundaries correspond to the inflow and outflow conditions, respectively. These should be chosen in a way which conforms with the concept of an open system, such that the boundaries, in particular the outflow boundary, do not obstruct the flow. Open boundaries are well defined for linear waves, requiring that waves are not reflected at the boundary, or, in mathematical terms, that for a hyperbolic system of differential equations all characteristics should be outgoing at the boundary, which guarantees that perturbations caused by the presence of the boundary do not propagate into the system. However, the incompressible dissipative MHD equations are of mixed hyperbolic parabolic type. In addition the equations are nonlinear and two-dimensional, and to date there is no rigorous method to determine whether for such a system a particular set of boundary conditions is admissible. (A detailed discussion of the boundary conditions for different MHD systems has been given by Forbes & Priest, 1987.)

Let us therefore apply a more practical procedure. While it seems to be sufficient to require continuity of ω and j at the boundaries in eq. (6.30), $\partial_n \omega = \partial_n j = 0$, the system is more sensitive to the boundary conditions for the potentials ψ in eq. (6.29) and ϕ in the Poisson equation $\nabla^2 \phi = \omega$, since an inappropriate choice may lead to singularities in ω and j, which would show up in the form of slow shocks in the vicinity of the boundaries. Any choice of the boundary conditions that does not give rise to such singularities should be regarded as acceptable, and numerical experience indicates that there is considerable freedom in this choice (Biskamp, 1986a). One may, for instance, specify $\phi(y)$ and $\partial_x \psi$,

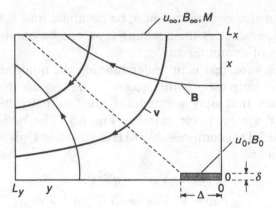

Fig. 6.8. Computational system and notation as used in numerical simulations of stationary driven reconnection.

i.e. $v_x(y)$ and $B_y(y)$, at the ingoing boundary $x = L_x$ and $\phi(x)$, i.e. $v_y(x)$, at the outgoing boundary $y = L_y$ and iterate $\partial_y \psi$, i.e. $B_x(x)$, at the latter, until the outflow configuration becomes smooth.

In order to describe the dependence of the configuration on the inflow conditions it is convenient to parametrize the inflow boundary functions, writing $v_x(y) = u_\infty f(y)$, $B_y(y) = B_\infty g(y)$. Then a stationary configuration is characterized by the inflow parameters u_∞, B_∞, or simply $u_\infty = M$ using the normalization $B_\infty = 1$, and the parameters u_0, B_0, Δ, δ describing the internal reconnecting current sheet, as illustrated in Fig. 6.8. While M is prescribed, the current sheet parameters depend on the internal dynamics and are functions of η, in particular. A homogeneous resistivity distribution is chosen to avoid additional complications arising from resistivity gradient effects. Reconnection is said to be independent of η if for fixed boundary conditions the configuration, in particular the width Δ of a macroscopic current sheet, does not depend on η (for sufficiently small η).

First consider the case of weak driving, $M = \partial_t \psi \lesssim \eta \ll 1$. We obtain an approximate stationary solution by expanding eqs (6.29), (6.30) in M, which to lowest order gives

$$\mathbf{B} \cdot \nabla j = 0 , \tag{6.31}$$

$$\mathbf{v} \cdot \nabla \psi = -M + \eta j . \tag{6.32}$$

If j vanishes asymptotically for $|x| \gg 1$, eq. (6.31) implies $j = 0$ everywhere in an X-point configuration with open field lines. There is a simple similarity solution of eqs (6.31), (6.32)

$$\psi = \tfrac{1}{2}(x^2 - y^2) , \tag{6.33}$$

$$\phi = \frac{M}{2} \ln \left| \frac{x+y}{x-y} \right| . \tag{6.34}$$

The solution is not valid, however, on the separatrix $x = \pm y$, where $\omega = 4Mxy/(x^2 - y^2)^2$ becomes singular invalidating the omission of the inertia term in eq. (6.31). Hence the current j does not vanish on the separatrix, in particular not in the X-point, which also makes the resistive term ηj in eq. (6.32) finite. In this case the magnetic configuration in the vicinity of the neutral point is fundamentally changed, as is discussed in more detail in section 6.4.1. However, for $M \lesssim \eta$, i.e. $M/\eta = MS = R_m \lesssim 1$, using normalizations $B_0 = v_A = 1$, $L = 1$ in the definitions of S and the magnetic Reynolds number R_m (see eq. (2.60)), the region affected is small such that the global configuration is still essentially described by eqs (6.33), (6.34).

Increasing M at constant η or decreasing η at constant M, such that $R_m > O(1)$, the configuration is modified by the formation of a current sheet of finite width Δ. A series of numerical simulations has been performed (Biskamp, 1986a) for identical boundary profile functions, but different values of M and η. The most conspicuous feature is the η-dependence of the width Δ of the internal current sheet. Figure 6.9 gives the flow pattern $\phi(x, y)$ and the magnetic configuration $\psi(x, y)$ for three cases differing only in the value of η. Obviously, Δ increases rapidly with decreasing η until reaching the system size $\Delta \sim L$, in contrast to a Petschek-like behavior $\Delta = O(\eta)$. The current has the properties of a Sweet-Parker sheet, satisfying in particular relations (6.6) and (6.7). Quantitatively one finds the following M, η scaling laws for the variables B_0, u_0, Δ, δ of the internal current sheet:

$$B_0 \propto M^2/\eta = MR_m , \tag{6.35}$$

$$u_0 = M/B_0 \propto R_m^{-1} , \tag{6.36}$$

$$\Delta \propto M^4/\eta^2 = (MR_m)^2 , \tag{6.37}$$

$$\delta \propto M\eta^0 . \tag{6.38}$$

Hence increasing the Reynolds number R_m leads to an increase of the field B_0 in front of the sheet (due to a pile-up of flux) and a corresponding decrease of the upstream velocity u_0 because of the stationarity condition $uB = M$. The sheet width Δ *increases* with decreasing η and the thickness δ is independent of η, contrary to expectation, and even *increases* with M because of the decrease of u_0, the deceleration of the upstream flow being apparent in Fig. 6.9c.

If the scaling laws (6.35)–(6.38) are universal and not just accidentally valid for the particular set of numerical simulations then they should

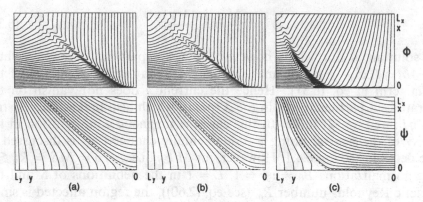

Fig. 6.9. Stream function $\phi(x, y)$ and flux function $\psi(x, y)$ for three stationary configurations of current sheet reconnection. (a) $\eta = \eta_0$; (b) $\eta = \eta_0/2$; (c) $\eta = \eta_0/4$ (from Biskamp, 1986a).

reflect the physics of the Sweet-Parker current sheet, an important feature of which is the acceleration along the current sheet. Consider the average force along the sheet. As is shown in section 6.4.1, the velocity v_y increases linearly $v_y \simeq B_0 y/\Delta$, such that we have

$$\overline{v_y \partial_y v_y} \simeq B_0^2/2\Delta \simeq M/2\delta \, , \qquad (6.39)$$

using mass conservation $u_0 \Delta = B_0 \delta$ and Ohm's law $u_0 B_0 = u_\infty B_\infty = M$ because of the normalization $B_\infty = 1$. Hence the scaling law (6.38) implies that the force along the sheet is invariant under changes of M and η, in particular it remains finite for $\eta \rightarrow 0$.

Finally we discuss the scaling of the energy dissipation rate, compared with the input power W_{in} considering only ohmic dissipation W_η, which is larger than viscous dissipation W_ν for $\nu \lesssim \eta$. Using eqs (6.35)–(6.38) we obtain the following estimate:

$$\begin{aligned}
\frac{W_\eta}{W_{in}} &\simeq \int \eta j^2 dF \Big/ \int u B^2 dl \\
&\simeq \eta \frac{B_0^2}{\delta^2} \Delta\delta \Big/ u_\infty B_\infty^2 L_y \\
&\propto M^6/\eta^3 \, , \qquad (6.40)
\end{aligned}$$

assuming that the dissipation is concentrated in the current sheet. W_η equals the magnetic field energy flux into the sheet, $W_\eta \simeq u_0 B_0^2 \Delta$. Relation (6.40) indicates in particular that the fraction of the input power which is dissipated increases with decreasing η, becoming of order unity for a macroscopic current sheet width.

When the current sheet width reaches the size of the global configuration $\Delta \simeq L$ as in Fig. 6.9c, the width cannot increase further and the scaling laws (6.35)–(6.38) are no longer valid. In this case the scaling behavior is directly determined by the properties of the Sweet-Parker current sheet of width $\Delta = L$. From

$$u_0 B_0 = M \simeq \eta B_0 / \delta$$

one obtains

$$u_0 = \eta / \delta , \quad B_0 = M \delta / \eta .$$

Inserting these results into the mass conservation equation $u_0 L = B_0 \delta$ yields the scaling relations

$$B_0 \propto (M^2 L / \eta)^{1/3} , \tag{6.41}$$

$$\delta \propto (\eta^2 L / M)^{1/3} , \tag{6.42}$$

$$A \propto (M L^2 / \eta^2)^{1/3} . \tag{6.43}$$

Examples of such current sheet reconnection processes are the coalescence of two magnetic islands and the nonlinear evolution of the resistive kink mode (section 6.6).

6.4 Current sheets: refined theory

We have seen in section 6.3 that the diffusion region in stationary reconnection has the form of a current sheet, which may reach macroscopic size and which has the characteristic properties of a Sweet-Parker sheet. In this section we present a more detailed theory of the diffusion layer, considering separately the central part and the edge region.

6.4.1 *Stationary solution in the vicinity of the neutral point*

The simplest way to investigate the solution of the resistive MHD equations in the vicinity of the neutral point $(x, y) = (0, 0)$ is to use a Taylor series expansion in x and y (Cowley, 1975; Shivamoggi, 1985). Assume a symmetric configuration as indicated in Fig. 6.10a, where the stagnation point of the flow coincides with the neutral point of the magnetic field:

$$\psi = \sum_{m,n} \psi_{2m,2n} \frac{x^{2m} y^{2n}}{(2m)! \, (2n)!} ,$$

$$\phi = \sum_{m,n} \phi_{2m+1,2n+1} \frac{x^{2m+1} y^{2n+1}}{(2m + 1)! \, (2n + 1)!} ,$$

$$f_{mn} = \partial_x^m \partial_y^n f |_{x,y=0} , \quad f = \psi, \phi .$$

For stationary conditions eqs (6.29), (6.30) read

$$\partial_x\phi\,\partial_y\psi - \partial_y\phi\,\partial_x\psi = \eta(\partial_x^2\psi + \partial_y^2\psi) - E \tag{6.44}$$

$$\partial_x\phi\,\partial_y\omega - \partial_y\phi\,\partial_x\omega - \partial_x\psi\,\partial_y j + \partial_y\psi\,\partial_x j = \nu(\partial_x^2\omega + \partial_y^2\omega)\,, \tag{6.45}$$

where $E \equiv \partial_t\psi = $ const. Since at the origin $\mathbf{B} = \mathbf{v} = 0$ and hence $\eta j = E$, eq. (6.44) gives

$$\eta\,(\psi_{20} + \psi_{02}) = E\,. \tag{6.46}$$

Differentiating eq. (6.44) twice with respect to x at the origin one obtains

$$2\phi_{11}\psi_{20} + \eta\,(\psi_{40} + \psi_{22}) = 0\,, \tag{6.47}$$

and twice with respect to y,

$$2\phi_{11}\psi_{02} - \eta\,(\psi_{22} + \psi_{04}) = 0\,. \tag{6.48}$$

Differentiating eq. (6.45) once with respect to x and y at the origin gives

$$-\psi_{20}\,(\psi_{22} + \psi_{04}) + \psi_{02}\,(\psi_{40} + \psi_{22}) = \nu\,(\phi_{51} + 2\phi_{33} + \phi_{15})\,, \tag{6.49}$$

which becomes by use of eqs (6.47), (6.48)

$$-\frac{4}{\eta}\phi_{11}\psi_{20}\psi_{02} = \nu\,(\phi_{51} + 2\phi_{33} + \phi_{15})\,. \tag{6.50}$$

First consider the inviscid case $\nu = 0$. Assuming $\phi_{11} \neq 0$, i.e. stream lines forming hyperbolae, either ψ_{20} or ψ_{02} (not both because of eq. (6.46)) must vanish. This implies that field lines are not hyperbolae, in particular that the separatrices do not intersect at a finite angle, but osculate as indicated in Fig. 6.10b, where we chose $\psi_{20} \neq 0$, $\psi_{02} = 0$. While in an X-point configuration B_x in the downstream cones increases linearly, $B_x = \psi_{02}y$, it is cubic in the osculating configuration, $B_x = (\psi_{04}/6)y^3$, whereas the velocity is in general linear, $v_y = \phi_{11}y$. This behavior indicates the inherent tendency to formation of current sheets in a resistive magnetized fluid.

In the case of finite viscosity these conclusions can no longer be drawn from eq. (6.50). In general ψ_{02} will be finite, its magnitude depending on the higher-order terms in the Taylor expansion of the streamfunction, in particular ϕ_{51}, the dominating term in an elongated configuration with $\partial_x \gg \partial_y$. Numerical simulations, however, show that even for $\nu \sim \eta$ the inviscid behavior $B_x \propto y^3$ is still nearly valid, implying ϕ_{51} to be small, $\phi_{51} \ll \phi_{11}/\delta^4$, where δ is the sheet thickness, i.e. the current gradient scale defined by $\psi_{40} \sim \psi_{20}/\delta^2$.

Since the configuration around a neutral point tends to be stretched out, one can make use of the quasi-one-dimensional character $\partial_x \gg \partial_y$ by performing a power expansion only in y,

$$\psi(x,y) \;=\; \psi_0(x) + y^2\psi_2(x)/2! + \cdots\,, \tag{6.51}$$

$$\phi(x,y) \;=\; y\phi_1(x) + y^3\phi_3(x)/3! + \cdots\,. \tag{6.52}$$

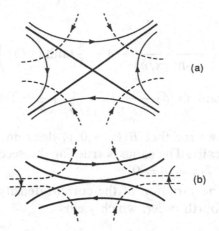

Fig. 6.10. Behavior of $\psi(x,y)$ (full lines) and $\phi(x,y)$ (dashed lines) in the vicinity of the neutral point. In the resistive case the separatrix branches cannot intersect at a finite angle, Fig. (a), but osculate, Fig. (b).

If the zeroth-order current distribution $j_0(x) = \psi_0''(x)$ is given then the other functions $\psi_2(x), \ldots, \phi_1(x), \ldots$ can be determined successively in terms of $j_0(x)$. Let us assume the profile

$$j_0(x) = \frac{j_m}{\cosh^2(x/\delta)}, \tag{6.53}$$

which is found in the simulations to fit the current density in the diffusion region surprisingly well, and calculate the first terms in the series (6.51), (6.52) explicitly. Integration of (6.53) gives

$$\psi_0(x) = j_m \delta^2 \ln\left[\cosh(x/\delta)\right]. \tag{6.54}$$

Inserting this into eq. (6.44) one obtains

$$\phi_1(x) = (\eta/\delta)\tanh(x/\delta) \tag{6.55}$$

and hence the lowest-order velocity components in the sheet:

$$v_x^{(1)} = -u_0 \tanh(x/\delta)$$
$$v_y^{(1)} = (u_0 y/\delta)/\cosh^2(x/\delta).$$

Here $u_0 = \eta/\delta$ is the upstream velocity and $v_0 = B_0 = j_m \delta$ is the downstream velocity, which defines the width Δ of the sheet using the mass conservation relation $v_0 \delta = u_0 \Delta$. In fact the inverse aspect ratio of the sheet $\delta/\Delta = u_0/B_0 = \eta/(j_m \delta^2) = M_0 \ll 1$ is the smallness parameter in the expansions (6.51), (6.52). The next order terms can be obtained by straightforward but somewhat tedious calculation using eqs (6.44), (6.45).

For $v = 0$ one finds

$$\psi_2(x) = -2M_0^2 j_m \left(\frac{1}{\cosh^2 (x/\delta)} - 1 + \frac{x}{\delta} \tanh (x/\delta) \right)$$

$$\phi_3(x) = M_0^3 \frac{j_m}{\delta} \left[\tanh (x/\delta) - \frac{3}{\cosh^2 (x/\delta)} \left(\frac{x}{\delta} - 2 \tanh (x/\delta) \right) \right] .$$

Since $\psi_2(x = 0) = 0$, we see that $B_x(x = 0, y)$ does not increase linearly in y but at most cubically. The same is true for the second-order current density contribution, $\psi_2''(x = 0) = 0$.

In order to obtain the variation of the current density along the y-axis one has to go to the fourth order, which yields

$$j(x = 0, y) = j_m \left[1 - \frac{4}{3} \left(M_0 \frac{y}{\delta} \right)^4 \right] . \tag{6.56}$$

This does not, however, contradict Syrovatskii's result, eq. (6.28), where the line current density $J(y) = \int j(x, y) dx$ varies parabolically, $J(y) \simeq 1 - ay^2$ for $y \ll b$. In fact eq. (6.56) is valid only along the centerline of the sheet, the y-axis, where j_2 vanishes. The x-integrated second-order contribution does not vanish; instead one obtains

$$J(y) = 2 j_m \delta \left[1 - \left(\frac{M_0 y}{\delta} \right)^2 \right] . \tag{6.57}$$

Since $B_x \propto y^3$, the Lorentz force along the sheet is small, $B_x j \propto y^3$, such that the plasma acceleration is caused mainly by the pressure force $-\partial_y p \simeq v_y \partial_y v_y \propto y$.

6.4.2 Current sheet edge region

Let us now discuss the behavior in the edge region of the current sheet $y \simeq \Delta$, where the Taylor series expansions (6.51), (6.52) break down, since $M_0 y/\delta$ becomes of order unity. Thus the only reliable information is obtained from direct numerical simulations. Earlier analytical treatments of the diffusion region (see e.g. Vasyliunas, 1975) assumed a smooth transition to the ideal exterior region of the downstream cone with the fluid continuing to flow at the upstream Alfvén velocity. However, such a highly super-Alfvénic flow (the local Alfvén velocity in the downstream cone is much smaller than in the upstream region) should be sensitive to shock formation, which would increase the field intensity and slow down the flow to sub-Alfvénic velocities. In fact, simulations show that the ideal downstream flow is clearly sub-Alfvénic and not related to the high speed reached *within* the diffusion region.

Closer inspection of the edge of the diffusion region reveals a complicated structure, as is illustrated in Figs 6.11 and 6.12. Figure 6.11 gives stereographic plots of the current distribution viewed from the upstream side (a) and the downstream side (b), which reveal the main features of the configuration, the diffusion layer represented by the central current sheet along the y-axis, the weaker sheet current along the separatrix, reminiscent of a Petschek slow shock, both joining in a region of rather complex behavior, the edge region of the diffusion layer. As is seen in Fig. 6.11b, the current density in the diffusion layer changes sign, i.e. the positive central part is followed by a negative part, terminated by a quasi-singular spike (numerically well resolved, however). Figure 6.12 gives contour plots of j, ϕ, ψ in the edge region of three stationary simulation states (the symmetric lower quadrant is added for clarity) with (a) $\eta = \eta_0$, (b) $\eta = \eta_0/\sqrt{2}$, (c) $\eta = \eta_0/2$, showing the rapid increase of complexity as η is reduced. The point where the current density of the diffusion layer changes sign (marked by the arrows in the j-plots) coincides with the location where the separatrix (the dashed line in the ψ-plots) branches off, as in Syrovatskii's current sheet configuration (Fig. 6.7a). The dynamics can most readily be interpreted when considering the ϕ-contours, high streamline density indicating high velocity. The flow, which is accelerated in the central current sheet up to the upstream Alfvén speed, is decelerated in the following part of reversed current density and is finally completely blocked and turned backwards at the current sheet end point singularity, the spike in Fig. 6.11b. The flow is subsequently accelerated again, forming two secondary current sheets parallel to the primary one, with the same characteristics, $j > 0$ part, $j < 0$ part and flow-blocking shock-like structure. In the limit $\eta \to 0$ a hierarchy of higher-order current sheets seems to be generated with a self-similar scaling behavior as drawn schematically in Fig. 6.13.

The properties of the diffusion layer as revealed by numerical simulation and outlined in this section are consistent with Syrovatskii's current sheet model, the multi-current sheet edge behavior representing the dynamically resolved singularity predicted in Syrovatskii's quasi-static theory. The picture indicated in Fig. 6.13, however, rests on a twofold idealization. One is the assumption of perfect symmetry, the other that of stationarity. As will be discussed in section 6.6, a dynamic current sheet, though more stable than a static one, becomes unstable if the aspect ratio $A = \Delta/\delta$ is sufficiently large, giving rise to a nonstationary behavior. In a less symmetric configuration such nonstationary behavior will be very complex. In fact we see in chapter 7 that the most probable dynamic state is that of fully developed turbulence, consisting of a statistical

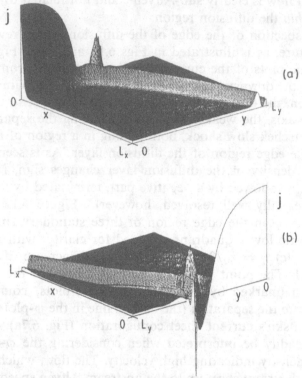

Fig. 6.11.   Stereographic plots of the current distribution of a stationary reconnection configuration, (a) viewed from the upstream side, (b) from the downstream side (from Biskamp, 1986a).

distribution of micro-current sheets. The stationary multiple current sheet configuration presented here gives a first indication about the complicated behavior to be expected at high Reynolds numbers.

6.5   Tearing instability of a Sweet-Parker current sheet

In a static sheet pinch, for which the tearing mode is usually considered (section 4.7.1), the instability condition $ka < 1$ implies that the configuration becomes unstable for a sheet width Δ of about two wavelengths of the marginal stable mode corresponding to an aspect ratio $A = \Delta/a \gtrsim 2 \cdot 2\pi/ka \sim 10$. The existence of apparently stable current sheets of considerably larger A, as observed in numerical simulations, indicates that the dynamics involved in a Sweet-Parker sheet has a strongly stabilizing effect. The stability properties of such a configuration with respect to the tearing mode were studied by Bulanov et al. (1979). Strictly

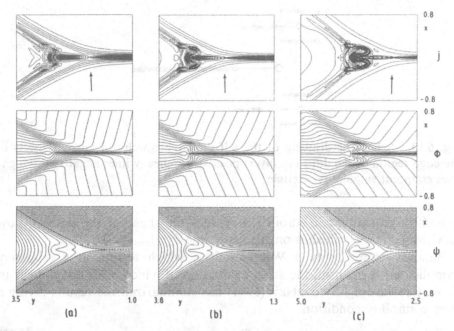

Fig. 6.12. Contours of j, ϕ, ψ in the edge region of the diffusion layer, (a) $\eta = \eta_0$; (b) $\eta = \eta_0/\sqrt{2}$; (c) $\eta = \eta_0/2$ (from Biskamp, 1986a).

speaking, a Sweet-Parker sheet is a two-dimensional system, where the inhomogeneity along the sheet arises because of the acceleration of the parallel flow as well as the increase of the normal magnetic field component. Since the latter is weak, $B_x \propto y^3$, as we have seen in section 6.4.1, the inhomogeneity of the parallel flow is the dominant effect,

$$v_y(y) = \Gamma y, \qquad (6.58)$$

where

$$\Gamma = v_A/\Delta \simeq u/\delta \simeq \eta/\delta^2, $$

using the properties of the Sweet-Parker sheet. Hence Γ^{-1} is just the resistive time τ_η of the current sheet and the aspect ratio equals the Lundquist number of the sheet,

$$\Delta/\delta = v_A\delta/\eta = \tau_\eta/\tau_A \equiv S_\delta, \qquad (6.59)$$

where the subscript δ should remind the reader that S_δ is to be distinguished from the Lundquist number $S = v_AL/\eta$ of the global configuration, $S_\delta \ll S$ usually. The normal component of the flow v_x only has the effect of balancing the resistive broadening of the sheet. Being zero at the sheet centerline $x = 0$, the resonant surface of the tearing mode, it does

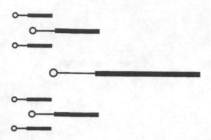

Fig. 6.13. Schematic drawing of the self-similar hierarchy of current sheets in the edge region of the diffusion layer: positive j = heavy line; negative j = light line; endpoint singularity = circle.

not affect tearing-mode stability, nor does the shear of the parallel flow $v_y(x)$, since it too vanishes on the centerline $\partial_x v_y|_{x=0} = 0$.

Bulanov et al. apply a WKB analysis which results in a relatively complicated formalism, the strict evaluation of which does not appear to be very rewarding. Syrovatskii (1981) has summarized the results giving a simple stability condition

$$\Gamma \gtrsim \gamma, \tag{6.60}$$

where γ is the tearing-mode growth rate for a one-dimensional static sheet. The result (6.60) can easily be understood. The tearing-mode corresponds to a local current condensation, which is counteracted by the wavelength stretching caused by the inhomogeneous parallel flow. More quantitatively, one may assume that the mode is effectively stabilized, if the relative change of the wavelength λ during one exponential growth time exceeds, say, $\frac{1}{4}$:

$$\left(v_y\left(y + \lambda\right) - v_y\left(y\right) \right)/\gamma\lambda = \Gamma/\gamma > \tfrac{1}{4}. \tag{6.61}$$

Hence the sheet is tearing-mode-stable, if $\Gamma > \gamma_{max}/4$. In the asymptotic limit of large S one has $\gamma_{max} \simeq 0.6 \left(\tau_A \tau_\eta\right)^{-1/2}$ (Furth et al., 1963; see also eq. (4.100)), valid for $k_m\delta \simeq S^{-1/4}$, which by use of eq. (6.59) gives the stability condition

$$\tau_\eta^{-1} \gtrsim 0.15 \left(\tau_\eta \tau_A\right)^{-1/2}$$

or

$$\delta/\Delta \gtrsim 2 \times 10^{-2}. \tag{6.62}$$

Since for relatively low S-values, $S_\delta \equiv \Delta/\delta < 10^2$, γ_{max} is somewhat smaller than predicted by the asymptotic formula, and because of the only semi-quantitative nature of the criterion (6.61), we may say that the tearing mode is unstable for current sheet aspect ratio $\Delta/\delta \gtrsim 10^2$. In a

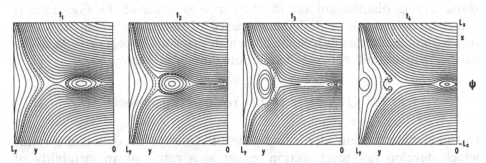

Fig. 6.14. Repetitive plasmoid generation in a tearing unstable Sweet-Parker current sheet. The figure shows ψ-contours at times $t_1 < t_2 < t_3 < t_4$ (from Biskamp, 1986a).

recent study, Phan & Sonnerup (1991) consider the case where the flow entering the sheet is diverted in the z-direction. Since there is no stabilizing parallel flow, their stability limit, $S_\delta \simeq 12.5$, is close to that of a static sheet.

In addition stability is expected to depend on the magnitude of the perturbation. Since the equilibrium flow, which constitutes the stabilizing effect, is affected by a finite perturbation amplitude, the current sheet may actually tear for smaller aspect ratios than predicted by the linear theory. In numerical simulations a current sheet configuration appears to be more prone to tearing when a poorer spatial resolution is used, since discreteness effects generate a higher noise level. Finally, the lack of stationarity, which is to be expected for general reconnection systems, leads to a higher sensitivity with respect to the tearing mode.

From the scaling laws for driven reconnection, eqs (6.37), (6.38), $\Delta/\delta \sim \eta^{-2}$, or (6.43), $\Delta/\delta \sim \eta^{-2/3}$, we see that the tearing mode will be unstable in a Sweet-Parker sheet for sufficiently small η. Let us therefore consider its nonlinear behavior. Though there are cases where a chain of several magnetic islands of comparable size is generated similar to the tearing mode in a one-dimensional static sheet pinch, more typically only one isolated island is generated in the most unstable center part of a dynamic current sheet. While growing in width and thickness in a self-similar way it is convected along the sheet and expelled into the downstream region. Such an isolated island is now generally called a plasmoid. Figure 6.14 illustrates the repetitive generation and evolution of plasmoids, showing a configuration of driven reconnection as given in Fig. 6.9 (for clarity the lower symmetric part is added in the plot). Changing the boundary conditions at $y = 0$ slighty from $\partial_y \psi = 0$ to $\psi(x, t) = \psi(x) + Mt$, where M is the externally imposed average reconnection rate and $\psi(x)$ is the stationary ψ distribution on the axis $y = 0$ observed in Fig. 6.9c, is sufficient to generate a periodic sequence of plasmoids. The properties of plasmoids and their acceleration, which plays an important role in the earth's magnetotail, are considered more closely in section 6.6.3.

6.6   Examples of 2-D reconnecting systems

In this section we consider in more detail some typical dynamic systems which develop fast reconnection, either as a result of an instability of the initial configuration in the cases of the coalescence instability and the resistive kink instability, or by a lack of equilibrium in the case of plasmoid acceleration.

6.6.1   *Coalescence of magnetic islands*

There are essentially two types of dynamic processes which may occur in an evolving two-dimensional magnetic configuration, tearing and coalescence. The natural plasma shape is that of circular cross-section, the cylindrical pinch. If such a system is forced into a strongly elongated shape, it will tear, i.e. break up into two or more fractions of roughly circular cross-section, which will subsequently coalesce to restore the original configuration, as illustrated in Fig. 6.15. Both processes are basically driven by the same physical effect, the attractive force between parallel currents. Tearing is usually a slow process. It corresponds to a local condensation or nucleation of the current density in a conducting medium, which from the beginning has to overcome the ideal flux conservation constraint by reconnection and hence cannot build up momentum to drive reconnection at a rate substantially faster than that of global resistive diffusion. By contrast, coalescence is a fast process. The currents flowing in the two flux tubes exert a finite attractive force on each other, which drives reconnection at a fast rate.

The coalescence process has been studied, starting from a periodically corrugated sheet pinch equilibrium consisting of a sequence of magnetic islands as shown in Fig. 3.7 (Pritchett & Wu, 1979; Biskamp & Welter, 1980). It belongs to the class of equilibria discussed in section 3.3.4,

$$\psi(x, y) = B_\infty a \ln \left(\cosh \left(x/a \right) + \varepsilon \cos \left(y/a \right) \right), \qquad (6.63)$$

where B_∞ is the asymptotic field for $|x| \to \infty$, and ε is a measure of the equilibrium island size w_0, given by the equation $\cosh \left(w_0/2a \right) = 1 + 2\varepsilon$, $w_0/a \simeq 4\sqrt{\varepsilon}$ for $w_0 \ll a$. It is interesting to note that the equilibrium eq. (6.63) corresponds to a finite amplitude tearing mode with wavenumber $ka = 1$, the marginal stable mode in an uncorrugated ($\varepsilon = 0$) sheet pinch.

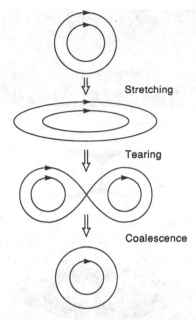

Fig. 6.15. A circular pinch forced into an elongated shape tends to restore its original shape by tearing and subsequent coalescence (schematic drawing).

It has been shown that the equilibrium (6.63) is ideally unstable with respect to pairwise island coalescence (Finn & Kaw, 1977) for any equilibrium island size $w_0 > 0$ (Pritchett & Wu, 1979). The nonlinear dynamic process has been investigated numerically, by solving the incompressible 2-D MHD equations (6.29), (6.30) (Biskamp & Welter, 1981). The process can be divided into an ideal MHD phase, where the islands are freely accelerated toward each other, leading to field compression and formation of a current sheet between the islands, and a quasi-stationary reconnection phase. For intermediate values of the normalized resistivity $\eta = S^{-1}$, typically 10^{-2} to 10^{-4}, a self-similar behavior is observed, with the following η-scaling laws for the upstream quantities u_0, B_0 taken just in front of the diffusion layer:

$$u_0 \propto \eta^{1/3} \,,$$
$$B_0 \propto \eta^{-1/3} \,, \tag{6.64}$$

and the width Δ and thickness δ of the layer:

$$\Delta \simeq w_0 \propto \eta^0 \,,$$
$$\delta \propto \eta^{2/3} \,. \tag{6.65}$$

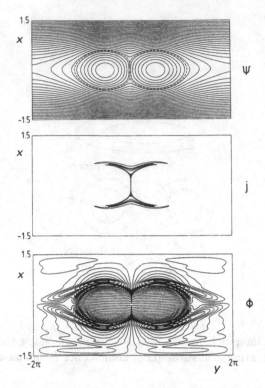

Fig. 6.16.   Contours of ψ, j, ϕ during island coalescence.

As a consequence the reconnection rate computed at the X-point, $\dot{\psi}_X = \eta j_X = u_0 B_0$, is independent of η. This does not mean, however, that the reconnection process can be associated with a Petschek-like behavior. In fact there is no relationship, since reconnection occurs in a macroscopic current sheet, $\Delta \simeq w_0$. Relations (6.64), (6.65) agree with the scaling laws for driven reconnection (6.41), (6.42), since the equivalent M produced by the coalescence instability is large enough for $\eta < 10^{-2}$ to generate a current sheet of the global system size, i.e. the equilibrium island width w_0. A typical state is illustrated in Fig. 6.16.

Obviously the equations (6.64) can be valid only in a certain η-range, since the value of the field B_0 cannot exceed the maximum value B_m, which would be obtained in the ideal case $\eta = 0$ for $u_0 = 0$, when the inward motion is reversed because of the repelling force produced by the compressed field. Hence the scaling law (6.64) breaks down, if B_0 approaches B_m. For smaller values of η, typically $\eta < 10^{-4}$, one finds a Sweet-Parker scaling law

$$\bar{u}_0 \propto \eta^{1/2}$$
$$\bar{B}_0 \simeq B_m \propto \eta^0 \,,$$

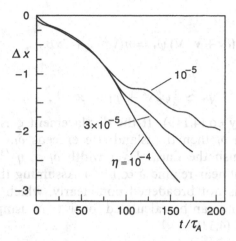

Fig. 6.17. Change of O-point position $\Delta x(t)$ during island coalescence for three different values of η. Ideal phase $t \lesssim 75\,\tau_A$, reconnection phase $t > 75\,\tau_A$.

$\bar{u}_0$, $\bar{B}_0$ indicating a time average, since in this regime the motion of the island plasma toward the current sheet is increasingly modulated by an internal oscillation or sloshing owing to the finite kinetic energy obtained in the first phase of the instability. The behavior is illustrated in Fig. 6.17, showing the change $\Delta x(t)$ of the position of the O-point during coalescence, $\Delta x = 0$ corresponding to the initial two-island equilibrium state, $\Delta x = -\pi$ to the final one-island state, for three values of the resistivity. While in the first phase $t \lesssim 90\,\tau_A$, the ideal process of island acceleration, Δx is independent of η, in the subsequent phase dominated by current sheet reconnection one finds in fact $d\Delta x/dt \simeq \text{const.} \propto \eta^{1/2}$.

6.6.2 Nonlinear evolution of the resistive kink mode

The linear theory of the $m = 1$ resistive kink mode given in section 4.7.2 indicates that in general this mode differs significantly from the $m \geq 2$ tearing mode. The physical reason is that the latter corresponds to an ideally stable mode, while the ideal $m = 1$ mode is usually close to marginal stability or even unstable. Hence the resistive mode is driven by the free energy of the ideal mode and a more rapid nonlinear evolution than the diffusive growth of the tearing mode can be expected.

First consider the geometric properties of the magnetic island produced by a finite amplitude $m = 1$ eigenmode. For simplicity we restrict ourselves to the case of marginal ideal stability $\lambda_H = 0$, eq. (4.117), corresponding to the reduced MHD approximation, where the eigenfunctions are given explicitly in eqs (4.118), (4.119). In the vicinity of the rational surface $r = r_1$, $x = r - r_1 \ll r_1$, the helical flux function ψ_*, eq. (5.11), following the equation

$$(\partial_t + \mathbf{v} \cdot \nabla) \psi_* = \eta (\nabla^2 \psi_* + 2\alpha B_0) ,$$

has the form

$$\psi_* \simeq \tfrac{1}{2}\psi_0'' x^2 + \psi_1(x) \cos\theta , \qquad (6.66)$$

where ψ_1 is given by eq.(4.119). If the displacement ξ is larger than the resistive layer width δ_l then the island size exceeds δ_l. (The subscript l is added to distinguish the linear layer width $\delta_l \propto \eta^{1/3}$ from the sheet thickness in the nonlinear regime $\delta \propto \eta^{1/2}$.) Assuming that the magnetic perturbation $\psi_1(x)$ is not broadened nonlinearly, which is shown below, the shape of the island can be calculated for $w \gg \delta_l$ using the asymptotic form ($\delta_l \to 0$) of eq. (4.119),

$$\psi_1(x) = \begin{cases} \psi_0'' \xi x & x < 0 \\ 0 & x > 0, \end{cases} \qquad (6.67)$$

where ξ is the uniform displacement of the plasma inside the $q = 1$ surface. The magnetic configuration is illustrated in Fig. 6.18. The condition $\nabla \psi_* = 0$ gives the O-point of the island at $x = -\xi$, $\theta = 0$. The two branches of the separatrix are obtained by considering the flux surfaces passing through the "X-point" at $x = 0$, $\theta = \pi$. They have the value $\psi_* = 0$, i.e.

$$x^2 + 2\xi x \cos\theta = 0 , \qquad x < 0 ,$$
$$x^2 = 0 , \qquad x > 0 .$$

Hence the outer separatrix is given by the concentric circle

$$x = 0 ,$$

while the inner separatrix is given by the equations

$$x = -2\xi \cos\theta , \qquad |\theta| < \pi/2 ,$$
$$x = 0 , \qquad |\theta| > \pi/2 ,$$

i.e. it consists of a shifted half-circle of the same radius as the outer separatrix for $|\theta| < \pi/2$ and coincides with the latter for $|\theta| > \pi/2$. This part represents a current sheet, the δ-function singularity in $\psi_1''(x)$ implied in eq. (6.67). While the current sheet is sustained for $|\theta| > \pi/2$ owing to the impinging plasma flow, it is smeared out over the island for $|\theta| < \pi/2$, since here the plasma flow is receding. Hence eq. (6.67) is strictly speaking only valid for $|\theta| > \pi/2$, while for $|\theta| < \pi/2$ the perturbation is smoothed across the island, the size of which is

$$w = 2\xi . \qquad (6.68)$$

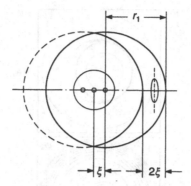

Fig. 6.18. Schematic drawing of the island shape caused by a central displacement ξ in the nonlinear resistive kink mode.

The current sheet, the magnetic configuration, and the flow pattern obtained from an exact simulation are given in Fig. 6.19. Hence the nonlinear reconnection process occurs in a quasi-stationary current sheet of width $\Delta \simeq \pi r_1/2$.

In order to compute the nonlinear evolution $w(t)$, we adjust the plane Sweet-Parker reconnection model outlined in section 6.1.1 to the geometry of the kink mode. In the laboratory frame the flow into the current sheet is not symmetric, but enters only from the interior region, as seen in Fig. 6.19c. Since the reconnection of the helical field is necessarily symmetric, involving equal positive and negative amounts of field, the current sheet itself is moving outward at a velocity $u/2$, where $u = \dot{\xi}$ is the plasma velocity in the laboratory frame, such that in the frame of the sheet the inflow velocity is $u_0 = u/2$. The continuity equation yields the relation

$$\int_0^{\pi/2} u_{0n} r_1 d\theta' = u_0 r_1 = v_0 \delta ,\tag{6.69}$$

where $v_\theta = v_0$ is the outflow speed, δ the sheet half-thickness and u_{0n} the normal component of the inflow velocity, $u_{0n} = u_0 \cos \theta'$, $\theta' = \pi - \theta$.

The outflow speed v_0 equals the upstream Alfvén velocity computed with the field component to be reconnected (see eq. (6.5)),

$$v_0 = B_* ,\tag{6.70}$$

where B_* is the helical field in front of the sheet. B_* can most easily be obtained from the behavior of the helical flux $\psi_*(r, \theta = \pi)$ shifted rigidly toward the sheet, as illustrated in Fig. 6.20,

$$B_* = |\partial_r \psi_*| = |\psi_0''| \, \xi/2 = |\psi_0''| \, w/4 .\tag{6.71}$$

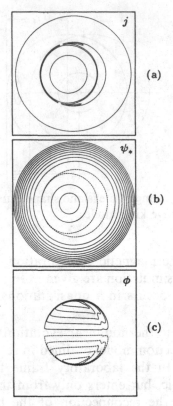

Fig. 6.19. Contour plots of (a) current density, (b) helical flux ψ_*, (c) stream function ϕ, for a simulation of the nonlinear resistive kink mode with $S = 10^7$ (from Biskamp, 1991).

The total change of the helical field across the sheet $[B_*] = 2B_*$ agrees with the expression obtained from eqs (6.66) and (6.67).

In contrast to a plane current sheet the inflow velocity u_{0n} is not homogeneous and hence the thickness δ is not constant but varies along the sheet, $\delta = \delta(\theta)$. This is determined by Ohm's law for stationary conditions across the sheet, generalizing eq. (6.2),

$$u_{0n}(\theta) \, B_*(\theta) = \eta \, j(\theta) \simeq \eta \, \frac{B_*(\theta)}{\delta(\theta)} \qquad (6.72)$$

and hence

$$\delta(\theta) = \eta/u_0 \cos\theta' \, . \qquad (6.73)$$

Evidently the current sheet cannot extend up to $\theta' = \pi/2$, but only to some value $\theta_0 < \pi/2$. A more detailed theory gives $\theta_0 \simeq 60°$ (Zakharov et al., 1993). Because of the broadening of the sheet the behavior in the edge region is rather smooth compared with the quasi-singular structure

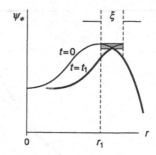

Fig. 6.20. Helical flux $\psi_*(r, \theta = \pi)$ (heavy line) resulting from a rigid shift of the interior branch which leads to a jump in the helical field B_*. The reconnected flux is indicated by the shaded area.

encountered in a plane current sheet (Figs 6.11, 6.12), the negative current density part being only weakly pronounced in Fig. 6.19a.

Inserting eqs (6.70), (6.71) and (6.73) with $\theta' = \theta_0$ into eq. (6.69) and using $u_0 = u/2 = \dot{w}/4$, one obtains the equation for the time evolution of the island,

$$\dot{w} = 2 \left(\eta \left| \psi_0'' \right| w / r_1 \cos \theta_0 \right)^{1/2}, \tag{6.74}$$

which yields the algebraic growth law for small but finite island size (Waelbroeck, 1989; Biskamp, 1991):

$$w = \eta \left| \psi_0'' / r_1 \cos \theta_0 \right| t^2 \tag{6.75}$$

or

$$u = \eta \left| \psi_0'' / r_1 \cos \theta_0 \right| t . \tag{6.76}$$

The nonlinear time scale obtained by setting $w \simeq r_1$ is $\tau = O(\eta^{-1/2})$. Though the flow velocity into the sheet is small, $u \lesssim O(\eta^{1/2})$ as seen from eq. (6.76), inertia is not negligible, since the velocity along the sheet v_0 is independent of η.

Equation (6.76) is in quasi-quantitative agreement with results obtained from numerical simulations. Figure 6.21 shows the velocity of the plasma center $u(t)$ (the center curve) for $\eta = 10^{-7}$. For reference the upper (dashed) curve represents continued exponential growth, while the lower curve represents the quasi-linear approximation, where u saturates, $u_{ql} \sim \gamma \delta_l = O(\eta^{2/3})$, in contrast to the exact behavior where $u(t)$ grows linearly reaching $u_{max} = O(\eta^{1/2})$. The resistive kink mode does not saturate at a finite island size but island growth proceeds until the entire helical flux originally inside the resonant surface $r = r_1$ is reconnected and the system has effectively returned to a symmetric state.

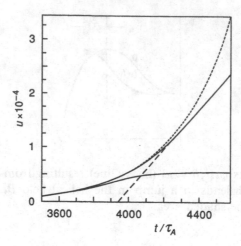

Fig. 6.21. Time evolution of the velocity u in the resistive kink mode (from Biskamp, 1991).

6.6.3   Plasmoids

As indicated in section 6.5, the nonlinear evolution of the tearing mode in an open, weakly two-dimensional current sheet differs significantly from that in periodic systems. In the latter there is a chain of islands, for instance an $(m, n) = (m_0, 1)$ mode in a cylindrical configuration (no island coalescence is possible because of $n = 1$ corresponding to the largest possible wavelength), which grow slowly on the global resistive time scale (section 5.4). Since inertia effects are negligible, the process constitutes a sequence of smooth equilibrium states. By contrast, in an open sheet configuration a single major plasmoid is usually generated, which moves rapidly along the sheet while growing in size; an example was given in Fig. 6.14. Inertia effects are important, since the system is not in MHD equilibrium, which allows much faster reconnection.

Let us consider this process in more detail. Plasmoids seem to be an important feature in various kinds of eruptive processes in astrophysical plasmas, notably magnetospheric substorms and solar flares. In fact in the numerical modelling of the earth's magnetotail the formation of plasmoids, which is believed to be the origin of the substorm phenomemon, has been investigated most intensively (Birn & Hones, 1981; Lee et al., 1985; Hautz & Scholer, 1987; Ugai, 1989; Otto et al., 1990; Kageyama et al., 1990).

The basic process is illustrated in Fig. 6.22. A weakly two-dimensional static equilibrium is assumed as initial state, modelling a typical magnetotail configuration. Equilibrium pressure variations along the tail midplane are usually chosen in such a way that the configuration has a distant X-point (measured from the earth, the left-hand boundary). This X-point

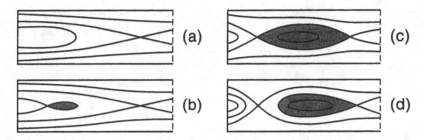

Fig. 6.22. Schematic drawing of the plasmoid evolution in a weakly two-dimensional magnetic configuration (from Otto et al., 1990).

is useful in order to define unambiguously the volume, mass and momentum of the plasmoid (Otto et al., 1990). Figure 6.23 gives a sequence of plasmoid states from a two-dimensional compressible MHD simulation. (Because of compressibility the flow pattern cannot be visualized by contour plots of a streamfunction ϕ, but requires a vector representation.) For finite resistivity the initial configuration is unstable to tearing. By applying a field perturbation or by locally increasing η one initiates reconnection at a particular position y_0 ($y_0 = 25$ in Fig. 6.23a), creating an X-point. The resulting plasmoid is accelerated along the sheet in the direction of decreasing pressure and field intensity. Since this motion leads plasma away from the X-point, it has to be replenished by flows from above and below into the X-point region, giving rise to a quasi-forced reconnection process on the Sweet-Parker time scale instead of the global resistive time scale in the case of a periodic tearing mode. In fact a long current sheet is created extending up to the receding plasmoid (Fig. 6.23b). In this phase the original distant X-point is hardly visible any more, but the plasmoid volume is well defined by the flow pattern, the velocity being large and nearly uniform over the plasmoid cross-section. Interaction with the downstream plasma still at rest leads to shock formation and a blunt leading plasmoid edge, which gives the plasmoid a drop-like shape and decreases the plasmoid acceleration. When the trailing current sheet is long enough, it becomes tearing-unstable itself, producing secondary plasmoids, which are strongly accelerated toward the primary one and eventually coalesce with the latter (Fig. 6.23c, d). Three-dimensional effects do not change the general picture qualitatively, as shown for instance by Kageyama et al. (1990), who present a global simulation of the magnetotail formation by the interaction of the dipole field with the solar wind. Tearing instability of the resistive tail gives rise to the continuous generation of plasmoids.

The question concerning the dominant force in the plasmoid acceleration has been investigated by Otto et al. (1990). It turns out that the

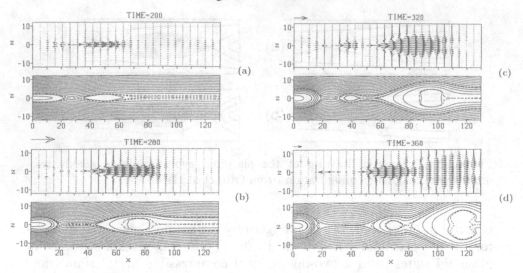

Fig. 6.23. Simulation of plasmoid generation in the geomagnetic tail. The arrows indicated above the charts of the flow pattern correspond to unit velocity (from Hautz & Scholer, 1987).

pressure force is significantly larger than the magnetic tension, which is similar to the plasma acceleration along a simple Sweet-Parker current sheet (section 6.4.1).

   While the plasmoid acceleration is essentially an ideal MHD process, the reconnection time scale τ and hence the plasmoid growth depend on the value of η, roughly $\tau \sim S^{1/2}$. This implies that plasmoid size decreases with η. In most magnetotail simulation studies this fact is concealed by the use of an anomalous resistivity model with $\eta \propto j$. It should be mentioned that the basic problem in the theory of magnetospheric substorms is indeed the identification of the relevant dissipation process in Ohm's law. Since the magnetotail plasma is collisionless, this has to be a collective process. Because of $T_i > T_e$, however, collisionless modes are in general strongly damped and cannot easily be excited which is consistent with the fact that the magnetotail is quiescent most of the time and substorms are rare events.

6.7 Magnetic reconnection in general three-dimensional systems

Up to this point consideration has been restricted to two dimensions, i.e. to systems with spatial symmetry, where the definition of magnetic reconnection in terms of a change of magnetic topology is unambiguous and the effect of reconnection is obvious on a simple inspection of the field configuration. Field line topology is determined by the flux function ψ with regions of different topology separated by separatrix surfaces ψ_s, connecting to X-type neutral points (or the generalizations thereof, such as osculating separatrices or current sheets with Y-points).

These concepts are no longer valid, however, in non-symmetric three-dimensional systems, which has led to certain misunderstandings. Following Schindler et al. (1988), we consider the process of plasmoid formation in a system of finite size in the z-direction with a superimposed weak B_z field, modelling the geomagnetic tail as illustrated in Fig. 6.24a. Obviously a separatrix, defined as a surface enclosing a region of field lines localized in x, does not exist, as all field lines finally connect to the earth, the left-hand tail edge. Nevertheless, field line reconnection does occur in the sense of a localized breakdown of the frozen-in condition and a resulting change of the field line connection, as is illustrated in Fig. 6.24b. Hence in three-dimensional systems one has to resort to the original physical meaning of the term reconnection as localized magnetic diffusion, which implies the presence of a finite parallel electric field $E_\parallel = \eta j_\parallel$ (or some other dissipation effect in Ohm's law). Intuitively speaking a general configuration will undergo reconnection if in a certain surrounding the projection perpendicular to a central field line has an X-point structure and flows have opposite directions in opposite quadrants.

Since the simple two-dimensional criteria based on flux surface topology cannot be applied, it is interesting to obtain a more general criterion to decide whether in a given plasma volume V magnetic reconnection takes place. To this purpose one may use the magnetic helicity (section 2.2),

$$H = \int_V \mathbf{A} \cdot \mathbf{B} \, d^3x \,.$$

However, H is a well-defined quantity only if gauge-invariant, which requires certain conditions on the surface S of the volume V to be satisfied. The time derivative of H is

$$\frac{dH}{dt} = -2 \int_V \mathbf{E} \cdot \mathbf{B} \, d^3x - \oint_S (\phi \mathbf{B} + \mathbf{E} \times \mathbf{A}) \cdot d\mathbf{F} \,, \tag{6.77}$$

using $\partial_t \mathbf{B} = -\nabla \times \mathbf{E}$ and $\mathbf{E} = -\nabla\phi - \partial_t \mathbf{A}$. If $B_n = E_t = 0$ on the boundary, which corresponds to a conducting wall, the gauge-dependent surface term vanishes. In this case $dH/dt \neq 0$ indicates $E_\parallel \neq 0$ somewhere in the system, i.e. reconnection occurs. In astrophysical applications, however, the case of an unbounded system is of more interest, where dynamical processes are limited to a finite region, while outside there is a static magnetic field and no electric field. Here one can define a more general helicity expression (Finn & Antonsen, 1985)

$$\bar{H} = \int_V (\mathbf{A} + \mathbf{A}_0) \cdot (\mathbf{B} - \mathbf{B}_0) \, d^3x \,, \tag{6.78}$$

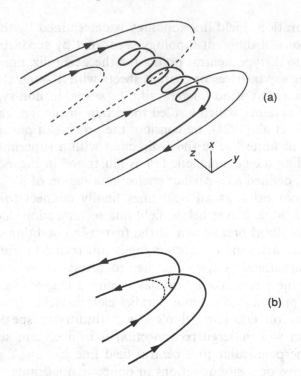

Fig. 6.24. (a) Schematic drawing of field lines in the geomagnetic tail with a finite component B_z, carrying a plasmoid. (b) Reconnection of two field lines in the process of plasmoid formation.

where $\mathbf{B}_0$, $\mathbf{A}_0$ are the field and vector potential at some reference time t_0. Analogously to eq. (6.77) one obtains

$$\frac{d\bar{H}}{dt} = -2\int_V \mathbf{E}\cdot\mathbf{B}\,d^3x - \oint_S [\phi\,(\mathbf{B}-\mathbf{B}_0) + \mathbf{E}\times(\mathbf{A}+\mathbf{A}_0)]\cdot d\mathbf{F}. \qquad (6.79)$$

For a static asymptotic field $\mathbf{B}=\mathbf{B}_0$ and vanishing asymptotic electric field $\mathbf{E}=0$ the surface integral vanishes. Hence $d\bar{H}/dt \neq 0$ implies $E_\parallel \neq 0$ and thus guarantees the presence of reconnection.

A different more explicit approach to the 3-D reconnection problem has been developed by Lau & Finn (1990). Given a regular magnetic field $\mathbf{B}(\mathbf{x},t)$, how can one decide whether reconnection is involved and where the reconnection layers are located in space? Here reconnection is defined in the following way. The ideal form of Ohm's law $\mathbf{E}+\mathbf{v}\times\mathbf{B}=0$, called the frozen-in condition, implies a perpendicular plasma flow

$$\mathbf{v}_\perp = \mathbf{E}\times\mathbf{B}/B^2, \qquad (6.80)$$

called the field line velocity. The electric field satisfies the condition

$\mathbf{E} \cdot \mathbf{B} = 0$, from which follows the equation for the electrostatic potential,

$$\mathbf{B} \cdot \nabla \phi = -\mathbf{B} \cdot \partial_t \mathbf{A}. \tag{6.81}$$

Only if a smooth solution ϕ exists in the volume considered can the plasma flow be regular and ideal MHD valid throughout the volume. If ϕ exhibits singularities, this implies a violation of the frozen-in condition. In this case magnetic reconnection is said to occur. Flow regularity requires that in the vicinity of such ideal singularities the ideal Ohm's law is replaced by

$$\mathbf{E} + \mathbf{v} \times \mathbf{B} = \mathbf{R},$$

where $\mathbf{R} = \eta \mathbf{j}$ or some equivalent process, hence $E_\parallel \neq 0$.

In two-dimensional systems singularities of $\phi(x, y)$ are located on the separatrix and in particular in the X-point. As an example consider the vector potential

$$\mathbf{A}(x, y, t) = -(\psi(x, y) + E_0 t)\mathbf{e}_z \tag{6.82}$$

with

$$\psi(x, y) = \tfrac{1}{2}(x^2 - y^2), \tag{6.83}$$

discussed in section 6.3, corresponding to the stationary magnetic field $\mathbf{B} = (B_x, B_y, B_z) = (y, x, B_0)$. Equation (6.81) becomes

$$\mathbf{B} \cdot \nabla \phi = B_0 E_0, \tag{6.84}$$

which has the solution

$$\phi = \tfrac{1}{2} B_0 E_0 \ln \left| \frac{x + y}{x - y} \right|, \tag{6.85}$$

essentially identical with eq. (6.34). Since in the present context ϕ is the electric potential, not the streamfunction of an incompressible flow (note that the field line velocity (6.80) is not assumed to be incompressible), the axial field B_0 appears in eq. (6.85). ϕ is singular on the separatrix $x = \pm y$, and hence reconnection occurs for any $E_0 \neq 0$. It is interesting to compute the field line velocity, eq. (6.80),

$$\begin{aligned}
\mathbf{v}_\perp &= \frac{(-\nabla \phi + E_0 \mathbf{e}_z) \times (\mathbf{e}_z \times \nabla \psi + B_0 \mathbf{e}_z)}{B^2} \\
&= \frac{B_0 \mathbf{e}_z \times \nabla \phi - E_0 \nabla \psi}{B^2} - \frac{\mathbf{e}_z \nabla \phi \cdot \nabla \psi}{B^2}.
\end{aligned} \tag{6.86}$$

In this expression the first term is in the poloidal plane, the second in the axial direction. $\mathbf{v}_\perp$ can be expressed by two components v_1, v_2:

$$\mathbf{v}_\perp = v_1 \nabla \psi / |\nabla \psi| + v_2 (\mathbf{B} \times \nabla \psi) / |\mathbf{B} \times \nabla \psi|,$$

where

$$v_1 = \mathbf{v}_\perp \cdot \nabla\psi/|\nabla\psi| = -E_0\big/\sqrt{x^2 + y^2}\,, \qquad (6.87)$$

$$v_2 = \mathbf{v}_\perp \cdot (\mathbf{B} \times \nabla\psi)\,/|\mathbf{B} \times \nabla\psi|$$

$$= \nabla\phi \cdot \nabla\psi/|\mathbf{B} \times \nabla\psi|$$

$$= -2B_0E_0\,xy\big/\left[(x^2 - y^2)\sqrt{x^2 + y^2}\sqrt{B_0^2 + x^2 + y^2}\right].\quad (6.88)$$

The component v_1 describes the flow in the poloidal plane toward or away from the X-point. It represents the convection of the flux surfaces ψ, which is singular only at the X-point. The component v_2 describes the field line flow in the flux surface $\psi =$ const. and has singularities along the separatrix surfaces $x = \pm y$. Only in the limit $B_0 = 0$ does one have $v_2 = 0$. (Note that the field line velocity gives only the plasma motion $\mathbf{v}_\perp$ perpendicular to the field line. The parallel component has to be obtained from an additional equation determined by the plasma dynamics, for instance from $\nabla \cdot \mathbf{v} = 0$ together with appropriate boundary conditions.)

On generalizing these concepts to three dimensions a new feature arises, the existence of isolated nulls of the magnetic field, i.e. points where $\mathbf{B} = 0$. Field nulls are structurally stable, meaning that they persist when the system is weakly perturbed, their position being only slightly shifted, in contrast to the separatrix in a two-dimensional system, which vanishes when the system becomes weakly three-dimensional. In the vicinity of a null (assumed to be located at $\mathbf{x} = 0$) the Taylor expansions of $\mathbf{B}$ and $\mathbf{j}$ are

$$B_i = \beta_{ij}x_j \qquad (6.89)$$

$$j_i = \varepsilon_{ijk}\beta_{jk}\,. \qquad (6.90)$$

Because of $\nabla \cdot \mathbf{B} = 0$ the trace of the real matrix β_{ij} vanishes. The matrix has three eigenvalues, which are used for a classification of nulls (Fukao et al., 1975; Greene, 1988). If all eigenvalues are real, either one is positive and two negative (A-type) or two are positive and one negative (B-type), since their sum must vanish. The eigenvectors of the two eigenvalues of the same sign locally form a plane, which can be continued into a global surface Σ by following the field lines located in this plane. Following the field lines along the third eigenvector defines a curve γ in space. The field lines near an A-type null are shown in Fig. 6.25, the surface Σ_A dividing space into two regions not connected by field lines, and all field lines in one region converging to form a bundle around γ_A. A B-type null has the same topology with the field line directions reversed. A- or B-type nulls are the generalization of a two-dimensional X-point, into which they degenerate if one of the two eigenvalues of equal sign becomes zero. The correspondence to X-points can also be seen from the fact that at

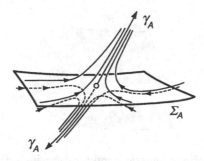

Fig. 6.25.   Field lines near an A-type magnetic null (from Lau & Finn, 1990).

A- or *B*-type nulls the current density (6.90) may vanish, since β_{ij} can be symmetric. These nulls are therefore expected to be important in the context of three-dimensional reconnection.

Nulls with one real and two conjugate complex eigenvalues (*S*-type) are generalizations of a two-dimensional *O*-point, to which they degenerate, if the real eigenvalue vanishes (in this case the remaining two are purely imaginary). As in the case of an *O*-point in the two-dimensional limit the current density cannot vanish at an *S*-type null, since β_{ij} is necessarily nonsymmetric.

A magnetic configuration with two nulls, one *A*-type and one *B*-type, is of special interest. The relative positions of the surfaces Σ_A, Σ_B and the curves γ_A, γ_B are illustrated in Fig. 6.26. Since in Σ_B all field lines are directed away from the point *B* and are collected in the bundle around γ_A, the line γ_A bounds Σ_B, and since all field lines in Σ_A must originate in the bundle around γ_B, the line γ_B bounds Σ_A. Hence Σ_A, Σ_B are semi-infinite half-sheets intersecting in a line connecting *A* and *B*, called a null–null line. A topologically equivalent configuration arises when superimposing a constant field on a dipole field (e.g. the earth's magnetic field in the presence of an interplanetary field). Since the surfaces Σ_A, Σ_B are not pierced by field lines, they serve as separators, the three-dimensional generalization of the two-dimensional separatrix.

Lau and Finn calculated the field line velocity, eq. (6.80), for a three-dimensional field configuration, consisting of the two-dimensional configuration, eq. (6.83), periodically modulated in the *z*-direction. If the modulation is weak, such that B_z does not change sign, the behavior of $\mathbf{v}_\perp$ is qualitatively the same as in the unmodulated case. For sufficiently strong modulation B_z becomes zero at certain values of *z* and field nulls appear, forming an alternating sequence of *A*- and *B*-type nulls. The singularities arising on the separator surfaces are of a more complicated type than in the case of no nulls and in addition essential singularities appear on the null–null lines between each pair of nulls.

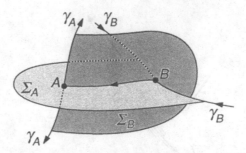

Fig. 6.26. Topology of a magnetic configuration including one A-type null (A) and one B-type null (B) (from Lau & Finn, 1990).

This kinematic approach as developed by Lau & Finn (1990) is, however, valid only in the limit of a vanishingly small reconnection rate $E_0 = O(\eta)$, since the effect of the singular flow dynamics on the magnetic field is neglected, **B** being prescribed as a smooth vector field. For larger values of E_0 the magnetic configuration is changed, presumably on a macroscopic scale. In two dimensions Syrovatskii has shown that, whenever $E_0 \neq 0$ at an X-point, a current sheet of finite length appears, as discussed in section 6.2.3. The three-dimensional generalization in the presence of field nulls has still to be worked out.

Finally we consider three-dimensional toroidal configurations, where field lines are endless. Though field nulls do in general not occur, the new feature compared with the axisymmetric case is the appearance of regions of stochastic field line behavior. As discussed in section 3.5 this means that such field lines are not localized to a particular flux surface but fill a finite volume. It appears that plasma flows resulting kinematically from the frozen-in condition $\mathbf{E} + \mathbf{v} \times \mathbf{B} = 0$ are singular in such a volume. Taking finite resistivity into account and allowing plasma dynamics to modify the field locally will probably lead to turbulent motions. MHD turbulence arising in regions of stochastic field lines has been invoked as a model for tokamak disruptions (section 8.2).

6.8 Turbulent reconnection

As stated at the beginning of this chapter, the fundamental problem in the theory of magnetic reconnection is to explain the observed fast times. In explosive magnetic processes such as tokamak disruptions or solar flares time scales seem to be essentially independent of the value of the collisional resistivity η, and primarily connected with typical Alfvén times $\tau_A = L/v_A$. Here L is the size of the region where the magnetic field is affected, for instance the minor radius of a tokamak plasma or the length of a magnetic loop or arcade in the solar atmosphere.

As a consequence, stationary current sheet reconnection, which appears to be the dominant mechanism for intermediate values of S, becomes too slow for "realistic" S-values. A possible solution of the problem is to consider further nonideal effects in Ohm's law such as (anomalous) electron viscosity or electron inertia, which in a weakly collisional plasma may be more important than resistivity, as is discussed in section 8.1.5. However, a more effective modification of the reconnection process at smaller η seems to be caused by a nonstationary behavior in the reconnection region. A first indication of such behavior is provided by the occasional generation and ejection of plasmoids. If η is further decreased, plasmoids are generated more frequently and in a more irregular way, with plasmoid coalescence becoming an important reconnection process, as can be seen in Fig. 6.23. Thus secondary small current sheets are generated with lifetimes that are small compared with the global time scale. It is not difficult to visualize a gradual transition to a state of fully developed MHD turbulence, in particular for nonsymmetric systems. In a region of stochastic field lines in a toroidal plasma a certain level of small-scale turbulence is expected even under macroscopically quiescent conditions.

As is discussed in section 7.6 turbulent dissipation in MHD systems is very efficient, with energy dissipation rates becoming essentially independent of η for large Reynolds numbers already in 2-D (in contrast to 2-D Navier-Stokes turbulence). This property is even more true in 3-D. In a reconnecting system turbulence is not spread uniformly over the entire magnetic configuration but predominantly confined to regions of relatively small extent, excited by strong field gradients. The regions of strong turbulence are probably not fixed in space but fluctuate, appearing here and there, which gives rise to a burst-like global behavior.

Small-scale MHD turbulence gives rise to an effective resistivity η_{eff} (sections 7.3.2 and 8.2.5) and hence an effective reconnection rate independent of the value of the collisional resistivity η. An explicit expression for η_{eff} in terms of the fluctuation amplitudes $\delta \mathbf{B}$, $\delta \mathbf{v}$ has been derived only under special assumptions. An example is given by eq. (8.68) for two-scale turbulence. The fact that turbulence is excited mainly in regions of high current density suggests a linear relation

$$\eta_{eff} = \eta + \alpha |j - j_0| , \qquad (6.91)$$

where j_0 is some threshold value. Such a relation is often used as a simple anomalous resistivity model (e.g. Sato & Hayashi, 1979). Note that this expression is motivated by properties of MHD turbulence and hence remains within the framework of MHD theory, so that no ad hoc assumption about the excitation of current-driven micro-instabilities is required, though the latter can of course also contribute to η_{eff}.

The arbitrariness in using expression (6.91) is above all connected with the choice of the threshold value j_0. Consider the process of island coalescence, section 6.6.1, in the limit of $\eta \rightarrow 0_+$, where an η-independent turbulent behavior is to be expected. If in modelling this process j_0 is assumed to be rather low, the resulting process resembles the reconnection by a quasi-stationary current sheet at relatively high η. Choosing a somewhat higher value of j_0 would allow the system to develop nonstationary features, connected with plasmoid generation, but still suppress further fine-scale effects. Hence the value of j_0 determines the level up to which the reconnection dynamics is resolved. The question remains, to what extent a low-level model using a simple relation such as (6.91) can describe the global features of the full turbulent ($\eta \rightarrow 0_+$) system.

Though a quantitative theory of turbulent reconnection is certainly very complicated, the existence of turbulent dissipation rates independent of η in principle solves the problem of explaining sufficiently fast reconnection time scales, a problem which arises only if one extrapolates stationary current sheet reconnection rates to the limit of small η. However, for practical applications the general statement that reconnection becomes independent of η for sufficiently small values is often not very helpful, and even misleading. An example is given by the sawtooth collapse in a hot tokamak plasma, which is treated in section 8.1. Though typical S-values appear to be very large, $S \sim 10^8$, experiments indicate that reconnection is considerably slower than the temperature collapse time scale, since little helical flux appears to be reconnected.

7

MHD turbulence

Ordinary nonmagnetic fluids are known to become turbulent at sufficiently high Reynolds numbers and a similar behavior is expected for electrically conducting magnetized fluids, though direct experimental evidence is scarce. Some confusion may arise, however, owing to the convention, widespread in the fusion research community, of calling the Lundquist number $S = Lv_A/\eta$ the magnetic Reynolds number, the latter being correctly defined by $R_m = Lv/\eta$, where v is some average fluid velocity. $S \gg 1$ simply means that the resistivity is small, while the system may well be nonturbulent, or even static corresponding to $R_m \simeq 0$. S is an important theoretical parameter characterizing growth rates of possible resistive instabilities. But only when large fluid velocities are generated in the nonlinear phase of an instability or by some external stirring R_m can become large, making the system prone to turbulence. MHD turbulence can thus be expected only in strongly dynamic systems, e.g. disruptive processes in tokamaks or flares in the solar atmosphere.

Though the behavior at Reynolds numbers close to the critical value, where the transition from laminar flow to turbulence occurs, has recently attracted much attention, the strongest interest is in the high-Reynolds-number regime, where turbulence is fully developed, which is characteristic of most turbulent fluids in nature. Let us give a convenient, geometry-independent definition of the Reynolds numbers. The dynamic state of the turbulence is characterized by the energy content per unit volume $E \sim v_0^2 \sim B_0^2$, the energy dissipation rate $\varepsilon = -dE/dt$, which equals the energy injection rate for stationary turbulence, and the values of the dissipation coefficients v and η. Hence a typical velocity is $E^{1/2}$ and a characteristic global or integral scale is defined by

$$L = E^{3/2}/\varepsilon \,, \tag{7.1}$$

175

such that

$$Re = \frac{E^2}{\nu\varepsilon}, \quad R_m = \frac{E^2}{\eta\varepsilon}. \tag{7.2}$$

Since consideration will be restricted to magnetic Prandtl numbers $Pr_m = \nu/\eta \sim 1$, $Re \sim R_m$, we often denote the Reynolds number by Re even if, strictly speaking, R_m is meant.

The most conspicuous feature of high-Reynolds-number turbulence is the simultaneous presence of a continuum of spatial scales ranging from the global scale of the order of the system size down to very small scales, where dissipation occurs. In order to separate individual macroscopic effects connected with the geometry of the particular system and the mechanism of turbulence generation, e.g. some MHD instability, from the more universal behavior at medium and small scales, the concept of homogeneous and possibly isotropic turbulence is introduced in section 7.1. In this framework the theoretical formalism is significantly simplified and relevant numerical simulations and their interpretation are facilitated. This chapter is therefore confined to homogeneous turbulence.

Section 7.2 deals with the properties of ideal, i.e. nondissipative, systems, their invariants and equilibrium distributions. Though very different from real turbulence, which constitutes a driven, dissipative system, equilibrium distributions provide valuable indications about the nonlinear interaction processes, e.g. the cascade directions to be expected in real systems. These considerations indicate that 2-D and 3-D MHD turbulence are much more closely related than 2-D and 3-D hydrodynamic (Navier-Stokes) turbulence.

Self-organization is an important aspect of MHD turbulence, which is considered in section 7.3. It is intimately connected with selective dissipation of the ideal invariants, leading to large-scale quasi-static magnetic configurations and to aligned states $\mathbf{v} \parallel \mathbf{B}$.

Energy spectra play a crucial role in turbulence theory. Section 7.4 presents a phenomenological derivation, emphasizing the Alfvén effect and the influence of velocity magnetic field correlations.

Section 7.5 gives a brief introduction to more rigorous theoretical approaches, in particular two-point closure theory. We first outline the general closure problem and then introduce a particular set of closure equations, which has proved to be very convenient for practical numerical computations.

The question of turbulent dissipation is considered in section 7.6. We concentrate on 2-D turbulence, where high-Reynolds-number numerical simulations have shed some light on the problem of the Reynolds number scaling of the energy dissipation rate and the spatial structure of the dissipative eddies.

Section 7.7 gives an introduction to intermittency theory. We first discuss two models, the log-normal theory and the β-model with its generalizations, which have been developed for Navier-Stokes turbulence. The basic concepts are, however, rather general and can easily be extended to the MHD case. Some results obtained from 2-D MHD simulations are presented.

Finally, section 7.8 deals with turbulent convection of magnetic fields, which occurs in weakly magnetized conducting fluids such as the solar convection zone. In contrast to the case with a strong global magnetic field not only the current density but also the magnetic field is distributed intermittently in thin rope-like or sheet-like structures.

Throughout this chapter consideration is confined to incompressible turbulence, an assumption which is made in most theoretical investigations, and which is also a reasonable approximation for many real turbulent systems.

7.1 Homogeneous isotropic turbulence

It has long been anticipated and recently been shown rigorously in the framework of chaos theory that turbulent systems can be described only by statistical means. If the ergodic theorem is valid, at least in some weak sense, time averages, which are usually measured experimentally, can be identified with ensemble averages, the basic quantities in turbulence theory. If the system is sufficiently homogeneous, i.e. the system size is large in comparison with the dominant turbulent scales, it can be visualized as being composed of a large number of equivalent weakly coupled subsystems such that the spatial average, which is very convenient in numerical simulations, corresponds to an ensemble average.

Globally, turbulence is in general inhomogeneous, being driven by the gradient of some quantity such as velocity (turbulent shear flow), temperature (thermal convection) or, in the case of MHD, current density (pinch instabilities). In fact the gradient scale length is in practice often taken as the global scale L in the definition of the Reynolds number. In a small section in the bulk of the turbulent region away from boundary layers, however, variations of the average quantities are weak. Such a subsystem may be considered homogeneous in a statistical sense. For homogeneous turbulence, quantities such as the kinetic energy density*

$$E = \tfrac{1}{2}\langle \mathbf{v}(\mathbf{x}) \cdot \mathbf{v}(\mathbf{x}) \rangle$$

---

* Following the convention in turbulence theory, we denote the energy by E instead of the W used in the previous chapters.

or, more generally, the two-point correlation functions

$$C_{ij}(\mathbf{r}) = \langle\, v_i(\mathbf{x})\; v_j(\mathbf{x}+\mathbf{r})\, \rangle$$

are independent of $\mathbf{x}$. Since such a quasi-homogeneous subsystem is open, periodic boundary conditions are a very adequate choice. Hence fluctuating quantities can be written in Fourier representation

$$f(\mathbf{x}) = \sum_{\mathbf{k}} f_{\mathbf{k}} e^{i\mathbf{k}\cdot\mathbf{x}}\;,\quad f_{-\mathbf{k}} = f_{\mathbf{k}}^*\;, \tag{7.3}$$
$$k_i = 2\pi n_i/L,\; n_i = \pm 1,\; \pm 2,\; \ldots\,,$$

a cubic box $L_x = L_y = L_z = L$ being assumed. The Fourier components are called modes, meaning spatial modes, instead of dynamic eigenmodes. The configuration space representation and the Fourier or spectral representation supplement each other. Whereas previously turbulence theory based mainly on closure theory tended to emphasize the spectral aspect, it has more recently become evident that problems connected with spatial structures such as the dissipative eddies and their intermittent distribution are more conveniently discussed in $\mathbf{x}$-space.

While the dissipative terms become very simple in $\mathbf{k}$-space,

$$\nabla^2 f \to -k^2 f_{\mathbf{k}}\;,$$

the nonlinear terms obviously become more complicated,

$$f(\mathbf{x})\, g(\mathbf{x}) \to \sum_{\mathbf{k}'} f_{\mathbf{k}-\mathbf{k}'}\, g_{\mathbf{k}'}\;.$$

Since the nonlinear terms are the dominant ones, being larger than the dissipative terms by a factor of the order of the Reynolds number, one might be a little skeptical about the basic value of the Fourier representation for fluid turbulence. Its essential advantage is a conceptual one, viz. the direct representation of spatial scales. Since turbulent dynamics is considered as a transfer process of certain quantities, such as energy or helicity, between different scales, the Fourier representation, which directly yields the corresponding spectral densities, is very convenient.

Assuming isotropy in addition to homogeneity further simplifies the theoretical formalism and the interpretation of experimental or simulation results. While in hydrodynamic turbulence isotropy is a valid assumption in a coordinate system moving with the average flow velocity, in an electrically conducting turbulent medium there is often a large-scale magnetic field $\mathbf{B}_0$, which cannot be eliminated by a Galilean transformation. A strong magnetic field has a profound effect on the turbulent dynamics, making the flow highly anisotropic. While the motion perpendicular to $\mathbf{B}_0$ may develop small-scale structures giving rise to turbulent dissipation, spatial variations along $\mathbf{B}_0$ remain smooth, corresponding to weakly interacting Alfvén waves. In the presence of $\mathbf{B}_0$ isotropy can thus only be expected in the perpendicular plane, which would make general statistical equations for MHD turbulence formally much more complicated than in the fully isotropic hydrodynamic case. Practically speaking, however, two limiting cases are of particular interest, 3-D isotropic MHD turbulence with no mean field $\mathbf{B}_0 = \langle \mathbf{B} \rangle = 0$, and 2-D isotropic MHD turbulence representing the turbulent motions perpendicular to a strong field $\mathbf{B}_0$.

7.2   Properties of nondissipative MHD turbulence

7.2.1   Ideal invariants

Important information about the turbulent dynamics, in particular the spectral transfer processes, is obtained by considering the ideal invariants and the corresponding statistical equilibrium properties. We limit ourselves to the case of incompressibility with the density normalization $\rho = 1$. When periodic boundary conditions are applied, the boundary contributions in the conservation laws discussed in section 2.2 vanish. For reasons given subsequently, the dynamically important invariants are the quadratic ones, which in 3-D MHD are

$$E = \tfrac{1}{2} \int (v^2 + B^2) d^3x = \tfrac{1}{2} \sum_{\mathbf{k}} (|v_{\mathbf{k}}|^2 + |B_{\mathbf{k}}|^2) = E^V + E^M \, ,$$

$$K = \tfrac{1}{2} \int \mathbf{v} \cdot \mathbf{B} d^3x = \tfrac{1}{2} \sum_{\mathbf{k}} \mathbf{v}_{\mathbf{k}} \cdot \mathbf{B}_{-\mathbf{k}} \, ,^\dagger \qquad (7.4)$$

$$H = \tfrac{1}{2} \int \mathbf{A} \cdot \mathbf{B} d^3x = \tfrac{1}{2} \sum_{\mathbf{k}} \mathbf{A}_{\mathbf{k}} \cdot \mathbf{B}_{-\mathbf{k}} \, ,^\ddagger$$

the total energy, the cross-helicity and the magnetic helicity, respectively. In the limit $B_0 \gg \tilde{B}$ one has $\mathbf{A} \cdot \mathbf{B} \simeq \mathbf{A} \cdot \mathbf{B}_0$ such that the quadratic helicity invariant degenerates to a linear one.

For 2-D MHD, which is identical with the reduced equations (2.49), (2.51) for $\partial_z = 0$, one finds a very similar set,

$$E = \tfrac{1}{2} \sum_{\mathbf{k}} \left(|v_{\mathbf{k}}|^2 + |B_{\mathbf{k}}|^2 \right) \, ,$$

$$K = \tfrac{1}{2} \sum_{\mathbf{k}} \mathbf{v}_{\mathbf{k}} \cdot \mathbf{B}_{-\mathbf{k}} \, , \qquad (7.5)$$

---

† The factor of $1/2$ introduced in H and K as compared with the usual definitions in eqs (2.24), (2.27) is convenient in the following considerations.

‡ Occasionally the quantity $\int \mathbf{B} \cdot \nabla \times \mathbf{B} d^3x$ is called magnetic helicity, in analogy to the kinetic helicity. This quantity is, however, no ideal invariant.

$$A = \tfrac{1}{2} \sum_{\mathbf{k}} |\psi_{\mathbf{k}}|^2 ,$$

where the first two are formally identical with the corresponding 3-D expressions, and A, the mean square magnetic potential, replaces the magnetic helicity H. It is worth noting that neither K nor A is conserved in 3-D reduced MHD.

For comparison we also give the ideal quadratic invariants of the Navier-Stokes equations, which in 3-D are

$$E^V = \tfrac{1}{2} \sum_{\mathbf{k}} |v_{\mathbf{k}}|^2 ,$$

$$H^V = \tfrac{1}{2} \sum_{\mathbf{k}} \mathbf{v}_{\mathbf{k}} \cdot \boldsymbol{\omega}_{-\mathbf{k}} , \qquad (7.6)$$

and in 2-D

$$E^V = \tfrac{1}{2} \sum_{\mathbf{k}} |v_{\mathbf{k}}|^2 ,$$

$$\Omega = \tfrac{1}{2} \sum_{\mathbf{k}} |\omega_{\mathbf{k}}|^2 , \qquad (7.7)$$

where $\omega = \nabla \times \mathbf{v}$, H^V is the kinetic helicity, and Ω the mean square vorticity, also called enstrophy. It is interesting to note that in the limit $\mathbf{B} \to 0$ the MHD invariants, eqs (7.4), (7.5), do not go over into those of nonmagnetic hydrodynamics, eqs (7.6), (7.7). In 3-D neither K nor H goes over into H^V, whereas in 2-D neither K nor A reduces to the enstrophy Ω. Instead the invariants involving $\mathbf{B}$ become identically zero, while the expressions for the hydrodynamic invariants become constant as $\mathbf{B} \to 0$.

In order to apply the formalism of equilibrium statistical mechanics to continuum fluid turbulence, some discretization is convenient. This can be achieved in configuration space by, for instance, introducing a finite spatial grid or in Fourier space by limiting the number of Fourier modes. Both methods are used in numerical turbulence simulations. In turbulence theory Fourier space discretization is more convenient, truncating the Fourier series

$$\sum_{\mathbf{k}} \to \sum_{\mathbf{k}}' ,$$

where the prime indicates summation over all $\mathbf{k}$ with $k_{min} \leq k \leq k_{max}$. In general, ideal invariants of the continuum system are not strict invariants in the truncated system. For example, quantities such as $\int \psi^n d^3x$ with $n \neq 2$ in 2-D MHD are conserved only in the continuum limit. The quadratic invariants (7.4)–(7.7), however, and presumably only these, are sufficiently robust or "rugged" to survive truncation. The origin of this property is the validity of corresponding detailed balance relations for the elementary interaction between any triad of wave vectors $\mathbf{k}, \mathbf{p}, \mathbf{q}$ forming a triangle, i.e. satisfying $\mathbf{k} + \mathbf{p} + \mathbf{q} = 0$. In the presence of only three such modes the spectral mode energy $E_\mathbf{k}$ follows the equation

$$\dot{E}_\mathbf{k} = T\left(\mathbf{k}, \mathbf{p}, \mathbf{q}\right) , \tag{7.8}$$

where T is the nonlinear energy transfer function. The detailed balance relations are

$$\dot{E}_\mathbf{k} + \dot{E}_\mathbf{p} + \dot{E}_\mathbf{q} = T\left(\mathbf{k}, \mathbf{p}, \mathbf{q}\right) + T\left(\mathbf{p}, \mathbf{q}, \mathbf{k}\right) + T\left(\mathbf{q}, \mathbf{k}, \mathbf{p}\right) = 0 \tag{7.9}$$

and similarly for the other quantities in eqs (7.4)–(7.7). The validity of relations such as eq. (7.9) can be verified by direct calculation using the dynamic equations in Fourier representation. For instance, in the case of 2-D MHD these read

$$\dot{\psi}_\mathbf{k} = -\tfrac{1}{2} \sum_{\mathbf{p}, \mathbf{q}} \mathbf{e}_z \cdot \left(\mathbf{p} \times \mathbf{q}\right) \left(\psi_{-\mathbf{p}} \phi_{-\mathbf{q}} - \psi_{-\mathbf{q}} \phi_{-\mathbf{p}}\right) \delta_{\mathbf{k}+\mathbf{p}+\mathbf{q}} , \tag{7.10}$$

$$\dot{\omega}_\mathbf{k} = -\tfrac{1}{2} \sum_{\mathbf{p}, \mathbf{q}} \mathbf{e}_z \cdot \left(\mathbf{p} \times \mathbf{q}\right) \left(\frac{1}{p^2} - \frac{1}{q^2}\right) \tag{7.11}$$

$$\times \left(\omega_{-\mathbf{p}} \omega_{-\mathbf{q}} - j_{-\mathbf{p}} j_{-\mathbf{q}}\right) \delta_{\mathbf{k}+\mathbf{p}+\mathbf{q}} ,$$

$$\omega_\mathbf{k} = -k^2 \phi_\mathbf{k} , \quad j_\mathbf{k} = -k^2 \psi_\mathbf{k} .$$

Multiplication of eq. (7.10) by $j_{-\mathbf{k}}$ and of eq. (7.11) by $\phi_{-\mathbf{k}}$ and addition gives the explicit form of the transfer function in eq. (7.8) for 2-D MHD, from which eq. (7.9) is easily derived, and similarly the detailed balance relations for K and A.

7.2.2 *Absolute equilibrium distributions*

An ensemble of ideal truncated MHD flows is now assumed. According to classical statistical theory the equilibrium distribution in phase space is given by the Gibbs distribution

$$\rho = Z^{-1} \exp\left\{-\alpha E - \beta H - \gamma K\right\} , \tag{7.12}$$

where E, H, K are quadratic forms in the phase space variables, the real and imaginary parts of the two components of $\mathbf{v}_\mathbf{k}$ and $\mathbf{B}_\mathbf{k}$ in the plane perpendicular to $\mathbf{k}$, because of $\mathbf{k} \cdot \mathbf{v}_\mathbf{k} = \mathbf{k} \cdot \mathbf{B}_\mathbf{k} = 0$, and Z is a normalizing factor. (The formal basis is provided by the validity of a Liouville theorem, as discussed for instance by Kraichnan & Montgomery, 1980.) If the ensemble initially has a non-equilibrium distribution then it will relax to the distribution (7.12), where the parameters α, β, γ are determined by the values of the invariants E, H, K. The most interesting quantities are the spectral functions $E_\mathbf{k}^V = \tfrac{1}{2} \langle |v_\mathbf{k}|^2 \rangle$, $E_\mathbf{k}^M = \tfrac{1}{2} \langle |B_\mathbf{k}|^2 \rangle$, $H_\mathbf{k} = \tfrac{1}{2} \langle \mathbf{A}_\mathbf{k} \cdot \mathbf{B}_{-\mathbf{k}} \rangle$, $K_\mathbf{k} =$

$\frac{1}{2}\langle \mathbf{v_k} \cdot \mathbf{B_{-k}} \rangle$. Using the lemma that the multivariate Gaussian probability distribution for the variables χ_i,

$$\rho = Z^{-1}\exp\left\{ -\tfrac{1}{2} \sum_{i,j} A_{ij}\chi_i\chi_j \right\} , \qquad (7.13)$$

has the second-order moments

$$\langle \chi_i\chi_j \rangle = \left(A^{-1} \right)_{ij} , \qquad (7.14)$$

where $(A^{-1})_{ij}$ is the inverse of the matrix A_{ij} (see, for instance, Lumley 1970, chapter 2), the equilibrium spectra can be derived in a straightforward way. One obtains (Frisch et al., 1975)

$$E_{\mathbf{k}}^V = \frac{1}{\alpha}\left(1 + \frac{\tan^2\phi}{1 - k_c^2/k^2} \right) , \qquad (7.15)$$

$$E_{\mathbf{k}}^M = \frac{1}{\alpha}\frac{1}{\cos^2\phi}\frac{1}{1 - k_c^2/k^2} , \qquad (7.16)$$

$$H_{\mathbf{k}} = -\frac{1}{\alpha\cos^2\phi}\frac{k_c/k^2}{1 - k_c^2/k^2} = -\frac{k_c}{k^2}E_{\mathbf{k}}^M , \qquad (7.17)$$

$$K_{\mathbf{k}} = -\frac{\gamma}{\alpha^2\cos^2\phi}\frac{1}{1 - k_c^2/k^2} = -\frac{\gamma}{\alpha}E_{\mathbf{k}}^M , \qquad (7.18)$$

where $\sin\phi = \gamma/2\alpha$, $k_c = \beta/\alpha\cos^2\phi$. Since ρ must be normalizable, the quadratic form $\alpha E + \beta H + \gamma K$ must be positive definite; in particular, expressions (7.15), (7.16) must be positive for any $\mathbf{k}$ with $k_{min} \le k \le k_{max}$. It thus follows that

$$\alpha > 0 , \quad |\gamma| < 2\alpha , \quad k_c^2 < k_{min}^2 .$$

The ratio β/α measures the magnetic helicity, while γ/α measures the velocity magnetic field correlation. Note that the difference

$$E_{\mathbf{k}}^M - E_{\mathbf{k}}^V = \frac{1}{\alpha}\frac{k_c^2}{k^2 - k_c^2} \qquad (7.19)$$

is always positive, being maximal for $k = k_{min}$. Hence the energy spectrum is dominated by the longest-wavelength magnetic contribution.

In the case of 2-D MHD the appropriate phase space distribution is

$$\rho = Z^{-1}\exp\left\{ -\alpha E - \beta' A - \gamma K \right\} . \qquad (7.20)$$

The equilibrium spectra are derived as in the three-dimensional case (Fyfe & Montgomery, 1976),

$$E_{\mathbf{k}}^V = \frac{1}{\alpha}\left(1 + \frac{\tan^2\phi}{1 - a/k^2} \right) , \qquad (7.21)$$

$$E_{\mathbf{k}}^M = \frac{1}{\alpha} \frac{1}{\cos^2 \phi} \frac{1}{1 - a/k^2}, \tag{7.22}$$

$$A_{\mathbf{k}} = k^{-2} E_{\mathbf{k}}^M, \tag{7.23}$$

$$K_{\mathbf{k}} = -\frac{\gamma}{\alpha} E_{\mathbf{k}}^M, \tag{7.24}$$

where eqs (7.21), (7.22) have been written in a form analogous to the corresponding 3-D expressions (7.15), (7.16) with $\sin \phi = \gamma/2\alpha$ and $a = -\beta'/\alpha \cos^2 \phi$. While α is intrinsically positive, β' and γ may be both positive and negative, the sign of β' being particularly important because of its appearance in the denominator $1 - a/k^2$. β' is essentially determined by the ratio of the mean square potential and the total energy A/E. If A/E is sufficiently large – the maximum being given by $A/E = k_{min}^{-2}$, where all energy resides in the smallest k – then β' is negative, i.e. $a > 0$, in which case the energy spectra $E_{\mathbf{k}}^V$, $E_{\mathbf{k}}^M$ are identical in 2-D and 3-D.

By contrast the absolute equilibrium energy spectra in Navier-Stokes theory are different for 2-D and 3-D (Kraichnan, 1967 and 1973):

$$E_{\mathbf{k}}^V = \begin{cases} \alpha \left(\alpha^2 - \beta^2 k^2 \right)^{-1} & \text{3-D} \\ \left(\alpha + \beta' k^2 \right)^{-1} & \text{2-D}, \end{cases} \tag{7.25}$$

with $\alpha, \alpha^2 - \beta^2 k^2 > 0$ in 3-D and $\alpha + \beta' k^2 > 0$ in 2-D in the truncated k range $k_{min} \le k \le k_{max}$; α, β and α, β' are the parameters in the Gibbs distributions connected with the 3-D invariants E^V, H^V, eq. (7.6), and the 2-D invariants E^V, Ω, eq. (7.7), respectively. In 3-D the equilibrium energy spectrum is enhanced at *large k* in the presence of kinetic helicity $\beta \ne 0$, with the limiting case of equipartition for nonhelical flows $\beta = 0$. In 2-D, by contrast, the energy spectrum may be enhanced at *low k*, an effect which is particularly strong for "negative temperature" $\alpha < 0$, which is excluded in 3-D.

7.2.3 Cascade directions

The ideal invariants are conserved in the nonlinear interactions, their spectral components following detailed balance relations such as eq. (7.9). If such a quantity is excited, or injected, to use the customary expression in turbulence theory, in a certain spectral range $k \sim k_{in}$, it will be scattered into other regions of k-space. Since triad interactions are strongest for $\mathbf{k}_i$'s of the same magnitude $|\mathbf{k}_i - \mathbf{k}_j| = \Delta k \lesssim k_i$, transfer in k-space occurs in relatively small steps, such that many steps are required from the injection scale k_{in}^{-1} to the small dissipation scales k_d^{-1}. The transfer process of a conserved quantity is hence called cascade.

In a real dissipative system two extreme cases occur. (a) A *normal* or *direct* cascade corresponds to the flow from k_{in} to larger wavenumbers up to k_d. In 3-D Navier-Stokes turbulence energy and also kinetic helicity (Kraichnan, 1973) are believed to follow a direct cascade. (b) In an *inverse* cascade transfer proceeds from k_{in} to small wavenumbers. For finite system size this eventually leads to accumulation (condensation) in the lowest possible k, since there is no dissipation at long wavelengths, preventing the system from reaching a stationary state. In 2-D Navier-Stokes theory energy follows an inverse cascade, while enstrophy has a direct cascade.

Cascade directions can be obtained by investigating the triad inter-actions and the corresponding local transfer processes or, as will be briefly done here, by inspection of the nondissipative equilibrium spectra. Though ideal equilibrium states are very far from real turbulence, which is a driven-dissipative nonequilibrium system, their nonlinear dynamics are identical, so that the ideal distributions indicate the direction of the evolution in the actual dissipative system. Since for MHD in three as well as two dimensions equilibrium energies are essentially equipartitioned at high k, they indicate a tendency for energy to flow toward large k, i.e. a normal energy cascade as in the case of 3-D Navier-Stokes theory. By contrast, the purely magnetic invariants, the magnetic helicity H in 3-D and the mean square potential A in 2-D, have equilibrium distributions strongly peaked at low k. Since these quantities are conserved by the nonlinear interactions, they are expected to exhibit an inverse cascade toward small k (from the intermediate injection range k_{in}). This picture is corroborated by the dominance of the magnetic energy at small k as indicated by eq. (7.19). In 2-D, in particular, the relation between E and A in MHD is analogous to that of Ω and E^V in Navier-Stokes theory, the first quantity exhibiting a direct cascade, the second an inverse one. Since the cross-helicity equilibrium spectrum is similar to the energy spectrum, it probably exhibits a direct cascade such as the kinetic helicity in Navier-Stokes turbulence. Ideal invariants and their presumable cascade directions in MHD and Navier-Stokes theory are summarized in Table 7.1. Note that concerning these properties MHD turbulence is very similar in 2-D and 3-D, while Navier-Stokes turbulence in 2-D differs significantly from that in 3-D.

7.3 Self-organization and turbulence decay laws

Three-dimensional Navier-Stokes turbulence does not seem to exhibit self-organization, i.e. the formation of large-scale coherent spatial struc-

Table 7.1. Ideal invariants and cascade directions

		3-D		2-D
MHD	E_k	direct	E_k	direct
	K_k	direct	K_k	direct
	H_k	inverse	A_k	inverse
Navier-Stokes	E_k^V	direct	E_k^V	inverse
	H_k^V	direct	Ω_k	direct

tures. In two-dimensional Navier-Stokes turbulence, by contrast, self-organization is a dominant feature consisting of the build-up of coherent vortices for which the nonlinear distortion essentially vanishes and which grow in size due to vortex coalescence (McWilliams, 1984). This increase of energy at small wavenumbers is associated with the inverse energy cascade.

Since in MHD turbulence, too, there are conserved quantities for which an inverse cascade is predicted, viz. the magnetic helicity H in 3-D and the mean square magnetic potential A in 2-D, similar processes of self-organization are expected to occur, leading to the formation of large-scale magnetic structures. In decaying turbulence this effect is related to the phenomenon of selective decay, while in driven turbulence it gives rise to the turbulent dynamo effect. Since the large-scale magnetic fields are found to be essentially force-free, the formation of such a structure is associated with an increasing alignment of the magnetic field and current density, $\mathbf{j} \times \mathbf{B} \to 0$. A similar process of self-organization, that of dynamic alignment of the velocity and magnetic field, $\mathbf{v} \times \mathbf{B} \to 0$, is connected with the conservation of cross-helicity K. Since K does not exhibit an inverse cascade, the mechanism of this alignment process is more subtle.

7.3.1 Selective decay

Self-organization in turbulence is intimately connected with the difference in the dissipation rates of the ideal invariants[§]. In 3-D MHD these are

---

[§] This does not mean, however, that different decay rates automatically give rise to self-organization. In 3-D Navier-Stokes turbulence kinetic helicity decays more rapidly than energy, but no self-organization processes are known to occur.

$$\frac{dE}{dt} = -\eta \int j^2 d^3x - v \int \omega^2 d^3x , \tag{7.26}$$

$$\frac{dK}{dt} = -\frac{(\eta + v)}{2} \int \omega \cdot \mathbf{j} d^3x , \tag{7.27}$$

$$\frac{dH}{dt} = -\eta \int \mathbf{j} \cdot \mathbf{B} d^3x . \tag{7.28}$$

In 2-D the equations for dE/dt, dK/dt are identical with eqs (7.26), (7.27), while eq. (7.28) is replaced by

$$\frac{dA}{dt} = -\eta \int B^2 d^2x . \tag{7.29}$$

For comparison, we also give the decay rates for the ideal invariants in Navier-Stokes turbulence, which in 3-D are

$$\frac{dE^V}{dt} = -v \int \omega^2 d^3x , \tag{7.30}$$

$$\frac{dH^V}{dt} = -v \int \omega \cdot \nabla \times \omega d^3x , \tag{7.31}$$

and in 2-D

$$\frac{dE^V}{dt} = -v \int \omega^2 d^2x = -2v\Omega , \tag{7.32}$$

$$\frac{d\Omega}{dt} = -v \int (\nabla \omega)^2 d^2x . \tag{7.33}$$

Different decay rates, called selective dissipation, arise because the dissipation terms contain different orders of spatial derivatives and also because some are negative definite and others are not. The 2-D Navier-Stokes case is most easily understood. Since from eq. (7.33) one has $d\Omega/dt \leq 0$, it follows that

$$\begin{aligned} \frac{dE^V}{dt} &= O(v) , \\ \frac{d\Omega}{dt} &= O(v^\alpha) . \end{aligned} \tag{7.34}$$

Numerical simulations indicate that $\alpha \simeq \frac{1}{2}$ for a system consisting of only a few vorticity gradient sheets, while $\alpha \simeq 0$ for fully developed turbulence (Kida et al., 1988). Hence for high Reynolds number one has $E^V \simeq$ const., while Ω decays rapidly. Since $E^V = \sum_k \Omega_k k^{-2}$, conservation of E^V implies that during turbulence decay the energy spectrum is shifted to small k, which is consistent with an inverse cascade.

The behavior of 2-D MHD is analogous when replacing E^V by A and Ω by E:

$$\frac{dA}{dt} = O(\eta) \,,$$

$$\frac{dE}{dt} = O\left(\eta^{\alpha'}\right) \,,$$

(7.35)

where again in fully developed turbulence $\alpha' \simeq 0$ (section 7.6). For $R_m \gg 1$ one has $A = \text{const.}$, implying growth of A_k at small wave numbers, i.e. an inverse cascade of the magnetic potential. The process is clearly seen in numerical simulations of decaying 2-D MHD turbulence (Fig. 7.1). The final state in a closed system is hence determined by the variational principle (see, for instance, Hasegawa, 1985):

$$\delta[E - \mu^2 A] =$$

$$\delta\left[\tfrac{1}{2}\int((\nabla\psi)^2 + (\nabla\phi)^2)d^3x - \frac{\mu^2}{2}\int \psi^2 d^3x\right] = 0 \,,$$

(7.36)

which gives the minimum energy state for a fixed value of the mean square potential. Variation with respect to ψ and ϕ gives the Euler equations

$$\nabla^2\psi + \mu^2\psi = 0 \,,$$

(7.37)

$$\nabla^2\phi = 0 \,,$$

(7.38)

where μ^2 is determined by the value of A. Equation (7.37) is identical with eq. (3.51) for a two-dimensional constant-μ force-free equilibrium. In homogeneous turbulence eq. (7.38) implies $\mathbf{v} = 0$. Hence the system tends toward a static magnetic state, $E^V/E^M \to 0$. The origin of this asymmetry between $\mathbf{B}$ and $\mathbf{v}$ is the lack of a conserved kinetic quantity equivalent to A. Before reaching the final state E_{min} equations (7.37), (7.38) may already be satisfied locally in certain coherent regions, magnetic eddies of approximately circular shape generated during turbulence decay. Such eddies are clearly visible in Fig. 7.1. They are dynamically rather isolated from the turbulent environment and thus form effectively closed subsystems. Scatter plots of j vs. ψ, with points taken from the central areas of several major eddies, in fact show linear relations $j \propto \psi$ (Fig. 7.2). Similar coherent structures, vorticity eddies, are known in 2-D Navier-Stokes turbulence (see, for instance, Benzi et al., 1988).

While in 2-D MHD turbulence the quantities E and A are intimately related, $E \geq E^M \neq 0$, implying $A \neq 0$, the coupling between E and H in 3-D MHD is less tight. For any reflectionally symmetric state one has $H = 0$, however large the energy. If $H \neq 0$, comparison of eqs (7.28) and (7.26) indicates that $\dot{H}/H \ll \dot{E}/E$. Though less obvious, there are good arguments and numerical evidence that H_k exhibits an inverse cascade, i.e.

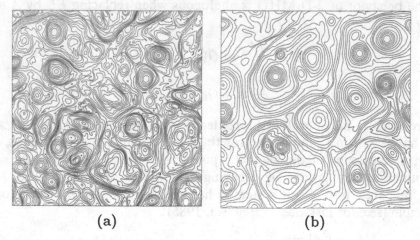

(a)   (b)

Fig. 7.1. Contours of ψ taken at times (a) $t = 2$, (b) $t = 6$ from a simulation of decaying MHD turbulence (from Biskamp & Welter, 1989a).

becomes more and more concentrated at small wave numbers. The system is expected to relax to a minimum-energy state under the constraint $H =$ const., satisfying the variational principle (Taylor, 1974)

$$\delta \left[E - \mu H \right] =$$
$$\delta \left[\tfrac{1}{2} \int (v^2 + B^2)d^3x - \frac{\mu}{2} \int \mathbf{A} \cdot \mathbf{B} d^3x \right] = 0 \; . \qquad (7.39)$$

Variation with respect to $\mathbf{A}$ gives the equation

$$\nabla \times \mathbf{B} - \mu \mathbf{B} = 0 \; , \qquad (7.40)$$

while variation with respect to $\mathbf{v}$ gives $\mathbf{v} = 0$. Hence the minimum-energy state is again a constant-μ, force-free magnetic field. Such states play a crucial role in the dynamics of the reversed-field pinch treated in chapter 9, where we return to Taylor's theory.

Hence one finds that because of selective decay turbulence in a closed system tends to relax to a large-scale, static, force-free magnetic configuration, corresponding to the alignment of $\mathbf{j}$ and $\mathbf{B}$.

For driven turbulence the inverse cascades of H_k in 3-D and A_k in 2-D have been demonstrated by numerical solution of the EDQNM closure equations (Pouquet et al., 1976; Pouquet, 1978), which are discussed in section 7.5.

7.3.2 *The α-term and turbulent dynamo theory*

The inverse cascade of the magnetic helicity is closely related to the turbulent dynamo effect, i.e. the generation of large-scale magnetic fields

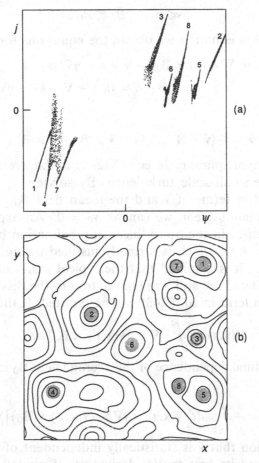

Fig. 7.2.   (a) Scatter plot of j vs. ψ for the central parts (shaded area) of the major magnetic eddies of the turbulent state shown in (b).

by turbulent fluid motions. Because of its application to the origin of cosmic magnetic fields, such as in planets, stars and galaxies, this topic has received widespread attention and several reviews exist (Moffat, 1978; Parker, 1979a; Krause & Rädler, 1980). Here we give only a brief introduction to the concepts of mean field electrodynamics and two-scale turbulence and derive the basic evolution equation for the mean magnetic field (Steenbeck et al., 1966).

One assumes that $\mathbf{v}$ and $\mathbf{B}$ can be decomposed into an average part varying only on the large scale and a weak, small-scale fluctuating part:

$$\mathbf{v} = \mathbf{v}_0 + \tilde{\mathbf{v}}, \quad \langle \tilde{\mathbf{v}} \rangle = 0,$$
$$\mathbf{B} = \mathbf{B}_0 + \tilde{\mathbf{B}}, \quad \langle \tilde{\mathbf{B}} \rangle = 0, \tag{7.41}$$

$$\tilde{v} \ll v_0 \,, \quad \tilde{B} \ll B_0 \,.$$

From the induction equation we obtain the equations for $\mathbf{B}_0$ and $\tilde{\mathbf{B}}$:

$$\partial_t \mathbf{B}_0 = \nabla \times (\mathbf{v}_0 \times \mathbf{B}_0) - \nabla \times \boldsymbol{\varepsilon} + \eta \nabla^2 \mathbf{B}_0 \,, \tag{7.42}$$

$$\partial_t \tilde{\mathbf{B}} = \nabla \times \left(\mathbf{v}_0 \times \tilde{\mathbf{B}} + \tilde{\mathbf{v}} \times \mathbf{B}_0 \right) + \nabla \times \tilde{\mathbf{G}} + \eta \nabla^2 \tilde{\mathbf{B}} \,, \tag{7.43}$$

where

$$\boldsymbol{\varepsilon} = -\langle \tilde{\mathbf{v}} \times \tilde{\mathbf{B}} \rangle \,, \quad \tilde{\mathbf{G}} = \tilde{\mathbf{v}} \times \tilde{\mathbf{B}} - \langle \tilde{\mathbf{v}} \times \tilde{\mathbf{B}} \rangle \,. \tag{7.44}$$

The most important quantity in eq. (7.42) is $\boldsymbol{\varepsilon}$, the average electric field generated by the small-scale turbulence. By solving eq. (7.43) this term can be expressed in terms of $\tilde{\mathbf{v}}$ and the mean field $\mathbf{B}_0$. Transforming to a moving coordinate system, we can set $\mathbf{v}_0 = 0$. An explicit solution of eq. (7.43) is obtained in the quasi-linear approximation by neglecting the nonlinear term $\nabla \times \tilde{\mathbf{G}}$, which is formally justified owing to the smallness of $\tilde{\mathbf{B}}$. In addition, it is assumed that the typical scales of the small-scale turbulence considered are still large in relation to the dissipative scales, so that the diffusion term in eq. (7.43) can be omitted. It thus follows that

$$\tilde{\mathbf{B}}(\mathbf{x}, t) = \int_{-\infty}^{t} \nabla \times \left(\tilde{\mathbf{v}}(\mathbf{x}, t') \times \mathbf{B}_0 \right) dt' \,, \tag{7.45}$$

where the space-time dependence of the average field $\mathbf{B}_0$ is weak. Insertion into $\boldsymbol{\varepsilon}$ gives

$$\boldsymbol{\varepsilon} = -\int_{-\infty}^{t} dt' \langle \tilde{\mathbf{v}}(\mathbf{x}, t) \times \left[\nabla \times \left(\tilde{\mathbf{v}}(\mathbf{x}, t') \times \mathbf{B}_0 \right) \right] \rangle \,. \tag{7.46}$$

On the assumption that $\tilde{\mathbf{v}}$ is statistically independent of $\mathbf{B}_0$, $\boldsymbol{\varepsilon}$ is a linear function of $\mathbf{B}_0$ and its first spatial derivative. If we take the turbulence field $\tilde{\mathbf{v}}$ to be isotropic then the most general expression of $\boldsymbol{\varepsilon}$ has the form

$$\boldsymbol{\varepsilon} = \alpha \mathbf{B}_0 + \beta \nabla \times \mathbf{B}_0 \,, \tag{7.47}$$

where α is a pseudo-scalar ($\alpha \to -\alpha$ for a reflection $\mathbf{x} \to -\mathbf{x}$), since $\mathbf{B}$ is an axial and $\boldsymbol{\varepsilon}$ a polar vector, and β is a scalar. In order to evaluate $\boldsymbol{\varepsilon}$ in eq. (7.46) consider a vector component decomposition $\boldsymbol{\varepsilon} = (\varepsilon_1, \varepsilon_2, \varepsilon_3)$. Because of eq. (7.47) ε_1 only contains B_1 and the derivatives $\partial_2 B_3, \partial_3 B_2$,

$$\varepsilon_1 = -\int_{-\infty}^{t} dt' \left(\langle \tilde{v}_2 \partial_1 \tilde{v}_3' \rangle - \langle \tilde{v}_3 \partial_1 \tilde{v}_2' \rangle \right) B_1$$

$$+ \int_{-\infty}^{t} dt' \langle \tilde{v}_2 \tilde{v}_2' \rangle \partial_2 B_3 - \int_{-\infty}^{t} dt' \langle \tilde{v}_3 \tilde{v}_3' \rangle \partial_3 B_2 \,. \tag{7.48}$$

Isotropy implies invariance under cyclic permutations,

$$\langle \tilde{v}_2 \partial_1 \tilde{v}_3' \rangle - \langle \tilde{v}_3 \partial_1 \tilde{v}_2' \rangle$$
$$= \langle \tilde{v}_2 \left(\partial_1 \tilde{v}_3' - \partial_3 \tilde{v}_1' \right) \rangle = -\tfrac{1}{3} \langle \tilde{\mathbf{v}} \cdot \nabla \times \tilde{\mathbf{v}}' \rangle \,,$$

and

$$\langle \tilde{v}_1 \tilde{v}_1' \rangle = \langle \tilde{v}_2 \tilde{v}_2' \rangle = \langle \tilde{v}_3 \tilde{v}_3' \rangle = \tfrac{1}{3} \langle \tilde{\mathbf{v}} \cdot \tilde{\mathbf{v}}' \rangle .$$

Comparison with eq. (7.47) gives the result

$$\begin{aligned} \alpha &= \tfrac{1}{3} \int_{-\infty}^{t} dt' \langle \tilde{\mathbf{v}} \cdot \nabla \times \tilde{\mathbf{v}}' \rangle \\ &= \frac{\tau}{3} \langle \tilde{\mathbf{v}} \cdot \nabla \times \tilde{\mathbf{v}} \rangle = \frac{\tau}{3} H^V , \end{aligned} \tag{7.49}$$

$$\begin{aligned} \beta &= \tfrac{1}{3} \int_{-\infty}^{t} dt' \langle \tilde{\mathbf{v}} \cdot \tilde{\mathbf{v}}' \rangle \\ &= \frac{\tau'}{3} \langle \tilde{\mathbf{v}}^2 \rangle , \end{aligned} \tag{7.50}$$

where $\tau \simeq \tau'$ is the velocity correlation time. Here β is essentially positive, but α may have both signs, depending on the sign of the kinetic helicity. Equation (7.42) now becomes

$$\partial_t \mathbf{B}_0 = -\nabla \times \alpha \mathbf{B}_0 + (\eta + \beta)\nabla^2 \mathbf{B}_0 . \tag{7.51}$$

While the β-term in the electromotive force (7.47) gives rise to an enhanced magnetic diffusivity (with $\beta \gg \eta$ in most cases of practical interest), the α-term may lead to an amplification of the large-scale field. Note that $\alpha \neq 0$ requires turbulent flows that, though isotropic, are not reflectionally symmetric. Such flows naturally arise in rotating systems. Since most astrophysical objects carrying large-scale magnetic fields consist of electrically conducting fluids in some state of rotation, the α-term has become the central point of dynamo theory. Let $\mathbf{B}$ be decomposed into a toroidal and a poloidal component with respect to the rotation axis as the main torus axis, $\mathbf{B} = \mathbf{B}_t + \mathbf{B}_p$. Differential rotation can easily generate a toroidal field $\mathbf{B}_t$ from a poloidal field $\mathbf{B}_p$. The reverse process is more complicated, involving twisting fluid motions. This process is described by the α-term, where a toroidal field with its poloidal current leads to amplification of $\mathbf{B}_p$ (for the right sign of α). A fully dynamic simulation of large-scale field amplification due to kinetic helicity injection has been performed by Meneguzzi et al. (1981).

Since the dynamics and statistics of the velocity field entering the coefficients α, β are assumed to be independent of $\mathbf{B}_0$, equation (7.51) is actually limited to weak magnetic fields, the so-called kinematic dynamo regime. In fact, for large Reynolds number the linear approximation, neglecting the Lorentz force in the equation of motion, becomes invalid already for very small $\mathbf{B}_0$, since this field acts as a guide field for the small-scale fluctuations $\tilde{\mathbf{v}}$, transforming them into Alfvén waves with $\tilde{\mathbf{v}} \times \tilde{\mathbf{B}} \simeq 0$, $\tilde{B} \simeq \tilde{v}$. This is called the *Alfvén effect* in MHD turbulence, which has a profound influence on the small-scale dynamics and is discussed in more detail in sections 7.3.3 and 7.4. It leads to rapid small-scale equipartition of magnetic and kinetic energy, however small the initial magnetic field $\tilde{\mathbf{B}}$. As a consequence the α-term is reduced, as is easily seen from the original form of ε, eq. (7.44). On the other hand, along with small-scale magnetic energy magnetic helicity H is also generated, which owing to its inverse cascade process leads to the build-up of large-scale magnetic fields. For finite field magnitude B_0 we thus have the nonlinear turbulent dynamo process (Pouquet et al., 1976)

$$H^V \overset{\text{Alfvén effect}}{\longrightarrow} H \overset{\text{inverse cascade}}{\longrightarrow} H_0 , E_0^M \tag{7.52}$$

instead of the linear process

$$H^V \overset{\alpha-\text{term}}{\longrightarrow} E_0^M .$$

In the case of large $\mathbf{B}_0$ eq. (7.47) has to be modified to account for the fact that the small-scale turbulence does not change the magnetic helicity, since $(dH/dt)/(dE/dt) = O(\eta^\nu), \nu < 1$. Hence we find from eq. (7.42)

$$\frac{dH_0}{dt} = -2 \int \mathbf{E} \cdot \mathbf{B} d^3 x \simeq -2 \int \varepsilon \cdot \mathbf{B}_0 d^3 x = 0 , \tag{7.53}$$

such that ε has the form

$$\varepsilon = (\mathbf{B}_0/B_0^2) \nabla \cdot \mathbf{h} , \tag{7.54}$$

where $\mathbf{h}$ is the helicity flux, which contains the effect of the small-scale turbulence. One can derive a more explicit form of $\mathbf{h}$. Inserting eq. (7.54) into the equation

$$\frac{dE}{dt} \simeq \frac{dE^M}{dt} = -\int \varepsilon \cdot \mathbf{j}_0 d^3 x < 0$$

gives

$$\int \mathbf{h} \cdot \nabla (\mathbf{j}_0 \cdot \mathbf{B}_0/B_0^2) d^3 x < 0 . \tag{7.55}$$

Since the turbulence and hence $\mathbf{h}$ is statistically independent at different locations, the integrand in eq. (7.55) itself should be negative, $\mathbf{h} \cdot \nabla \left(\mathbf{j}_0 \cdot \mathbf{B}_0/B_0^2 \right) < 0$, from which follows $\mathbf{h} = -D\nabla \left(\mathbf{j}_0 \cdot \mathbf{B}_0/B_0^2 \right), D > 0$. Hence ε becomes (Boozer, 1986; Bhattacharjee & Hameiri, 1986)

$$\varepsilon = -(\mathbf{B}_0/B_0^2) \nabla \cdot D \nabla (\mathbf{j}_0 \cdot \mathbf{B}_0/B_0^2) . \tag{7.56}$$

The result implies that in a magnetic configuration with a quasi-static mean field $\mathbf{B}_0 \gg \mathbf{v}_0$ the turbulent resistivity β vanishes, since a finite β would also lead to rapid helicity decay. Instead the effect of small-scale turbulence on the average fields, both the (nondissipative) dynamo effect and the turbulent dissipation, is described by a hyperresistivity, defined in eq. (2.57), $\eta_2 = D$ (apart from a factor B_0^{-1}).

7.3.3 Dynamic alignment of velocity and magnetic field

In addition to the magnetic self-organization leading to an alignment of **j** and **B**, which is connected with the quasi-constancy of the magnetic helicity, there is a further process of self-organization in a turbulent MHD system leading to an alignment of **v** and **B**. In fact, satellite observations of the solar wind, the only accessible system of quasi-stationary, fully developed MHD turbulence, often show states where velocity and magnetic field fluctuations are strongly correlated, $\mathbf{v} \simeq \pm\mathbf{B}$ (see e.g. Burlage & Turner, 1976). Comparing the decay rates of the energy E, eq. (7.26), and cross-helicity K, eq. (7.27), the latter is expected to be smaller than the former. This suggests that the relaxed state is the minimum energy state for a given value of the cross-helicity determined by the variational principle

$$\delta\left[E - 2\lambda K\right] =$$
$$\delta\left[\tfrac{1}{2}\int (v^2 + B^2)d^3x - \lambda\int \mathbf{v}\cdot\mathbf{B}d^3x\right] = 0. \qquad (7.57)$$

Variation with respect to **v** and **B** gives the equations

$$\mathbf{v} - \lambda\mathbf{B} = 0,$$
$$\mathbf{B} - \lambda\mathbf{v} = 0,$$

hence

$$\mathbf{v} = \pm\mathbf{B}, \qquad (7.58)$$

valid at each point in space. Note that no differential operators appear, in contrast to the equations following from the constancy of the magnetic helicity, eqs (7.37), (7.40). The relaxed state is a so called pure Alfvénic state, corresponding to a linearly polarized Alfvén wave.

Following Dobrowolny et al. (1980), we can show that this relaxation is in fact likely to occur dynamically. For this purpose it is convenient to write the MHD equations in terms of the Elsässer variables $\mathbf{z}^{\pm} = \mathbf{v} \pm \mathbf{B}$ (Elsässer, 1950; see eq. (4.14)):

$$\partial_t\mathbf{z}^{\pm} \mp \mathbf{v}_A\cdot\nabla\mathbf{z}^{\pm} + \mathbf{z}^{\mp}\cdot\nabla\mathbf{z}^{\pm}$$
$$= -\nabla\left(p + \frac{B^2}{2}\right) + \frac{\nu + \eta}{2}\nabla^2\mathbf{z}^{\pm} + \frac{\nu - \eta}{2}\nabla^2\mathbf{z}^{\mp}, \qquad (7.59)$$

where $\mathbf{v}_A = \mathbf{B}_0$ (in our normalization with $\rho = 1$) is the background magnetic field. Equation (7.59) indicates that the small-scale turbulence may be considered as an ensemble of Alfvén waves propagating along the average field $\mathbf{B}_0$, with interactions occurring only between modes traveling in opposite directions, which is the formal expression of the Alfvén effect in MHD turbulence mentioned in the preceding subsection. (A dynamical model of the Alfvén effect has been presented by Diamond & Craddock (1990).) We assume that these interactions are quasi-local in wavenumber space, meaning that modes of a certain scale l (or wavenumber $k \sim l^{-1}$) interact most efficiently with modes of similar scales. We distinguish two dynamic time scales, the Alfvén time

$$\tau_A = l/v_A \qquad (7.60)$$

and the time for distortion of an Alfvén wave packet or eddy $z_l^{\pm}$ (we give a more detailed introduction to these and similar concepts in section 7.4)

$$\tau_l^{\pm} = l/z_l^{\mp} \,, \qquad (7.61)$$

where usually $\tau_A \ll \tau_l^{\pm}$. Since the interaction time between two oppositely propagating eddies is τ_A, the change of amplitude occurring during a single collision of two eddies is small:

$$\frac{\delta z_l^{\pm}}{z_l^{\pm}} = \frac{\tau_A}{\tau_l^{\pm}} \ll 1 \,.$$

Because of the random nature of the process $N = \left(z^{\pm}/\delta z^{\pm}\right)^2$ elementary interactions are needed to produce a relative amplitude change of order unity. Hence the energy transfer time is

$$T_l^{\pm} = N \tau_A = (\tau_l^{\pm})^2 / \tau_A \,. \qquad (7.62)$$

Since both $E = E^+ + E^-$ and $K = (E^+ - E^-)/2$ are ideal invariants, so are E^+ and E^-, where $E^{\pm} = \int z^{\pm 2} d^3x/4$. Hence both $(z_l^+)^2$ and $(z_l^-)^2$ are cascading quantities with the energy fluxes given by

$$\begin{aligned} \Pi_l^{\pm} &= (z_l^{\pm})^2 / T_l^{\pm} \\ &= (z_l^+)^2 (z_l^-)^2 \tau_A / l^2 \,. \end{aligned} \qquad (7.63)$$

Since for a stationary state $\Pi^{\pm}(l) = \varepsilon^{\pm}$ are constant in the inertial range (for a definition see section 7.4.1) and equal to the injection rates, the symmetry in eq. (7.63) indicates that $\varepsilon^+ = \varepsilon^-$ and hence cross-helicity K is not dissipated,

$$- dK/dt = (\varepsilon^+ - \varepsilon^-)/4 = 0 \,. \qquad (7.64)$$

Strictly speaking, the result $\varepsilon^+ = \varepsilon^-$ holds only in the local approximation (in wavenumber space) and can be softened by considering interactions within a certain band of wavenumbers, as is discussed in section 7.4.3.

We now consider the consequences of eq. (7.64) on the dynamics of turbulence decaying from a state with $z_l^+ > z_l^-$. In this case the energy fluxes are time-dependent $\varepsilon^{\pm}(t)$ but preserve the property $\varepsilon^+ = \varepsilon^-$, which follows from the $\pm$ symmetry of eq. (7.63) and does not require (global) stationarity. The transfer times are, however, different. From eq. (7.62) it is found that

$$T_l^+/T_l^- = (z_l^+)^2/(z_l^-)^2 > 1 \,. \tag{7.65}$$

This means that transfer, and hence damping, of the minority field z^- is more rapid, which leads to a continuous increase of the ratio E^+/E^-, until practically all remaining energy is in the pure Alfvénic or aligned state z^+, i.e. $\mathbf{v} = \mathbf{B}$. (This state can, however, be reached only asymptotically, since in such a state the nonlinear interactions vanish.) Thus dynamic alignment of $\mathbf{v}$ and $\mathbf{B}$ is a direct consequence of the MHD equations. Since eqs (7.59) are formally identical in 2-D and 3-D, dynamic alignment should occur in both 2-D and 3-D systems. The process has been investigated quantitatively in the framework of closure theory (Grappin et al., 1982 and 1983) and in direct numerical simulations of decaying 2-D MHD turbulence (Pouquet et al. 1988, Biskamp & Welter 1989a).

7.3.4 *Energy decay laws*

In two-dimensional Navier-Stokes turbulence the constancy of the energy E^V at high Reynolds number, eq. (7.34), allows a simple similarity solution for the enstrophy decay, $\Omega \propto t^{-2}$ (Batchelor, 1969). In MHD turbulence the quantities A (in 2-D) and H (in 3-D) play a role similar to that of E^V in two-dimensional Navier-Stokes theory, so that Batchelor's argument can also be applied to the MHD case (Hatori, 1984). The theory uses the fact that a cascading quantity such as $A_\mathbf{k}$ exhibits a power-law inertial-range spectrum. Hatori chooses a Kolmogorov $k^{-5/3}$ energy spectrum, which is not justified for MHD turbulence, as is discussed in section 7.4. (In addition, the use of an inertial range spectrum in the argument is questionable, since it implicitly assumes that the inertial range extends up to the largest energy-containing scales.) However, it turns out that the results are independent of the particular spectral law, even independent of the existence of an inertial range, and uniquely determined by the global behavior. Here I present a simple heuristic derivation, which contains the essence of the physics involved. As in the 2-D Navier-Stokes case the essential feature is selective decay, i.e. certain quantities decay more slowly than others and can thus be considered constant. In 2-D MHD turbulence we have $A = \text{const.}$, while $E = E^M + E^V$ is decaying. Define a length scale L by

$$\begin{aligned} A &= E^M L^2 \\ &\simeq E L^2 \,, \end{aligned} \tag{7.66}$$

since $E \simeq E^M > E^V$ in decaying turbulence. L should be independent of the details of the turbulent motion and depend only on global quantities, i.e. E and the energy dissipation rate ε. Hence $L = E^{3/2}/\varepsilon$, which is the global scale defined in eq. (7.1). Since L could in principle also depend on K, the third ideal invariant, we restrict consideration to the case of weak correlation $K \ll E$. Inserting L into eq. (7.66), the constancy of A gives

$$E^2/\varepsilon = \text{const.} , \tag{7.67}$$

which has the similarity solution,

$$E \propto t^{-1} , \tag{7.68}$$

using $\varepsilon = -dE/dt$. Numerical simulations of decaying 2-D MHD turbulence confirm this decay law (Biskamp & Welter, 1989a). Since the Reynolds number is defined by $Re = E^2/\varepsilon v$, eq. (7.2), we find that

$$Re = \text{const.} \tag{7.69}$$

in decaying 2-D MHD turbulence.

The 3-D case is treated analogously, with the magnetic helicity H replacing the mean square potential A. Expressing H in terms of E and ε we obtain

$$\begin{aligned} H &= E^M L \simeq EL \\ &= E^{5/2}/\varepsilon = \text{const.} , \end{aligned} \tag{7.70}$$

which has the similarity solution

$$E \propto t^{-2/3} . \tag{7.71}$$

In contrast to the 2-D case the Reynolds number is not constant in decaying turbulence but increases, $Re \propto t^{1/3}$. Note that the derivation relies on $H \neq 0$. Hence the decay law eq. (7.71) may not be valid in the case of nonhelical turbulence $H = 0$.

7.4   Energy spectra

7.4.1   Inertial range spectra in MHD turbulence

A characteristic property of high-Reynolds-number turbulent fluids is the inertial-range energy spectrum, where the inertial range is defined by

$$k_{in} \ll k \ll k_d . \tag{7.72}$$

Here k_{in} represents the injection range, the large scales, where turbulence excitation occurs and which carry most of the turbulent energy, and k_d the dissipation range, the smallest scales present in the turbulence. Obviously, a clearly discernible inertial range exists only for sufficiently high ratio k_d/k_{in}, i.e. for large enough Reynolds numbers (see eqs (7.95), (7.96)). If the turbulent energy transfer is characterized by a local cascade process in

k-space, which implies that interactions between strongly disparate wave numbers are weak, the dynamics of the inertial range modes is expected to show a similarity or scaling behavior.

Since the case of isotropic turbulence is the most interesting, only the angle-integrated spectra are considered,

$$E_k = \int d\Omega_\mathbf{k} E_\mathbf{k} , \tag{7.73}$$

$$E = \int dk E_k .$$

Note that E_k is formally independent of the spatial dimensions.

The spectral dynamics of the energy is determined by the energy injection rate ε_{in} at $k \sim k_{in}$, the energy transfer rate or energy flux ε_t in the inertial range, and the energy dissipation rate ε_d at $k \sim k_d$. For stationary turbulence we have

$$\varepsilon_{in} = \varepsilon_t = \varepsilon_d = \varepsilon .$$

The relation, however, also holds when the injection rate varies in time as in decaying turbulence, since the rapid small-scale dynamics in the inertial and dissipation range adjusts quasi-instantaneously to the slower changes at the energy-carrying scales.

Using the locality of the transfer process, the inertial range energy spectrum can easily be derived in a heuristic way. First consider the case of Navier-Stokes turbulence. It is convenient to divide the inertial range into a discrete set of scales

$$k_0 < k_1 < \cdots < k_N ,$$

with $k_n = l_n^{-1}$. Consider a fluid element or eddy v_n of size l_n, where v_l is defined by the velocity difference between two points separated by a distance l

$$v_l = \Delta v(l) \equiv v(x + l) - v(x) . \tag{7.74}$$

(The vector character of $\mathbf{v}$ and $\mathbf{x}$ is not important in this context.) The energy transfer time between two eddies v_n, v_{n+1} is given by

$$\tau_n \simeq l_n/v_n ,$$

which is a typical distortion or turnover time of an eddy v_n. Since the energy flux is constant,

$$\varepsilon = E_n/\tau_n = \text{const.} \tag{7.75}$$
$$\simeq v_n^3/l_n ,$$

it follows that

$$v_n \simeq \varepsilon^{1/3} l_n^{1/3} . \tag{7.76}$$

With

$$v_n^2 = E_n \simeq \int_{k_n}^{k_{n+1}} E_k dk \simeq E_{k_n} k_n$$

we obtain

$$E_k = C_K \varepsilon^{2/3} k^{-5/3} , \qquad (7.77)$$

the famous Kolmogorov spectrum (Kolmogorov, 1941a; Obukhov, 1941). It is noteworthy that the scaling $v_l^2 \propto l^{2/3}$ or $E_k \propto k^{-5/3}$ already follows from dimensional analysis using the dimensions $[\varepsilon] = L^2 T^{-3}$, $[E_k] = L^3 T^{-2}$. The numerical factor C_K, the Kolmogorov-Obukhov constant, cannot be determined by scaling arguments but requires a dynamic theory such as the Lagrangian History Direct Interaction Approximation (Kraichnan, 1965a). The Kolmogorov spectrum has been observed in many different types of turbulent flows and appears to be rather independent of the fluid material and the mechanism of turbulence generation, though the Kolmogorov "constant" C_K varies somewhat for different experiments, $C_K \simeq 1.4$–2.

The derivation of the spectrum given above assumes a direct energy cascade and is therefore valid only for 3-D Navier-Stokes turbulence. In the two-dimensional case, where the enstrophy Ω_k has a direct cascade, while the energy exhibits an inverse cascade, the argument has to be modified. Replacing the energy flux ε in eq. (7.75) by that of the enstrophy η_Ω, we obtain

$$\eta_\Omega = \Omega_l / \tau_l \simeq (v_l/l)^2 / \tau_l$$
$$\simeq v_l^3 / l^3 , \qquad (7.78)$$

which yields the energy spectrum for 2-D Navier-Stokes turbulence

$$E_k \simeq \eta_\Omega^{2/3} k^{-3} , \qquad (7.79)$$

neglecting a weak logarithmic factor. The spectrum (7.79) has been verified in numerical simulations (e.g. Brachet et al., 1988; Kida et al., 1988). If the injection wavelength is much smaller than the system size, $k_{in}^{-1} \ll L$, the inverse energy cascade may become effective. In this case the energy spectrum for $k < k_{in}$ is again the Kolmogorov spectrum, since the cascade direction does not enter the scaling argument.

Let us now discuss how these scaling properties are modified in the MHD case. According to the Alfvén effect (see section 7.3.3), small-scale fluctuations are not independent of the global state, but are strongly influenced by the large-scale magnetic field, behaving approximately as Alfvén waves $z^\pm$. For the time being we restrict the discussion to weak velocity magnetic field correlation, such that $z_l^+ \simeq z_l^- \simeq v_l \simeq B_l$. The case of strong correlation will be treated in section 7.4.3. The interaction time of two Alfvén wave packets v_l is much shorter than the nonmagnetic eddy distortion time, $\tau_A \ll \tau_l$. Hence many interaction events are necessary to change the amplitude appreciably. As outlined in section 7.3.3, the energy transfer time becomes (see eq. (7.62))

$$\tau_l \to T_l \simeq \tau_l^2/\tau_A \ .$$

With this substitution in eq. (7.75) one obtains (eq. (7.63))

$$\varepsilon \simeq v_l^4 \tau_A/l^2 \ ,$$

hence

$$v_l^2 \simeq (\varepsilon v_A)^{1/2} l^{1/2} \ , \tag{7.80}$$

$$E_k = C_K'(\varepsilon v_A)^{1/2} k^{-3/2} \ , \tag{7.81}$$

which is the Iroshnikov-Kraichnan spectrum, the inertial-range energy spectrum for (uncorrelated) MHD turbulence (Iroshnikov, 1964; Kraichnan, 1965b). It is less steep than the Kolmogorov spectrum because the factor τ_l/τ_A, by which the transfer time is longer, increases with decreasing l, and therefore increasingly higher amplitudes are required compared with the Navier-Stokes case to produce the same energy transfer. In contrast to the Kolmogorov spectrum, which depends only on the inertial-range quantity ε, the MHD energy spectrum also depends explicitly on the large-scale quantity $v_A = B_0$ and hence cannot be derived by dimensional analysis.

It is interesting to note that formally the same spectrum as eq. (7.81), with v_A replaced by the large-scale fluid velocity v_0, results from Kraichnan's original Direct Interaction Approximation for Navier-Stokes turbulence (Kraichnan, 1959). This is, however, a spurious effect due to insufficient discrimination between eddy convection and distortion. The small-scale dynamics should not depend on the large-scale velocity, which can be eliminated by a Galilean transformation, in contrast to a large-scale magnetic field which has a significant effect on small scales by coupling velocity and magnetic field fluctuations.

The MHD energy spectrum (7.81) should be valid in both 3-D and 2-D since there is a direct energy cascade in both cases. Only the constant C_K' may depend on the spatial dimensions. C_K' also depends on the precise definition of the average field B_0 and hence on the geometry of the large-scale eddies, i.e. C_K' is a priori less universal than the Kolmogorov-Obukhov constant C_K.

The Alfvén effect determines not only the total spectral energy E_k, but also the individual contributions E_k^V and E_k^M. Since for Alfvén waves $|\mathbf{v}| = |\mathbf{B}|$, one has

$$E_k^V = E_k^M \ . \tag{7.82}$$

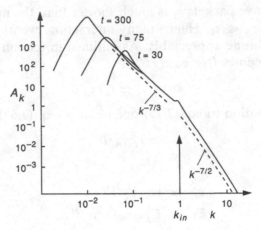

Fig. 7.3. Inverse cascade of the mean square potential A_k for $k < k_{in}$. The right-hand part of the spectrum ($k > k_{in}$) corresponds to the direct energy cascade, $A_k = k^{-2}E_k \propto k^{-7/2}$. (From Pouquet, 1978.)

This result is, however, only approximately valid since it considers only the interaction of a mode k with the large-scale field B_0, neglecting the weak but finite interactions between Alfvén modes of about equal scale. The deviation from strict equipartition (7.82) is derived in section 7.4.3.

Finally, let us consider the consequences of the inverse cascade of the magnetic helicity and the mean square potential in 3-D and 2-D turbulence, respectively, in the small-wavenumber range $k < k_{in}$. Since for long-wavelength modes the Alfvén effect is weak, a Kolmogorov-type argument should be appropriate. Assuming a constant helicity flux η_H,

$$\eta_H \simeq H_l/\tau_l = \text{const.} ,$$

$$H_l \simeq B_l^2 l , \quad \tau_l \simeq l/B_l ,$$

gives

$$H_k \simeq \eta_H^{2/3} k^{-2} , \tag{7.83}$$

and analogously for a constant mean square potential flux η_A,

$$A_k \simeq \eta_A^{2/3} k^{-7/3} . \tag{7.84}$$

The helicity spectrum (7.83) and the potential spectrum (7.84) are recovered in numerical solutions of the closure equations (section 7.5). Figure 7.3 illustrates the numerical result in 2-D, showing the spectrum eq. (7.84) for $k < k_{in}$ and the Iroshnikov–Kraichnan spectrum for $k > k_{in}$. The spectrum (7.84) has also been found in 2-D direct numerical simulations of the inverse cascade process (Biskamp & Bremer, 1994).

7.4.2 *Dissipation scales*

Though dissipation is negligible in the inertial range, the cascade process, i.e. the unidirectional flow in k-space, implies the presence of a sink at some wavenumber k_d where dissipation dominates. Let us first consider Navier-Stokes turbulence. The dissipation scale $l_d = k_d^{-1}$ is determined by the condition that the nonlinear transfer rate equals the dissipation rate,

$$\tau_l^{-1} = v_l/l = v/l^2 , \qquad (7.85)$$

which with eq. (7.76) yields

$$l_d = (v^3/\varepsilon)^{1/4} = l_K , \qquad (7.86)$$

where l_K is called the Kolmogorov micro-scale (Kolmogorov, 1941b). The dissipation scale can also be considered as the scale where the local Reynolds number $v_l l/v$ becomes unity. For MHD turbulence the nonlinear energy transfer is weakened by the Alfvén effect, such that condition (7.85) becomes

$$(\tau_A/\tau_l)/\tau_l = v/l^2 . \qquad (7.87)$$

Using eq. (7.80) one obtains

$$l_d = (v^2 v_A/\varepsilon)^{1/3} = l_K' , \qquad (7.88)$$

the modified Kolmogorov micro-scale. In the derivation we assume $v_l \simeq B_l$ and $v \simeq \eta$, which implies equal dissipation contributions $\varepsilon_\eta \simeq \varepsilon_v$, $\varepsilon = \varepsilon_\eta + \varepsilon_v$,

$$\varepsilon_\eta = \eta \int j^2 d^3x , \quad \varepsilon_v = v \int \omega^2 d^3x . \qquad (7.89)$$

The energy spectrum for (uncorrelated) MHD turbulence can now be written in a form valid in both the inertial range and the dissipation range:

$$E_k = v_A v \, \hat{E}(\hat{k}) , \quad \hat{k} = kl_K' . \qquad (7.90)$$

Here $\hat{E}(\hat{k})$ is the nondimensional or normalized spectrum. In the inertial range $\hat{E}(\hat{k})$ has the limiting form

$$\hat{E}(\hat{k}) = C_K' \hat{k}^{-3/2} , \quad \hat{k} \ll 1 , \qquad (7.91)$$

the Iroshnikov-Kraichnan spectrum eq. (7.81). In the dissipation range $\hat{k} > 1$ the spectrum is expected to fall off rapidly. Numerical simulations show an exponential decay

$$\hat{E}(\hat{k}) \propto e^{-\alpha \hat{k}} , \quad \alpha \simeq 4.8 , \quad \hat{k} \gg 1 . \qquad (7.92)$$

The normalized spectrum $\hat{E}(\hat{k})$ is independent of the numerical values of the parameters v_A, ε, v and should be universal, if the spectrum depends only on these parameters. Before giving some evidence for such behavior, we introduce the Taylor micro-scale λ, an intermediate spatial scale between the global scale $L = E^{3/2}/\varepsilon$ of the energy carrying eddies and the dissipation scale l_d. In Navier-Stokes turbulence λ is defined by

$$\lambda^2 = \langle v^2 \rangle / \langle \omega^2 \rangle = E/\Omega \qquad (7.93)$$
$$= Ev/\varepsilon \, .$$

While L depends only on the macroscopic quantities E and ε (here in the sense of the energy injection rate) and the dissipation scale l_K depends only on the microscopic quantities v and ε (in the sense of the energy dissipation rate), the Taylor micro-scale depends on all three quantities E, ε, v, and is typically located in the inertial range. With this scale one defines the Taylor micro-scale Reynolds number

$$R_\lambda = \frac{\lambda v}{v} = \frac{E}{(v\varepsilon)^{1/2}} = Re^{1/2} \, . \P \qquad (7.94)$$

R_λ and λ are often used to characterize the inherent dynamic turbulence properties. The three characteristic scales L, λ, l_K are related to the Reynolds numbers:

$$\lambda/l_K = R_\lambda^{1/2} \, ,$$
$$L/l_K = R_\lambda^{3/2} = Re^{3/4} \, . \qquad (7.95)$$

The lower definition of λ in eq. (7.93) and that of R_λ, eq. (7.94), are also used in MHD turbulence for $\eta \simeq v$. Here, however, the dissipation scale l'_K, eq. (7.88), depends also on the macroscopic state, $v_A \simeq E^{1/2}$, such that relations (7.95) are somewhat modified, since $l'_K/l_K = R_\lambda^{1/6}$:

$$\lambda/l'_K = R_\lambda^{1/3} \, ,$$
$$L/l'_K = Re^{2/3} \, . \qquad (7.96)$$

Since the Reynolds number remains constant in 2-D MHD turbulence, eq. (7.69), R_λ can be used to characterize a system of decaying turbulence.

¶ In turbulence theory the Taylor micro-scale is usually defined in a slightly different way as the curvature radius of the longitudinal velocity correlation function $f(r) = \langle v_x(x+r)v_x(x) \rangle / \langle v_x^2 \rangle$, $\lambda^{-2} = -f''(0) = \langle (\partial_x v_x)^2 \rangle / \langle v_x^2 \rangle$, such that in 3-D λ is larger by a factor $\sqrt{10}$ than in our definition eq. (7.93). $f(r)$ also defines a global or integral scale $L_0 = \int_0^\infty f(r)dr$, which gives a direct measure of the size of the energy carrying eddies. Since in general $L_0 < L$, the corresponding integral Reynolds number is smaller than Re defined in eq. (7.2), while the Taylor micro-scale Reynolds number is somewhat larger than in our definition eq. (7.94), $R_\lambda = u\lambda/v = \sqrt{20/3}\, E/(v\varepsilon)^{1/2}$, with $E = (3/2)u^2$; for more details see Monin & Yaglom (1975), vol. 2.

If the spectrum depends only on $E\left(\simeq E^M \simeq B_0^2 = v_A^2\right)$, ε and $v\,(\simeq \eta)$, the normalized spectrum should be invariant during turbulence decay. Results from numerical simulations of freely decaying 2-D turbulence are consistent with these properties. Figure 7.4 shows time-averaged normalized spectra for five simulation runs with different values of R_λ, indicating the existence of a universal energy spectrum. Increasing R_λ leads only to a lengthening of the horizontal part, the inertial range, which agrees quite well with a $k^{-3/2}$ law rather than a $k^{-5/3}$ one. The result must, however, be considered with some caution. The formation of large-scale, force-free eddies (see Fig. 7.2) has the effect of enhancing the energy spectrum at small k, a process not included in the scaling behavior on which the $k^{-3/2}$ spectral law is based. Hence in decaying turbulence a $k^{-3/2}$ law can be expected only during the first period before force-free eddies are formed, while at later times such coherent structures lead to an effective steepening of the energy spectrum. A similar even stronger steepening has been found in high resolution simulations of 2-D Navier-Stokes turbulence (Santangelo et al., 1989). It should also be mentioned that the energy spectrum measured in the solar wind is often found to be close to $k^{-5/3}$ (see e.g. Matthaeus et al., 1982), from which it has sometimes been concluded that the Alfvénic turbulence in the solar wind has a Kolmogorov instead of an Iroshnikov-Kraichnan spectral behavior. However, only the *magnetic* energy spectrum is measured which is somewhat steeper since E_k^M slightly exceeds E_k^V (see the following subsection). Numerical simulations indeed give E_k^M close to $k^{-5/3}$.

The clearest indication of the Alfvén effect in MHD turbulence comes from the high-k region, where the effect is strongest. The universality of the spectrum in Fig. 7.4 is most convincing in the dissipation range $\hat{k} = kl_K' \gtrsim 1$. The Alfvén effect modifies the dissipation scale $l_K \to l_K'$. Though the R_λ-dependence introduced by this modification is rather weak, $l_K'/l_K = R_\lambda^{1/6}$, Fig. 7.4 shows beyond doubt that the dissipation scale is given by l_K'. Normalization of k with l_K instead of l_K' would lead to systematic horizontal shift of the spectra with increasing R_λ, giving rise to a total spread of the five curves in Fig. 7.4 of a factor of $(160/21)^{1/6} \simeq 1.4$ (Biskamp & Welter, 1989a, Fig. 16).

Finally we consider the modifications induced by using more general dissipation operators,

$$v\nabla^2 \to (-1)^{j-1}\, v_j\, \nabla^{2j}\,,$$

and similarly η_j, where v_j, η_j have the dimension $L^{2j}T^{-1}$. η_2 is called hyperresistivity, eq. (2.57). Instead of eq. (7.88) one obtains the generalized dissipation scale

$$l_{Kj}' = (v_j^2\, v_A/\varepsilon)^{1/(4j-1)}\,. \tag{7.97}$$

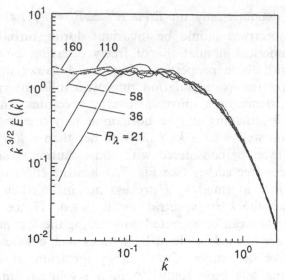

Fig. 7.4. Plots of $\hat{k}^{3/2}\hat{E}(\hat{k})$, $\hat{E}(\hat{k})$ = normalized energy spectrum, for five simulation runs of decaying 2-D MHD turbulence with different values of R_λ (from Biskamp & Welter, 1989a).

The energy spectrum assumes the form

$$E_k = (\varepsilon^{2j-2} v_A^{2j+1} v_j^3)^{1/(4j-1)} \hat{E}_j(\hat{k}) , \qquad (7.98)$$

$$\hat{k} = k \, l'_{Kj} ,$$

where $\hat{E}_j(\hat{k})$ has the small-argument behavior (7.91) for any j since the inertial-range is independent of the dissipation, while in the dissipation range numerical simulations indicate an exponential asymptotic law

$$\hat{E}_j \propto e^{-\alpha_j \hat{k}}, \qquad \hat{k} \gg 1 , \qquad (7.99)$$

where the coefficient α_j increases with j (Biskamp and Welter, 1989a),

$$\alpha_j \simeq 1.6(1 + 2j) .$$

The result that the energy spectrum is purely exponential in the far dissipation range is consistent with theoretical arguments by Frisch and Morf (1981).

Figure 7.5 illustrates how a higher order j of the diffusion operator results in a more abrupt onset of the dissipative spectrum fall-off and hence a widening of the inertial-range (the horizontal part) for the same numerical resolution given by the ratio k_{max}/k_{min}. It should, however, be noted that the order j cannot be chosen arbitrarily large without affecting the inertial-range behavior, since the limit $j \to \infty$ corresponds to a truncated nondissipative system, which relaxes to an absolute equilibrium distribu-

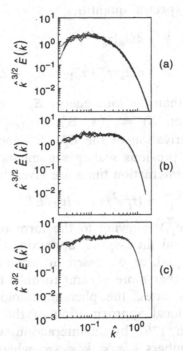

Fig. 7.5. Scatter plots of the normalized energy spectrum for three simulation runs of 2-D decaying MHD turbulence illustrating the effect of generalized diffusion operators (a) $\eta_1 = 1.25 \times 10^{-3}$, (b) $\eta_2 = 6 \times 10^{-8}$, (c) $\eta_3 = 4 \times 10^{-12}$ (from Biskamp & Welter, 1989a).

tion (section 7.2). With increasing j a growing fraction of the energy flow to large k seems to be reflected at $\hat{k} \simeq 1$, i.e. instead of a unidirectional cascade process $l_1 \to l_2 \to \cdots \to l_n$ one has $l_1 \overrightarrow{\leftarrow} l_2 \overrightarrow{\leftarrow} \cdots \overrightarrow{\leftarrow} l_n$, which leads to an increasingly flatter inertial-range spectrum. (The behavior is analogous to the reflection of an electromagnetic wave at a metallic surface, where the WKB approximation breaks down.) Only for $j \lesssim 2$ does the effect seem to be negligible.

7.4.3 *Energy spectra in highly aligned systems*

In the derivation of the energy spectrum (7.81) the velocity magnetic field correlations are not considered. In fact, it turns out that the $k^{-3/2}$ spectrum is valid only for weak correlation $\rho = 2K/E \ll 1^{\|}$. In the case of finite ρ the Elsässer variables $\mathbf{z}^\pm$, eq. (4.14), are more fundamental than

---

$^{\|}$ In this chapter ρ does not denote the mass density which is set equal to unity.

v, B, or in terms of the spectral quantities,

$$E_k^\pm = \tfrac{1}{4} \int d\Omega_k |z_k^\pm|^2 , \qquad (7.100)$$

$$E_k^R = \tfrac{1}{2} \int d\Omega_k \, \mathbf{z}_k^+ \cdot \mathbf{z}_{-k}^- = E_k^V - E_k^M , \qquad (7.101)$$

are more fundamental than the total energy $E_k = E_k^V + E_k^M = E_k^+ + E_k^-$ and the cross-correlation $K_k = \tfrac{1}{2}\langle \mathbf{v} \cdot \mathbf{B} \rangle_k = (E_k^+ - E_k^-)/2$. We give a phenomenological derivation of the E_k^+, E_k^- spectra (Grappin et al., 1983), generalizing the previous scaling arguments to finite ρ. For the fields $z_l^\pm$, the nonlinear interaction times are given by eq. (7.62),

$$T_l^\pm \simeq (\tau_l^\pm)^2/\tau_A = l v_A/E_l^\mp , \qquad (7.102)$$

since $\tau_l^\pm \simeq l/z_l^\mp = l/(E_l^\mp)^{1/2}$ owing to the form of the nonlinear term in eq. (7.59). In this local (in k-space) approximation the energy fluxes $\varepsilon^\pm$ would be exactly equal, as discussed in section 7.3.3. Closure theory, however, indicates that more general turbulent states with $\varepsilon^+ \neq \varepsilon^-$ exist. To allow such states, the phenomenological theory is modified, by replacing the local approximation of the mode interaction in the expression $\tau_l^\pm \propto (E_l^\mp)^{-1/2}$ by an interaction with modes in a certain band of wave numbers $k/\alpha < k < k\alpha$, where α is some yet undetermined parameter $\alpha > 1$. Hence eq. (7.63) is replaced by (with $k = l^{-1}$, $E_l = kE_k$, $\tau_A = (kv_A)^{-1}$, $\tau_k^\pm = (k^3 E_k^\mp)^{-1/2}$)

$$\varepsilon^\pm = k E_k^\pm/T_k^\pm = k E_k^\pm \tau_A \langle (\tau_k^\pm)^{-2} \rangle , \qquad (7.103)$$

where the brackets indicate the average over the wavenumber band. Assuming a spectral law $E_k^\pm = C^\pm k^{-m_\pm}$, one can easily evaluate the average, e.g.

$$\langle (\tau_k^+)^{-2} \rangle = C^- \left(\int_{k/\alpha}^{k\alpha} k^{3-m} dk \Big/ \int_{k/\alpha}^{k\alpha} dk \right) = F^+ E_k^- k^3 ,$$

where $F^+ \simeq \alpha^{3-m_-}/(4 - m_-)$ is independent of k. One therefore obtains

$$\varepsilon^\pm = F^\pm E_k^+ E_k^- k^3/v_A . \qquad (7.104)$$

Since in general $F^+ \neq F^-$, the symmetry is broken. The only constraint on the spectral indices $m_\pm$, following from eq. (7.104), is

$$m_+ + m_- = 3 . \qquad (7.105)$$

To determine $F^\pm$ (or the band parameter α), one uses the condition that the energy injection rates $\varepsilon^\pm$ equal the dissipation rates for stationary turbulence, which for $v = \eta$ reads

$$\varepsilon^\pm = 2v \int_{k_{in}}^{k_\pm} dk\, k^2 E_k^\pm , \qquad (7.106)$$

where k_{in} is the injection wave number and $k_{\pm}$ the dissipation wave numbers of the $E^{\pm}$ fields. From the balance relation

$$T_{k_{\pm}}^{\pm} = 1/\nu\, k_{\pm}^2$$

one obtains, using (7.102),

$$E_{k_+}^- = E_{k_-}^+ = \nu v_A \,. \tag{7.107}$$

If k_+, k_- were strongly disparate, e.g. $k_- \ll k_+$, there would be an intermediate wave number k, $k_- < k < k_+$, where E_k^- is negligible because of the exponential fall-off in the dissipation range. Hence the E^+ transfer rate would be negligible for $k > k_-$, and E_k^+ would rapidly decay, too. Both dissipation wave numbers must therefore be of about equal magnitude,

$$k_+ \simeq k_- \simeq k_d \,, \tag{7.108}$$

and the $E_k^{\pm}$ spectra become

$$E_k^{\pm} \simeq \nu\, v_A \left(\frac{k}{k_d}\right)^{-m_{\pm}} \,. \tag{7.109}$$

The quantitative difference between ε^+ and ε^- in eq. (7.104) arises from the different behavior in the *inertial range*. With eq. (7.109) evaluation of the dissipation integrals (7.106) provides a relation between the injection rates $\varepsilon^{\pm}$ and the spectral indices $m_{\pm}$,

$$\frac{\varepsilon^+}{\varepsilon^-} \simeq \frac{3 - m_-}{3 - m_+} = \frac{m_+}{m_-} \tag{7.110}$$

where eq. (7.105) has been used. Comparison with eq. (7.104) indicates that $F^{\pm} \simeq 1/m_{\pm}$.

Thus, only in the uncorrelated case $E_k^+ = E_k^-$, implying $\varepsilon^+ = \varepsilon^-$, one finds the Iroshnikov-Kraichnan spectrum (7.81), since only for $m_+ = m_- = 3/2$ the total energy $E_k = E_k^+ + E_k^-$ also follows a $k^{-3/2}$ law. In the general case of finite correlation ρ, however, $m_+ \neq m_-$, and E_k does not obey a power law. Only for strong correlation $\rho \simeq 1$, where $\mathbf{v}$ and $\mathbf{B}$ are almost aligned so that, say, $E^+ \gg E^-$, E_k again assumes an approximate power law, that of the dominant component $m \simeq m_+$. Let us also calculate the ratio of the total energies $E^{\pm} = \int_{k_{in}}^{k_d} E_k^{\pm} dk$ using eq. (7.108),

$$E^+/E^- \simeq (k_d/k_{in})^{n_+ - n_-} \,, \quad n_{\pm} = \sup(1, m_{\pm}) \,. \tag{7.111}$$

Thus for given injection rates $\varepsilon^{\pm}$ the velocity magnetic field correlation $\rho = (E^+ - E^-)/(E^+ + E^-)$ depends on the Reynolds number, $k_d/k_{in} \simeq Re^{2/3}$, eq. (7.96), ρ approaching unity for $Re \to \infty$. Since the spectral intensities are equal at the dissipation scale, eq. (7.107), the dominant component, say E^+ for $m_+ > m_-$, will completely outrun the smaller one at the energy-containing scale k_{in}^{-1}.

The phenomenological theory has been verified by numerical solution of the EDQNM closure equations (Grappin et al., 1983) as well as 2-D simulations (Pouquet et al., 1988; Politano et al., 1990).

In addition to $E_k^\pm$ the reduced energy spectrum $E_k^R = E_k^V - E_k^M$ is found to follow a power law. While one has $E_k^R \simeq 0$ for high k owing to the Alfvén effect, i.e. the interaction with the large-scale field B_0, local interactions between Alfvén waves of similar wavenumbers become increasingly important for smaller k, making E_k^R finite. The evolution equation for E_k^R can be written in the following way (Grappin et al., 1983):

$$\partial_t E_k^R = T_k^{\mathrm{nonloc}} + T_k^{\mathrm{loc}} , \tag{7.112}$$

dissipation terms being ommitted. T_k^{nonloc} represents the Alfvén effect,

$$T_k^{\mathrm{nonloc}} \simeq -E_k^R/\tau_A = -E_k^R k v_A , \tag{7.113}$$

which would lead to decay of E_k^R within one Alfvén time if the local interactions T_k^{loc} were absent. From dimensional arguments the latter assumes the form

$$T_k^{\mathrm{loc}} \simeq \varepsilon/k \simeq k^3 E_k^+ E_k^- / k v_A \simeq C^+ C^- / k v_A \tag{7.114}$$

since $E_k^\pm = C^\pm k^{-m_\pm}$ and $m_+ + m_- = 3$. Hence for stationary conditions, $T_k^{\mathrm{nonloc}} + T_k^{\mathrm{loc}} \simeq 0$, one obtains

$$E_k^R \simeq C^+ C^- / (k v_A)^2 . \tag{7.115}$$

Note that the relative magnitude E_k^R/E_k is small for large k, in particular for weak correlation $m_+ \simeq m_- \simeq 3/2$, where $E_k^R/E_k \propto k^{-1/2}$. The sign of E_k^R is not determined by these phenomenological considerations. Numerical solution of the EDQNM closure equations (Grappin et al. 1983) as well as 2-D simulations (Biskamp & Welter, 1989a) have verified the validity of the k^{-2} spectral law. These calculations also show that there is an excess of magnetic energy, $E_k^R < 0$, a tendency which is already indicated by the behavior of the absolute equilibrium spectra, eq. (7.19). As is to be expected from the phenomenological derivation, the residual spectrum is found to be rather independent of the velocity magnetic field correlation ρ.

7.5 Closure theory for MHD turbulence

In the preceding section spectral properties were derived in a phenomenological way, using dimensional and scaling arguments. Though such arguments are very powerful and reliable, representing basic physical principles, they are restricted to a few quantities and even then cannot predict proportionality constants, e.g. the value of the Kolmogorov constant C_K or the sign of the residual spectrum E_k^R. In addition, they provide little insight into the turbulence *dynamics*, in particular for nonstationary systems. Such properties should be described by a statistical theory derived from the basic fluid equations. In the past turbulence theory has mainly been confined to two-point closure models on the lines initiated by the quasi-normal approximation (Proudman & Reid, 1954). Recently, different approaches have been introduced, notably the renormalization group method, which has been rather successful in predicting several parameters in hydrodynamic turbulence (Yakhot & Orszag, 1986) and which has also been applied to MHD turbulence (Longcope and Sudan, 1991; Camargo and Tasso, 1992). The actual analysis, however, still contains a considerable amount of arbitrariness. Hence the discussion given here is restricted to closure theory.

7.5.1 *The problem of closure*

Let us write the dynamic equations in the general form valid for many types of nonlinear systems

$$\frac{du_i}{dt} + v_i u_i = \sum_{j,m} M_{ijm} u_j u_m , \qquad (7.116)$$

where in the incompressible MHD case $\{u_i\} = \{\mathbf{v_k}, \mathbf{B_k}\}$ and $\{v_i\} = \{v\, k^2, \eta k^2\}$. Multiplying eq. (7.116) by $u_j, u_j u_l, \ldots$ and averaging over an ensemble of equivalent systems, one obtains a system of equations for the moments or correlation functions $\langle u_i u_j \rangle, \langle u_i u_j u_l \rangle, \ldots$,

$$\frac{d}{dt}\langle u_i u_j \rangle + (v_i + v_j)\langle u_i u_j \rangle$$
$$= \sum_{m,n} \left[M_{imn}\langle u_j u_m u_n \rangle + M_{jmn}\langle u_i u_m u_n \rangle \right] , \qquad (7.117)$$

$$\frac{d}{dt}\langle u_i u_j u_l \rangle + (v_i + v_j + v_l)\langle u_i u_j u_l \rangle$$
$$= \sum_{m,n} \left[M_{imn}\langle u_j u_l u_m u_n \rangle + M_{jmn}\langle u_i u_l u_m u_n \rangle \right.$$
$$\left. + M_{lmn}\langle u_i u_j u_m u_n \rangle \right] , \qquad (7.118)$$
$$\text{etc.}$$

The hierarchy of moments forms an infinite set of statistical equations replacing the single dynamic equation (7.116). This apparent incongruity is due to the fact that in reality the solution of the hierarchy is equivalent to an infinite set of dynamic solutions with different initial conditions, or in the case of stationary turbulence a solution over infinitely long times.

While the solution of the "simple" dynamic equation is very complicated, with rapid spatial and temporal variations, the solution of the statistical equations can be stationary and is usually smooth.

To be useful, the hierarchy must be closed, leading to a set of approximate equations for a finite number of functions. Since in the most interesting case of large Reynolds number the nonlinear term in eq. (7.116) is much larger than the linear one, finite series approximations, neglecting for instance correlations of order $n > N$, are not useful. Instead approximation must be guided by physical reasoning. A natural starting point is the experimental and numerical finding that the components of the velocity and the magnetic field have Gaussian or normal probability distributions. This property is to be expected in a fluid with random motions as a consequence of the central limit theorem, much in the same way the molecules in a gas assume a Maxwellian distribution.

The properties of a random variable χ become particularly clear when considering its cumulants. The nth cumulant $C^{(n)}$ is defined as the nth derivative of the logarithm of the characteristic function $\varphi(s) = \langle e^{is\chi} \rangle$ at the origin,

$$C^{(n)} = \left(d^n/ds^n \right) \ln \varphi(s)|_{s=0} , \tag{7.119}$$

which can be expressed in terms of the moments $\langle \chi^\nu \rangle$, $\nu \le n$ (see, for instance, Lumley 1970, chapter 2). In the case of zero mean the first four cumulants are

$$C^{(1)} = 0 , \quad C^{(2)} = \langle \chi^2 \rangle , \quad C^{(3)} = \langle \chi^3 \rangle ,$$

$$C^{(4)} = \langle \chi^4 \rangle - 3\langle \chi^2 \rangle^2 .$$

For a set of random variables $\chi_1, \ldots, \chi_n$ with a multivariate Gaussian distribution, eq. (7.13), the characteristic function $\varphi(s_1, \ldots, s_n) = \langle \exp i \sum_j s_j \chi_j \rangle$ can easily be computed. Its logarithm is

$$\ln \varphi (s_1, \ldots, s_n) = -\tfrac{1}{2} \sum A_{ij}^{-1} s_i s_j . \tag{7.120}$$

Hence all cumulants of order $n > 2$ vanish, so that in particular one has

$$\langle \chi_i \chi_j \chi_m \rangle = 0 \tag{7.121}$$

and

$$\langle \chi_i \chi_j \chi_m \chi_n \rangle = \langle \chi_i \chi_j \rangle \langle \chi_m \chi_n \rangle + \langle \chi_i \chi_m \rangle \langle \chi_j \chi_n \rangle$$
$$+ \langle \chi_i \chi_n \rangle \langle \chi_j \chi_m \rangle . \tag{7.122}$$

However, even if the probability distributions of the velocity components considered at some point in space, i.e. the one-point distributions, are normal, this cannot exactly be the case for the joint probability distribution involving different space arguments, since it would imply the vanishing of the three-point correlations, eq. (7.121), which would eliminate all nonlinear interactions in the statistical equations (7.117). Instead one can assume that the joint probability distributions are *close* to normal, such that all cumulants of order $n > 3$ vanish, which implies in particular eq. (7.122), and compute the third-order moments from the evolution equation (7.118), using eq. (7.122). This closure of the hierarchy is called the quasi-normal or cumulant-discard approximation (Proudman & Reid, 1954). Though inconsistent in its elementary or naked form, as explained below, this approximation serves as the formal skeleton for most momentum closure models proposed to date. Integrating eq. (7.118) and inserting the resulting expression into eq. (7.117) gives the nonlinear evolution equation for the mode intensity $U_k = \langle |u_k|^2 \rangle$, which for homogeneous isotropic turbulence can be written in the general form

$$\frac{dU_k}{dt} + 2v_k U_k = \int_{\Delta_k} dp\,dq \int_{-\infty}^{t} dt'\, G(t,t') \left[A U_k(t') U_q(t') \right.$$
$$\left. + B U_p(t') U_q(t') \right] . \tag{7.123}$$

Here Δ_k indicates integration over the subdomain of the p,q-plane such that $\mathbf{k}, \mathbf{p}, \mathbf{q}$ form a triangle. A, B are coefficients depending on k, p, q and the cosines of the angles in the triangle formed by the three wave vectors, and $G = G_{kpq}(t,t')$ is called the linear response function, which in the quasi-normal approximation is simply

$$G(t,t') = G^{(0)}(t - t') = \exp\left\{ -(v_k + v_p + v_q)(t - t') \right\} . \tag{7.124}$$

(The precise form of the quasi-normal theory for Navier-Stokes turbulence can be found in Leslie's book (1973), which gives a detailed review of closure theory.)

By solving the quasi-normal equations numerically it is found that U_k, which by definition should be nonnegative, becomes negative after some time, in essential regions of k-space and for typical initial conditions. This unacceptable feature is due primarily to the unphysically long memory time $\sim v_k^{-1}$ in the time integral in eq. (7.123), which for a mode in the inertial range $kl_d \ll 1$ is much longer than a typical eddy interaction time. Hence the viscous damping v_k should be replaced by an effective eddy damping $\tilde{v}_k$. A self-consistent theory for Navier-Stokes turbulence accounting for this effect is the Direct Interaction Approximation (DIA) (Kraichnan, 1959), or its Lagrangian variant (Kraichnan, 1965a), which corrects the inconsistency of the former Eulerian version mentioned in section 7.4.1. Formally, the DIA amounts to supplementing eq. (7.123) by a second nonlinear equation for the response function G, instead of using the quasi-normal expression (7.124). The Lagrangian DIA seems to be the only fundamental turbulence theory developed to date with no adjustable parameters which gives good agreement with experimental measurements of the energy spectrum. However, the equations are too complicated for practical numerical evaluations. In addition, the theory has not been generalized to the MHD case. We shall therefore present a more phenomenological approach which satisfies important consistency requirements and allows numerical solutions for sufficiently high Reynolds numbers.

7.5.2 Eddy-damped quasi-normal Markovian approximation

Since the quantity of primary interest is the modal intensity $U_k(t)$, we turn our attention to eq. (7.123) and try to model the response function G, using the following guidelines:

(a) Introduction of an eddy viscosity. The time scale involved in $G(t, t')$, which is the relaxation time of the triple correlations in eq. (7.118), should be determined by the eddy damping rate $\tilde{v}_k$. In Navier-Stokes turbulence dimensional arguments give

$$\tilde{v}_k \simeq \varepsilon^{1/3} k^{2/3} \simeq E_k^{1/2} k^{3/2} , \tag{7.125}$$

using the Kolmogorov spectrum (7.77) connecting ε and E_k. A more specific expression valid for the MHD case is given below. Replacing the collisional viscous damping $v_k = v\,k^2$ in expression (7.124) by the eddy damping rate $\tilde{v}_k$ gives the modified or renormalized[**] response function

$$\tilde{G}(t, t') = \exp\left\{-\int_{t'}^{t} d\tau \left[\tilde{v}_k(\tau) + \tilde{v}_p(\tau) + \tilde{v}_q(\tau)\right]\right\} . \tag{7.126}$$

There is of course a considerable amount of arbitrariness about the way in which the quasi-normal expression (7.124) can be modified. Expression (7.126) represents the simplest and most straightforward form. An important property is the symmetry of G in k, p, q, which guarantees a detailed balance relation for each triad interaction and hence conservation of the ideal quadratic invariants of the dynamic equations.

(b) Markovization. The solution of eq. (7.123) should be realizable in the sense that it corresponds to the statistical average with some nonnegative probability distribution. This implies in particular that U_k remains nonnegative if it is so initially. Violation of this condition led to rejection of the original quasi-normal approximation. A simple way to satisfy this condition is to replace the time argument t' in the

---

[**] Borrowed from quantum field theory, this term indicates renormalization of the bare viscosity coefficient by including the effect of mode interactions.

functions U_k under the time integral in eq. (7.123) by the outer time t, a process called markovization. Since the effective time domain in the integral is the eddy interaction time $\tilde{v}_k^{-1}$, which is the shortest time scale on which the mode intensity U_k can change, markovization is not expected to change the results qualitatively.

Equation (7.123) now takes the form (indices and coefficients being suppressed for clarity)

$$\frac{dU}{dt} + 2\nu U = \theta(t)U(t)U(t) , \tag{7.127}$$

where $\theta(t) = \int_0^t \tilde{G}(t,t')dt'$ is basically the triple correlation or triad relaxation time. Realizability of the Markovian equation (7.127) can be proved directly by showing that it gives the *exact* ensemble average of the mode intensity of a model dynamic system. The model amplitude equation has the form of a Langevin equation,

$$\frac{d}{dt}\hat{u}_k(t) = -\alpha_k(t)\hat{u}_k(t) + \hat{q}_k(t) , \tag{7.128}$$

where $\hat{u}_k$ is a stochastic variable, which is *not* identical with u_k but has the same variance $\langle \hat{u}_k^2 \rangle = U_k$, $\hat{q}_k(t)$ is a white-noise stochastic variable essentially independent of $\hat{u}_k$, and $\alpha_k(t)$ is a nonstochastic damping rate. Both $\hat{q}_k$ and α_k can be chosen such that the equation for the average mode intensity derived from eq. (7.128) coincides with eq. (7.127). Spectral equations of this type with G given by eq. (7.126) are called eddy-damped quasi-normal Markovian approximations (EDQNM) (see, for instance, Orszag, 1977).

In the case of MHD turbulence the following expression for the eddy damping rate was chosen (Pouquet et al., 1976):

$$\tilde{v}_k = C_s \left[\int_0^k q^2 (E_q^M + E_q^V) dq \right]^{1/2}$$

$$+ C_A k \left[\int_0^k E_q^M dq \right]^{1/2} + (\nu + \eta) k^2 . \tag{7.129}$$

The first term corresponds to the self-distortion or scrambling of the flow as in Navier–Stokes turbulence and is an obvious generalization of the scaling expression (7.125). The second term $\sim k B_0$ represents the Alfvén effect, usually the dominant contribution in the inertial range. C_s, C_A are free parameters of order unity. The last term dominates in the dissipation range.

Two sets of EDQNM equations for 3-D MHD turbulence have been derived and evaluated. In the first case (Pouquet et al., 1976) these are equations for the kinetic and magnetic energies E_k^V, E_k^M and the kinetic and magnetic helicities H_k^V, H_k^M (=H_k in our notation) assuming zero cross-correlation, $K_k = 0$. In steady state the model gives a Iroshnikov-Kraichnan energy spectrum $E_k \propto k^{-3/2}$ and a reduced spectrum $E_k^R = E_k^V - E_k^M \propto k^{-2}$ with $E_k^R < 0$. If magnetic helicity is injected, numerical integration in time shows the inverse H cascade. The second model (Grappin et al., 1982, 1983) deals with the effect of velocity magnetic field alignment. Equations are written in Elsässer variables describing the evolution of E_k^+, E_k^- and E_k^R, assuming $H_k^M = 0$. Stationary spectra are consistent with the phenomenological theory given in section 7.4.3, and if cross-correlation is injected the system develops toward an aligned state. The EDQNM theories have thus proved to be a valuable instrument to describe dynamic properties such as the evolution of the spectrum in Reynolds number regimes which are still outside the feasibility of direct numerical simulation. The latter, however, can provide more general information concerning in particular the spatial structure of the turbulence, e.g. that of the dissipative eddies.

7.6   Energy dissipation in 2-D MHD turbulence

The dissipation properties in 2-D MHD turbulence have been elucidated by a series of numerical simulations with relatively high Reynolds numbers. Let us consider these properties in some detail.

7.6.1   *Spontaneous excitation of small-scale turbulence*

It is intuitively taken for granted that a fluid of high Reynolds number develops small-scale turbulence spontaneously from a smooth initial flow. In conventional turbulence theory, however, this issue is not clear, since stationary turbulence has usually been assumed to be generated by a random external stirring force, the statistics of which may significantly influence the statistical properties of the turbulence. In numerical simulations a random forcing often seems to create a pseudo-turbulence in systems, which may otherwise behave in a rather nonturbulent way. This is particularly true in two-dimensional systems, where the strong tendency to large-scale self-organization might affect or even suppress generation of small-scale dissipative turbulence. In fact, typical two-dimensional MHD simulations of instability evolution, such as the $m = 1$ resistive kink mode or steady-state reconnection simulations, discussed in chapter 6, do not develop small-scale turbulence even at what seem to be rather small values of η and ν, i.e. high Lundquist numbers (but not necessarily high Reynolds numbers).

The question was re-examined in numerical simulation studies of freely decaying homogeneous turbulence (Biskamp & Welter, 1989a; Politano et al., 1989). It is found that for sufficiently high Reynolds number truly turbulent states are generated spontaneously. Figure 7.6 illustrates the evolution of a system starting from a smooth initial state at $t = 0$. After a period of the order of an eddy turnover time, $t = 1.5$, extended current sheets have developed. The increase of current density in the sheets (and the corresponding decrease of sheet thickness) is found to be exponential in time (Frisch et al., 1983), until dissipation prevents further thinning, which implies that in the limit $Re \to \infty$ exponential growth would continue indefinitely. This seems to rule out the possibility of a finite time singularity in ideal two-dimensional MHD flows, a point which has attracted considerable attention in turbulence theory (Pouquet, 1978). A similarity solution showing the exponential flattening of the configuration into a current sheet near an X-point was derived by Sulem et al. (1985). The question is, however, more a matter of principle. Practically speaking, for finite Reynolds number, no matter how huge, dissipative saturation of the current density in the sheet occurs in a finite time t_0, depending only very weakly on Re, $t_0 \propto \ln Re$.

Subsequently, the extended (macro-)current sheets tend to disintegrate as illustrated in Fig. 7.7, which shows essentially the same system as in Fig. 7.6 at a somewhat later time. We can distinguish two processes. In regions of weak magnetic field extending around the X-points of the major magnetic eddies a random dynamic folding process occurs, in particular in the upper left part of the figure, which is similar to the folding of vorticity gradient sheets in 2-D Navier-Stokes turbulence. The second process is the tearing instability clearly recognizable in the ψ contours. It occurs in regions of stronger magnetic field, in which the extended sheets are imbedded. As discussed in section 6.5, a dynamic current sheet is accompanied by strong inhomogeneous flows along the sheet, which have a considerable stabilizing effect on the tearing mode, such that a dynamic sheet becomes unstable only for aspect ratio $A = \Delta/\delta \gtrsim 10^2$. Hence the tearing mode only sets in at Reynolds numbers exceeding some threshold value. Both processes generate small-scale turbulence, with dissipative eddies consisting of micro-current sheets and their associated vorticity sheets. Since small-scale structures are due mainly to flows perpendicular to the large-scale magnetic field, one can expect current sheets, viz. two-dimensional structures, to dominate dissipation also in 3-D MHD turbulence in strongly magnetized plasmas. By contrast, in 3-D Navier-Stokes turbulence dissipation seems to be localized in vorticity ropes, i.e. quasi-one-dimensional structures (e.g. Vincent & Meneguzzi, 1991).

Figure 7.8 illustrates the distribution of micro-current sheets for three cases of 2-D MHD turbulence with nearly identical macro-states but

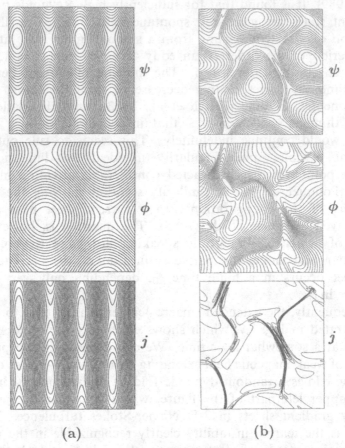

Fig. 7.6.   Generation of current sheets from an initially smooth state. (a) $t = 0$, (b) $t = 1.5$.

different Reynolds numbers. With increasing Reynolds number current sheets become smaller and more numerous.

7.6.2   *Energy dissipation rates*

The process of small-scale turbulence generation is reflected in the evolution of the energy dissipation rate $\varepsilon = -dE/dt$. In Fig. 7.9 $\varepsilon(t)$ is plotted for five simulation runs with identical smooth initial conditions but different values of the dissipation coefficients $\eta = v$. The upper four curves correspond to ordinary diffusion with $\eta = \eta_1 = 2.5 \times 10^{-3}$, 1.25×10^{-3}, 6.25×10^{-4}, 3.12×10^{-4}. While in the smooth initial state we have $\varepsilon \propto \eta$, energy dissipation increases exponentially owing to the formation and thinning of macro-current sheets. Saturation of this exponential

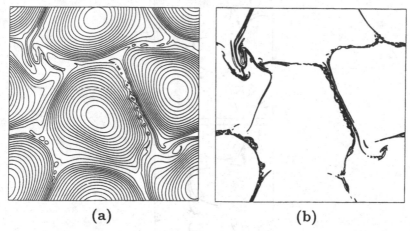

(a)   (b)

Fig. 7.7. Filamentation of macro-current sheets at high Reynolds number. From a simulation similar to the one shown in Fig. 7.7, $t = 2.1$. (a) ψ, (b) j.

growth is reached at $t \simeq 1.5$, where dissipation rates scale as $\varepsilon \propto \eta^{1/2}$, the Sweet-Parker scaling expected for such current sheet configurations (chapter 6.1.1). As these current sheets break up, ε further increases, an effect which is the more pronounced the smaller η. The dashed curve in Fig. 7.9 corresponds to a case with higher-order diffusion operators, $\eta_2 = \nu_2 = 10^{-8}$, and hence much smaller initial dissipation. The time evolution, however, shows the same features as the upper cases, with which it finally merges. Thus we find that for fully developed turbulence the energy dissipation rate becomes essentially independent of the Reynolds number. This result is rather remarkable, since it is usually expected to hold only for 3-D turbulence. A corollary is the observed invariance of the constant C'_K in the inertial range spectrum, $C'_K \simeq 2$, as seen in Fig. 7.4, which does not depend on R_λ.

7.7 Intermittency

In the derivation of the energy spectra in section 7.4 it was tacitly assumed that the energy transfer or dissipation rate ε is a nonfluctuating quantity distributed uniformly in space. The results of the last section (see in particular Fig. 7.8), however, show that the distribution of the dissipative eddies is far from uniform. This spottiness of the dissipative eddies is a special feature of what is now believed to be a general property of fully developed turbulence that with decreasing scale turbulent fluctuations become less and less space-filling, i.e. are concentrated in regions of smaller and smaller volume but increasingly complicated shape. This

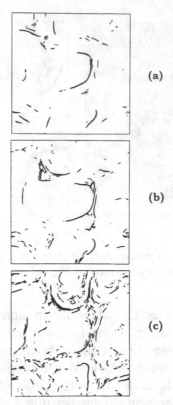

Fig. 7.8. Distribution of micro-current sheets in 2-D MHD turbulence for three different Reynolds numbers, (a) $R_\lambda = 110$, (b) $R_\lambda = 160$, (c) $R_{\lambda eff} \simeq 340$, for essentially identical macro states (from Biskamp & Welter, 1989a).

phenomenon is called intermittency, which is a central topic in actual turbulence research. In sections 7.7.1 and 7.7.2 two intermittency models are outlined which have been developed for Navier-Stokes turbulence. In section 7.7.3 the probability distribution functions of the velocity fluctuations v_l are discussed in some detail. Finally, section 7.7.4 presents some results concerning intermittency effects in 2-D MHD turbulence.

7.7.1 *The log-normal theory*

The first important intermittency model was presented by Obukhov (1962) and Kolmogorov (1962), refining Kolmogorov's 1941 theory (K-41). The theory deals with the statistical properties of the local dissipation rate $\varepsilon(\mathbf{x}, t) = v/2 \sum_{i,j} (\partial_i v_j + \partial_j v_i)^2$. The dynamics of a fluid element v_l depends on the values of $\varepsilon(\mathbf{x}, t)$ in the volume $V_l \sim l^3$ and the time interval τ_l, and the simplest assumption is that it depends only on the average value in

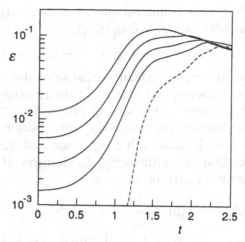

Fig. 7.9. Time evolution of the energy dissipation rate for five simulations with identical initial conditions but different values of the dissipation coefficients. $t = 0$: $\varepsilon \propto \eta$; $t \simeq 1.4$: $\varepsilon \propto \eta^{1/2}$; $t > 2$: $\varepsilon \propto \eta^0$ (from Biskamp & Welter, 1989a).

this volume

$$\varepsilon_l = \langle \varepsilon(x,t) \rangle_{V_l, \tau_l} ,^{\dagger\dagger} \tag{7.130}$$

such that instead of eq. (7.76) we now have

$$v_l^3 \simeq \varepsilon_l \, l , \tag{7.131}$$

where ε_l is now a random variable with a probability distribution $P(\varepsilon_l)$. To obtain an approximate expression for $P(\varepsilon_l)$ one assumes a simple model of the self-similar eddy fragmentation or cascade process. Following Yaglom (1966) we consider a cube of length l_0 and mean dissipation rate ε_0, where l_0 is of the order of the integral length L and $\varepsilon_0 = \bar{\varepsilon}$, and divide it into j cubes of length $l_1 = l_0 j^{-1/3}$. In each of these the (space-averaged) dissipation rate ε_1 is a random variable with identical probability distributions and mean value $\overline{\varepsilon_1} = \bar{\varepsilon}$. The l_1-cubes are further subdivided, each into j cubes of length $l_2 = l_1 j^{-1/3} = l_0 j^{-2/3}$, and the process is continued N times until the dissipation scale l_d is reached, $l_N = l_d$. Consider a scale l_n in the inertial range,

$$l_0 \gg l_n = l_0 j^{-n/3} \gg l_d . \tag{7.132}$$

The random variable ε_n can be written as a product

$$\varepsilon_n = \varepsilon_0 \frac{\varepsilon_1}{\varepsilon_0} \cdots \frac{\varepsilon_n}{\varepsilon_{n-1}} , \tag{7.133}$$

---

†† In this section $\langle f \rangle$ indicates the space time average of a particular realization, while $\bar{f}$ denotes the statistical average.

where $\varepsilon_i/\varepsilon_{i-1}$, $1 \le i \le n$, is a dimensionless random variable. Hence $\ln \varepsilon_n$ is a sum of random variables $\chi_i = \ln(\varepsilon_i/\varepsilon_{i-1})$,

$$\ln \varepsilon_n = \ln \varepsilon_0 + \chi_1 + \cdots + \chi_n . \qquad (7.134)$$

The χ_i are assumed to be independent, because of the random character of the fragmentation process, and identically distributed with mean value m and variance σ^2, because of the scaling property in the inertial range. By the central limit theorem the probability distribution of $\ln \varepsilon_n$ becomes Gaussian for $n \gg 1$ with mean value $m_n = nm$ and variance $\sigma_n^2 = n\sigma^2$. One now concludes that the probability distribution of ε_n tends toward the so-called log-normal distribution

$$P(\varepsilon_n) = P_0 (\ln \varepsilon_n) \frac{1}{\varepsilon_n}$$

$$= \frac{1}{\sqrt{2\pi}\sigma_n} \frac{1}{\varepsilon_n} \exp\left\{-\frac{(\ln \varepsilon_n - m_n)^2}{2\sigma_n^2}\right\} . \qquad (7.135)$$

The requirement that the average dissipation rate does not depend on the level n of subdivision, $\overline{\varepsilon_n} = \bar\varepsilon$, gives the relation $m_n = -\sigma_n^2/2$.

Since n is proportional to $\ln(L/l_n)$, we find that the σ_n^2 can be written as $\sigma_n^2 = \mu \ln(L/l_n)$, introducing the parameter μ. Hence the distribution depends weakly on the global scale L. From the distribution eq. (7.135) the moments of ε_l can be computed (replacing l_n by the continuous scale variable l)

$$\overline{\varepsilon_l^q} = \bar\varepsilon^q \exp\left\{\tfrac{1}{2}q(q-1)\sigma_l^2\right\}$$

$$= \bar\varepsilon^q \left(\frac{L}{l}\right)^{\mu_q} , \quad \mu_q = \mu q(q-1)/2 . \qquad (7.136)$$

Since in the nonintermittent K-41 theory $\overline{\varepsilon_l^q} = \bar\varepsilon^q$, i.e. $\mu = 0$, μ is called the intermittency parameter, the essential free parameter in the log-normal theory, which must be determined by comparison with experiments. Experimentally it is, however, more convenient to measure the velocity fluctuations $\Delta v(l)$ as defined in eq. (7.74) rather than ε_l, most importantly the moments

$$F_p(l) = \overline{\Delta v(l)^p} \qquad (7.137)$$

which are called structure functions. In the inertial range self-similarity requires a scaling law

$$F_p(l) \propto l^{\zeta_p} , \qquad (7.138)$$

or for $f_p(l)$, the dimensionless structure functions,

$$f_p(l) = \frac{F_p(l)}{(F_2(l))^{p/2}} \propto l^{\xi_p} , \qquad (7.139)$$

$$\xi_p = \zeta_p - p\zeta_2/2 \ . \tag{7.140}$$

To connect the statistics of ε_l with that of $\Delta v(l)$ one uses eq. (7.131), writing $\Delta v(l)$ explicitly for clarity,

$$\Delta v(l) \simeq \varepsilon_l^{1/3} l^{1/3} \ . \tag{7.141}$$

Note that ε_l and $\Delta v(l)$ are random quantities. Using relation (7.141) together with eq. (7.136), we easily obtain the structure functions

$$F_p(l) \simeq \bar{\varepsilon}^{p/3} l^{p/3} (L/l)^{p(p-3)\mu/18} \ , \tag{7.142}$$

i.e.

$$\begin{aligned}
\zeta_p &= \frac{p}{3} - \mu \frac{p(p-3)}{18} \\
&= (q - \mu_q)|_{q=p/3} \ .
\end{aligned} \tag{7.143}$$

In particular for $p = 2$

$$\begin{aligned}
F_2(l) &\simeq \bar{\varepsilon}^{2/3} l^{2/3} (L/l)^{-\mu/9} \ , \\
E_k &\simeq \bar{\varepsilon}^{2/3} k^{-5/3} (kL)^{-\mu/9} \ .
\end{aligned} \tag{7.144}$$

While the μ correction to the 5/3 law seems to be too small to be reliably measurable, the corrections to the higher-order moments become substantial for large p. The intermittency parameter μ is usually associated with the correlation function of the energy dissipation

$$\overline{\varepsilon(\mathbf{x})\varepsilon(\mathbf{x} + \mathbf{l})} \simeq \overline{\varepsilon_l^2} \propto l^{-\mu} \ , \tag{7.145}$$

using eq. (7.136), which corresponds to the structure function F_6, hence

$$\mu = 2 - \zeta_6 \ . \tag{7.146}$$

Both experiments (e.g. Anselmet et al., 1984) and numerical simulations (e.g. Vincent & Meneguzzi, 1991) give a value $\mu \simeq 0.2$. With this value the prediction of ζ_p, eq. (7.143), is quite good up to $p \simeq 16$, which corresponds to the maximum value of ζ_p for $\mu = 0.2$. The subsequent decrease of ζ_p for larger p is not observed, however. In fact Novikov (1970) derives from a fairly general model that μ_q increases at most linearly, $\mu_q \leq q + \mu - 2$, $q > 2$, in the 1-D case considered by Novikov, or $\mu_q \leq q + (\mu - 2)D$ for general spatial dimensions D, in contrast to the quadratic dependence predicted by the log-normal theory, eq. (7.136). On the other hand, it has been found, both experimentally (see e.g. Monin & Yaglom, 1975, Chapter 7, section 25) and numerically (e.g. Biskamp et al., 1990), that the distribution $P(\varepsilon_l)$ is rather well approximated by the log-normal distribution (apart from a slight enhancement at small and a slight deficiency at large values). The origin of this somewhat paradoxical behavior can be traced to the weak convergence of $P(\varepsilon_l)$ to a log-normal distribution as R_λ is increased, such that the moments do not tend to those of the log-normal distribution. This is related to the mathematical property that the distribution function is not uniquely determined by the value of all the moments, if the latter grow too rapidly (see e.g. Feller (1966), chapter 7; also Paladin & Vulpiani (1987), appendix A). Hence it is more practical to consider models predicting the moments directly as we do in the following.

7.7.2   *The β-model and its generalizations*

A further objection regards the use of the space-averaged dissipation rate ε_l, eq. (7.130), as a fundamental inertial range quantity in the log-normal theory, since it mixes inertial range and dissipation range properties, as is evident from eq. (7.141). In fact this equation is based on little more than dimensional arguments[‡‡]. When discussing inertial range properties it is therefore preferable to use the inertial range quantities $\Delta v(l)$ directly. The simplest intermittency theory following this concept is the β-model by Frisch et al. (1978), which was in fact first proposed by Novikov & Stewart (1964). The model uses the discrete sequence of scales introduced above, $l_n = 2^{-n}L$, with corresponding wave numbers $k_n = l_n^{-1}$. The basic property of intermittency, that smaller scales are less space filling, is described in the β-model in the following simple way (Fig. 7.10a). In each step of the cascade an eddy v_n of scale l_n splits into $2^D\beta$ eddies of scale $l_{n+1} = l_n/2$, D = spatial dimension, where β is a fixed parameter with $0 < \beta \leq 1$, i.e. only a fraction β of the 2^D volume elements are filled. From the construction it is clear that in the limit $n \to \infty$ the spatial distribution of eddies is a fractal of dimension $D_F = D - \delta$, $\beta = 2^{-\delta}$.[§§] The spatial average energy E_n of eddies of scale l_n is smaller than v_n^2,

$$E_n = \langle v_n^2 \rangle \simeq \beta^n v_n^2 \,, \tag{7.147}$$

while the transfer time, i.e. the time for $l_n \to l_{n+1}$ fragmentation, is determined by the actual velocity, $\tau_n = l_n/v_n$. Hence the energy transfer rate is

$$\varepsilon = E_n/\tau_n = \beta^n v_n^3/l_n \,, \tag{7.148}$$

modifying eq. (7.75). Note the conceptual difference between eqs (7.131) and (7.148). While in the former v_l is modified with respect to the K-41 theory by the random character of the averaged dissipation rate ε_l, the

---

[‡‡] Relation (7.141) is called Kolmogorov refined similarity hypothesis (see Monin & Yaglom, 1975, vol. 2). Recently, this relation has been verified to a certain extent, both experimentally (Praskovsky, 1992) and numerically (Chen et al., 1993).

[§§] To visualize the concept of fractal dimension consider special cases; for instance $D_F = 2$ in 3-D corresponds to sheet-like structures.

latter is modified by the geometric effect of the increasing sparseness of the distribution of the dynamic eddies. From eq. (7.148) one finds the spectrum

$$E_n = \langle v_n^2 \rangle = \varepsilon^{2/3} l_n^{2/3} \beta^{-2n/3} \beta^n , \qquad (7.149)$$

$$E_k = \varepsilon^{2/3} k^{-5/3} (kL)^{-\delta/3} , \qquad (7.150)$$

using $\beta^n = (l_n/L)^\delta$. As in eq. (7.148), the last factor β^n in eq. (7.149) accounts for the spatial average. The scaling of the structure functions F_p in the β-model is obtained directly from eq. (7.148),

$$F_p(l) = \langle v_l^p \rangle = \varepsilon^{p/3} l^{p/3} (L/l)^{\delta(p-3)/3} ,$$

i.e.

$$\zeta_p = \frac{p}{3} - \frac{\delta}{3}(p - 3) , \qquad (7.151)$$

which is a linear function of p. If one again determines δ from F_6, one obtains $\delta = 2 - \zeta_6$, identical with μ, eq. (7.146). The skewness factor S_3 and the flatness factor (or kurtosis) S_4, usually defined in the form

$$S_3 = \langle (\partial_x v_x)^3 \rangle / \langle (\partial_x v_x)^2 \rangle^{3/2} , \qquad (7.152)$$

$$S_4 = \langle (\partial_x v_x)^4 \rangle / \langle (\partial_x v_x)^2 \rangle^2 , \qquad (7.153)$$

are particularly important quantities in turbulence theory, which for non-intermittent turbulence should have the values $S_3 = 0, S_4 = 3$, corresponding to a Gaussian probability distribution for $\partial_x v_x$ (section 7.7.3). Experiments and numerical simulation show substantially different values $S_3 \simeq -0.5$, $S_4 \simeq 2.4 R_\lambda^{0.18}$ (Kerr, 1985; Vincent & Meneguzzi, 1991). While for odd-order moments such as S_3 the β-model can only give an upper limit, it predicts a definite scaling of the even-order moments. Since the dissipation scale l_d is the smallest scale excited, we have $\partial_x v_x \simeq \Delta v(l_d)/l_d$. Hence we obtain

$$S_4 \simeq f_4(l_d) \propto (L/l_d)^\delta = R_\lambda^{3\delta/2} , \qquad (7.154)$$

using $L/l_d = R_\lambda^{3/2}$, eq. (7.95), and $l_d = l_K$, ignoring the weak δ-correction to the dissipation scale. With $\delta \simeq 0.2$ we have $S_4 \propto R_\lambda^{0.3}$, which is a stronger R_λ-dependence than observed. The discrepancy is still larger for higher order moments. The β-model outlined above is too simple to describe the intermittency of Navier-Stokes turbulence satisfactorily. Because of the linear p-dependence of ζ_p, eq. (7.151), the quantitative agreement of ζ_p with experiments and simulations in the range $p \lesssim 16$ is less good than for the quadratic law (7.144) of the log-normal theory. Asymptotically for large p, however, the log-normal theory is unacceptable, since ζ_p becomes negative, while the β-model remains qualitatively correct. For instance the

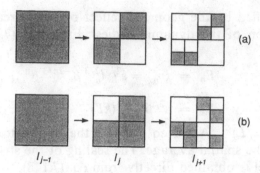

Fig. 7.10. Schematic representation of the increasingly intermittent distribution of smaller scales. (a) β-model with $\beta = \beta_0 = 0.5$; (b) random β-model with $\bar{\beta} = 0.5$.

scaling exponents μ_q of the ε_l-moments obtained from ζ_p, eq. (7.151), by use of the relation (7.141) is linear in q, satisfying Novikov's condition.

An obvious generalization of the β-model is to regard β as a random variable in analogy to ε_l in the log-normal model. This random β-model (Benzi et al., 1984) is an example of a broad class of multifractal models which have been developed for various different physical systems exhibiting a scaling behavior; see for instance Paladin & Vulpiani (1987). Instead of

$$v_n \propto l_n^{1/3} \beta^{-n/3}$$

from eq. (7.148), v_n is now a random quantity,

$$v_n \propto l_n^{1/3} \prod_{i=1}^{n} \beta_i^{-1/3} , \tag{7.155}$$

depending on the random variables β_i. As illustrated in Fig. 7.10b, the fraction of the volume occupied by the eddies v_n, which is determined by the sequence $\{\beta_1, \ldots, \beta_n\}$, is $\prod_{i=1}^{n} \beta_i$ (instead of β^n). The structure function $F_p(l)$ is the β-average

$$\langle \Delta v(l_n)^p \rangle \propto l_n^{p/3} \int \prod_{i=1}^{n} d\beta_i \beta_i^{1-p/3} P(\beta_1, \ldots, \beta_n)$$

$$\simeq l_n^{p/3} \left[\int d\beta \rho(\beta) \beta^{1-p/3} \right]^n , \tag{7.156}$$

assuming the fragmentation processes between subsequent levels to be independent, $P(\beta_1, \ldots, \beta_n) = \prod_{i=1}^{n} \rho(\beta_i)$. With $l_n/L = 2^{-n}$ one obtains

$$\zeta_p = \frac{p}{3} - \log_2 \overline{\beta^{1-p/3}} . \tag{7.157}$$

For $\rho(\beta) = \delta(\beta - \beta_0)$ we regain the β-model result eq. (7.151) with $\delta = -\log_2 \beta_0$. Knowledge of all structure constants ζ_p allows one in principle to determine $\rho(\beta)$. In practice, however, only a few ζ_p, $p \lesssim 20$, are known with sufficient accuracy, which leaves substantial freedom in the choice of $\rho(\beta)$.

As an example assume that there are only two different fragmentation processes characterized by values β_1 and β_2, $1 \geq \beta_1 \geq \beta_2$, occuring with probabilities α and $1 - \alpha$, respectively,

$$\rho(\beta) = \alpha \, \delta(\beta - \beta_1) + (1 - \alpha) \, \delta \, (\beta - \beta_2) \,, \tag{7.158}$$

hence

$$\overline{\beta^{1-p/3}} = \alpha \beta_1^{1-p/3} + (1 - \alpha)\beta_2^{1-p/3} \tag{7.159}$$

$$\simeq (1 - \alpha)\beta_2^{-p/3} \quad \text{for} \quad p \gg 1 \,.$$

The asymptotic behavior of ζ_p is linear as in the (non-random) β-model,

$$\zeta_p \simeq p \left(\tfrac{1}{3} - c \right) \,, \quad c = -\tfrac{1}{3}\log_2 \beta_2 \,. \tag{7.160}$$

Thus for $p \gg 1$ the intermittency effect may be much larger than the intermittency correction δ to the energy spectrum in eq. (7.150). For $p = 2$ eq. (7.157) gives an effective δ in E_k, $\delta = -3\log_2 \overline{\beta^{1/3}}$. The asymptotic, linear behavior of ζ_p extrapolated from numerical simulations (Vincent & Meneguzzi, 1991) suggests a value $\beta_2 \simeq 0.6$. One can easily verify that eq. (7.160) satisfies Novikov's condition only for $c < \tfrac{1}{3}$, i.e. $\beta_2 \geq \tfrac{1}{2}$, $\delta = D - D_F \leq 1$, which would limit the degree of sparseness generated by the fragmentation process, allowing sheet-like but not rope-like structures in 3-D.

7.7.3 *Probability distribution functions*

The basic assumption underlying most closure approximations of the hierarchy of moments eq. (7.117) is that the probability distribution of the set $\{\mathbf{v}(\mathbf{x}_i)\}$ of random variables is close to multivariate Gaussian, eq. (7.13), which implies that quantities such as $v_x(x + l) - v_x(l) = \Delta v(l) \equiv v_l$ are approximately normally distributed. While this is found to be true for sufficiently large separation l, where $\mathbf{v}(\mathbf{x})$ and $\mathbf{v}(\mathbf{x} + \mathbf{l})$ are essentially independent, for small separation experiments (e.g. by Anselmet et al., 1984) as well as simulations (e.g. by Vincent & Meneguzzi, 1991) show that $P(v_l)$ differs significantly from Gaussian, being reduced for small values of v_l and enhanced for large values. Hence the statistical properties are dominated by sparsely distributed large fluctuation amplitudes, a direct evidence of spatial intermittency. In fact for small l the behavior of v_l is determined by the local deterministic dynamics rather than by some random process. For a dynamical equation of the form

$$\partial_t v = v^2 , \quad \partial_t v \simeq \partial_x v , \tag{7.161}$$

a Gaussian distribution for v results in an exponential distribution for $\partial_x v$. Since for any monotonic function $s = s(v)$ one has

$$P(s) = \int dv P_0(v) \, \delta(s - s(v)) = P_0(v(s))/|ds/dv| , \tag{7.162}$$

we immediately find for $s = v^2$

$$P(s) \propto e^{-|s|}/|s|^{1/2} . \tag{7.163}$$

She (1991) and She & Orszag (1991) use the idea, introduced by Kraichnan (1990), of deterministic mapping of the probability distribution function. They compute $P(v_l) = \lim_{t \to \infty} P(v_l, t)$ from an initial Gaussian distribution by solving a simple evolution equation which accounts in a phenomenological way for the important dynamical effects of vortex stretching and dissipation. Surprisingly good global agreement of $P(v_l)$ with simulation results is obtained by adjusting a free parameter h, which characterizes the asymptotic behavior of $P(v_l)$, the wings of the distribution,

$$P(v_l) \propto \exp\left[-c|v_l|^{1+h}\right] , \tag{7.164}$$

where the value of h decreases with increasing R_λ and decreasing l.

The probability distribution function can also be discussed in the random β-model (Benzi et al., 1991). The theory relates the statistics of the small-scale velocity fluctuations v_l to that of the Gaussian statistics of the large-scale velocity difference $v_0 = v(x + l_0) - v(x)$. Using eqs (7.155) and (7.162) the probability distribution $P(v_n)$ can be written in the form

$$P(v_n) = \int dv_0 P_0(v_0) \int \delta \left(v_n - v_0 l_n^{1/3} \prod_{j=1}^n \beta_j^{-1/3} \right) \prod_{j=1}^n \rho(\beta_j) \beta_j d\beta_j / \overline{\beta}^n , \tag{7.165}$$

where

$$P_0(v_0) = \frac{1}{\sqrt{2\pi} V_0} \exp[-v_0^2/2V_0^2] ,$$

and the normalization $l_0 = 1$ is used to simplify the notation. Hence $P(v_n)$ is determined by the probability density $\rho(\beta)$ of the filling factors β_j. Consider the simple model eq. (7.158). The case $\alpha = 1$, $\beta_1 = 1$ corresponds to the K-41 theory, where $P(v_n)$ is Gaussian,

$$P(v_n) = \frac{1}{\sqrt{2\pi} V_0 l_n^{1/3}} \exp[-v_n^2 / 2V_0^2 l_n^{2/3}] . \tag{7.166}$$

The choice $\alpha = 1$, $\beta_1 < 1$ corresponds to the nonrandom β-model, which again gives a Gaussian $P(v_n)$. In the general case $0 < \alpha < 1$, $\beta_1 = 1$,

$\beta_2 = \beta_0 < 1$, eq. (7.165) gives $P(v_n)$ as a sum of Gaussians, since each factor in the product consists of two terms, which leads to the binomial expression

$$P(v_n) = \frac{1}{\sqrt{2\pi} V_0 \overline{\beta}^n} \sum_{j=0}^{n} \binom{n}{j} \alpha^{n-j}(1-\alpha)^j \beta_0^{4j/3} \frac{1}{l_n^{1/3}}$$

$$\times \exp\left[-\left(\beta_0^{2j/3} \big/ 2V_0^2 l_n^{2/3}\right) v_n^2\right], \qquad (7.167)$$

There is no simple asymptotic behavior, because in general no single term dominates in this expression. Numerical evaluation shows that for large n, i.e. small-scale l_n, the distribution becomes curved upward on a logarithmic plot (such as for $h < 0$ in eq. (7.164)). Thus the random β-model describes in a simple, natural way the increase of the deviation of $P(v_n)$ from Gaussian with decreasing scale l_n.

The probability distribution $P(s)$ of the velocity derivative s, for instance $s = \partial_x v_x$ or $\partial_y v_x$, is of particular significance both in turbulence experiments and in numerical simulations. We have $s \simeq v_n/l_n$, where $l_n = l_d$ is the dissipation scale defined by $v_n l_n = v$ (see section 7.4.2), which limits the inertial range. Note, however, that here l_d is not constant but depends on the random variable v_n. For an inertial range scaling law $v_n = v_0 l_n^h$ one finds by eliminating l_n, $l_n^2 = v/|s|$, that $v_0^2 \simeq v^{1-h}|s|^{1+h}$ and hence by use of eq. (7.162)

$$P(s) \propto \left(\frac{v}{|s|}\right)^{(1-h)/2} \exp\left[-cv^{1-h}|s|^{1+h}\right], \qquad (7.168)$$

assuming $P(v_0)$ to be Gaussian. (The singularity at $s = 0$ should not be taken seriously since consideration is restricted to the inertial range $l < l_0$, such that s is finite.) In the K-41 theory we have $h = \frac{1}{3}$ (see eq. (7.76)). Thus even in the absence of intermittency $P(s)$ is not Gaussian. This somewhat surprising result is due to the random character of the dissipation scale which is the larger the smaller the value of v_n. In the nonrandom β-model the scaling exponent h is reduced compared with the K-41 value $\frac{1}{3}$, $h = (1-\delta)/3$ with $2^{-\delta} = \beta_0$, see eq. (7.148). Since Novikov's condition requires $\beta_0 \geq 1/2$, i.e. $\delta < 1$, $\log P(s)$ cannot become curved upward. In the simple random β-model (7.158) $P(s)$ is a sum of distributions of the type (7.168). From the definitions $s = v_n/l_n$, $l_n = v/v_n$, $l_n^2 = 2^{-2n} = v/|s|$ we see that the number of steps n is a function of the value of s, $n = n(s) = \frac{1}{2}\log_2(|s|/v)$. With $\beta_1 = 1$, $\beta_2 = \beta_0 = 2^{-\delta}$, $k_j = j\delta/n$, we easily find

$$P(s) = P(v_n)dv_n/ds = P(v_n)l_n$$

$$\propto \sum_{j=0}^{n} \binom{n}{j} \alpha^{n-j} (1-\alpha)^j \left(\frac{v}{|s|}\right)^{(1+2k_j)/3}$$

$$\times \exp\left[-v^{(2+k_j)/3}|s|^{(4-k_j)/3}/2V_0^2\right] \tag{7.169}$$

Choosing $\alpha = \frac{7}{8}$, $\delta = 1$, i.e. $\beta_0 = \frac{1}{2}$, which gives structures coefficients ζ_p, eq. (7.157), consistent with experimental results, one obtains qualitative agreement of $P(s)$ with the probability distribution of the transverse derivative $s = \partial_x v_y$ found in the simulations of Vincent & Meneguzzi. The agreement should, however, not be overemphasized. $\beta_0 = \frac{1}{2}$ describes sheet-like structures, while simulations show that in 3-D Navier-Stokes turbulence small-scale structures are rope-like. More importantly the intermittency models discussed above assume that turbulent eddies have roughly isotropic shapes, while in real turbulence small eddies tend to be much longer than wide. In 3-D Navier-Stokes turbulence simulations the dissipative ropes are found to have diameters of the order of the mean dissipation scale, but lengths which extend far into the inertial range possibly up to the integral scale. Hence correlation functions, in particular $\overline{\varepsilon(\mathbf{x})\varepsilon(\mathbf{x}+\mathbf{l})} \simeq \varepsilon_l^2$, may be more strongly determined by the length of the dissipative structures than by their spatial distribution.

7.7.4 *Intermittency in MHD turbulence*

Since scaling properties are somewhat different in MHD turbulence because of the Alfvén effect, the intermittency models given above have to be modified accordingly. This can easily be performed for the β-model. While eq. (7.147) remains valid, eq. (7.148) is changed replacing $\tau_n^{-1} \rightarrow \tau_A/\tau_n^2$, $\tau_n = l_n/v_n$, as discussed in section 7.4.1,

$$\varepsilon = E_n \tau_A/\tau_n^2 = \beta^n v_n^4/v_A l_n , \tag{7.170}$$

which gives

$$v_n = (\varepsilon v_A)^{1/4} l_n^{1/4} \beta^{-n/4} , \tag{7.171}$$

$$E_n = (\varepsilon v_A)^{1/2} l_n^{1/2} \beta^{n/2} ,$$

$$E_k = (\varepsilon v_A)^{1/2} k^{-3/2} (kL)^{-\delta/2} . \tag{7.172}$$

Hence the intermittency correction of the energy spectrum is slightly stronger than in the hydrodynamic case, eq. (7.150). From eq. (7.170) we obtain the scaling of the structure function for v_l (the model does not distinguish between v_l and B_l):

$$F_p(l) = \langle v_l^p \rangle = (\varepsilon v_A)^{p/4} l^{p/4} (L/l)^{\delta(p-4)/4} ,$$

i.e.

$$\zeta_p = \frac{p}{4} - \frac{\delta}{4}(p-4) . \tag{7.173}$$

Generalization of these results to the random β-model proceeds as in the hydrodynamic case. Instead of eq. (7.171) we now have

$$v_n = v_0 l_n^{1/4} \prod_{j=1}^{n} \beta_j^{-1/4} , \tag{7.174}$$

hence

$$\zeta_p = \frac{p}{4} - \log_2 \overline{\beta^{1-p/4}} . \tag{7.175}$$

The probability distribution function is

$$P(v_n) = \int dv_0 P_0(v_0) \int \delta\left(v_n - v_0 l_n^{1/4} \prod_{k=1}^{n} \beta_k^{-1/4}\right) \prod_{j=1}^{n} d\beta_j \beta_j \rho(\beta_j)/\overline{\beta}^n ,$$

which in the case of the simple model eq. (7.158) with $\beta_1 = 1$, $\beta_2 = \beta_0 < 1$ becomes

$$P(v_n) \propto \sum_{j=0}^{n} \binom{n}{j} \alpha^{n-j}(1-\alpha)^j \beta_0^{5j/4} \frac{1}{l_n^{1/4}} \exp\left[-\beta_0^{j/2} v_n^2/2V_0^2 l_n^{1/2}\right] . \tag{7.176}$$

To derive the probability distribution $P(s)$ for the derivative $s \simeq v_n/l_n$, where $l_n = l_d$, the dissipative scale, is defined by eq. (7.87),

$$\tau_A/\tau_n^2 = v/l_n^2 ,$$

$$l_n = v v_A/v_n^2$$

instead of $l_n = v/v_n$, we express l_n in terms of s:

$$l_n^3 = v v_A/s^2 . \tag{7.177}$$

From eq. (7.176) one obtains

$$P(s) \propto \frac{1}{\overline{\beta}^n} \sum_{j=0}^{n} \binom{n}{j} \alpha^{n-j}(1-\alpha)^j \left(\frac{v v_A}{s^2}\right)^{(1+5k_j/3)/4}$$

$$\times \exp\left[-(v v_A)^{(1+k_j/3)/2}|s|^{1-k_j/3}\right] , \tag{7.178}$$

where $n = \frac{1}{3}\log_2(s^2/v v_A)$ and, as before, $k_j = j\delta/n$, $\beta_0 = 2^{-\delta}$. It is interesting to note that in the absence of intermittency $\delta = 0$, $k_j = 0$, $P(s)$ is purely exponential, $P(s) \propto e^{-c|s|}$, in contrast to the K-41 case in Navier-Stokes turbulence where $P(s) \propto \exp[-c|s|^{4/3}]$, which is caused by the Alfvén effect in MHD turbulence. Hence finite intermittency always leads to a upwardly curved behavior of $\log P(s)$.

Simulation studies of intermittency in MHD turbulence have been limited to 2-D systems. Figure 7.11 illustrates the behavior of the probability distributions $P(B_l)$, $B_l = B_x(x + l) - B_x(x)$, observed in a high resolution simulation. While for $l \sim L$ the probability distribution is Gaussian (Fig. 7.11.c), it becomes increasingly non-Gaussian with decreasing l, Fig. 7.11a corresponding to $P(s)$. In general $\log P(s)$ seems to be more strongly curved upward than in the 3-D Navier-Stokes case, which may be due to the influence of the Alfvén effect mentioned above, but also to the two-dimensionality of the system.

There is, however, some indication from the simulations that in 2-D MHD turbulence the wings of the probability distribution $P(s)$ reach an asymptotic shape for $R_\lambda \gg 10^2$, instead of increasing with R_λ, as predicted by the random β-model. For large R_λ only the central part changes, becoming more and more peaked. Such behavior is also reflected by the R_λ-dependence of the moments of the probability distribution, notably the flatness factor S_4, which seems to approach some constant value. Let us consider a simple model assuming that dissipation is concentrated in a set of current sheets with mean thickness l_d, the dissipation scale given in eq. (7.88), width l_w, and separation r such that the number density of sheets is r^{-2}. Energy balance requires that

$$\varepsilon \simeq \eta \langle j^2 \rangle , \qquad (7.179)$$

since for $\nu \sim \eta$ resistive dissipation dominates. The current sheets are generated by magnetic fluctuations of scale l_w, $j \simeq \delta B(l_w)/l_d$, hence

$$\langle j^2 \rangle \simeq \left(\frac{\delta B(l_w)}{l_d} \right)^2 \frac{l_d l_w}{r^2} .$$

Insertion into eq. (7.179) and use of the inertial range scaling $\delta B(l)^2 \simeq (\varepsilon v_A)^{1/2} l^{1/2}$, eq. (7.80), gives the relation

$$l_d^{1/2} l_w^{3/2} / r^2 \simeq 1 . \qquad (7.180)$$

The flatness factor can now be estimated in the following crude way. Assuming that $\partial_x B_x \simeq \partial_x B_y \simeq j$ is concentrated in the current sheets, S_4 is essentially determined by the current sheet filling factor $l_d l_w / r^2$, the fraction of the area covered by current sheets,

$$S_4 \propto r^2 / l_d l_w \simeq (l_w/l_d)^{1/2} , \qquad (7.181)$$

using eq. (7.180). If the width l_w reaches up to the integral scale, $l_w \lesssim L$, as simulations indicate for relatively low R_λ, we will have the scaling $S_4 \propto (L/l_d)^{1/2} = R_\lambda^{2/3}$. However, tearing mode stability limits the current sheet aspect ratio, $l_w/l_d < 10^2$, as discussed in section 6.5. Hence this simple model in fact predicts that $S_4 \to$ const. for sufficiently high R_λ. This is also consistent with the numerically observed current sheet distribution

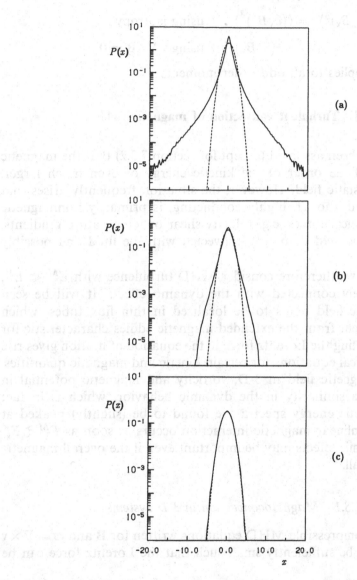

Fig. 7.11. Probability distribution function of the longitudinal difference $\Delta B_x(l) \equiv B_x(x+l) - B_x(x)$ in 2-D MHD turbulence, obtained from a numerical simulation with 2048^2 modes: (a) $l = 0.012$ in the dissipation range; (b) $l = 0.14$ in the inertial range; (c) $l = 1.57$ in the global scale range (the system size is 2π).

such as shown in Fig. 7.8. It should be mentioned that the skewness factor
S_3 vanishes for 2-D isotropic turbulence:

$$\langle (\partial_x B_x)^3 \rangle = \langle (\partial_y B_y)^3 \rangle \qquad \text{using isotropy}$$
$$= -\langle (\partial_y B_y)^3 \rangle \quad \text{using} \quad \nabla \cdot \mathbf{B} = 0 \,.$$

The argument applies to all odd-order moments.

7.8 Turbulent convection of magnetic fields

Up to now it has been assumed (except for section 7.3.2) that the magnetic
field energy is of the order of the kinetic energy or even much larger,
forming a quasi-static field. However, the situation frequently arises that
the turbulent fluid, though highly conducting, is primarily nonmagnetic
stirred by mechanical forces, e.g. velocity shear or temperature gradients.
Here the magnetic field is mainly convected with the fluid and possibly
amplified.

   In this section we therefore consider MHD turbulence with $E^M \ll E^V$,
which is intimately connected with the dynamo effect. It will be seen
that the magnetic field tends to be localized in thin flux tubes, which
is basically different from the extended magnetic eddies characteristic for
$E^M \gtrsim E^V$. Neglecting the Lorentz force in the equation of motion gives rise
to formally identical equations for certain kinetic and magnetic quantities,
vorticity and magnetic field in 3-D, vorticity and magnetic potential in
2-D, suggesting a similarity in the dynamic behavior, which is in fact
observed. Magnetic energy spectra are found to be (slightly) peaked at
large k. Since nonlinear magnetic interaction occurs as soon as $E_k^M \gtrsim E_k^V$
for some k, dynamic effects may be important even if the overall magnetic
energy is still small.

7.8.1 *Magnetoconvection in 3-D systems*

Consider the incompressible MHD equations written for $\mathbf{B}$ and $\omega = \nabla \times \mathbf{v}$
and assume $\mathbf{B}$ to be sufficiently small such that the Lorentz force can be
neglected:

$$\begin{aligned}
\partial_t \mathbf{B} + \mathbf{v} \cdot \nabla \mathbf{B} &= \mathbf{B} \cdot \nabla \mathbf{v} + \eta \nabla^2 \mathbf{B} \,, \\
\partial_t \omega + \mathbf{v} \cdot \nabla \omega &= \omega \cdot \nabla \mathbf{v} + \nu \nabla^2 \omega \,.
\end{aligned} \qquad (7.182)$$

Since $\mathbf{v}$ is now independent of $\mathbf{B}$, the induction equation is linear in $\mathbf{B}$,
describing the passive convection of magnetic field.

   For magnetic Prandtl number $Pr_m = \nu/\eta = 1$, the equations for ω
and $\mathbf{B}$ are identical and the solutions differ only because of different

initial conditions, where ω and $\mathbf{v}$ are strongly correlated, while $\mathbf{B}$ may be independent of $\mathbf{v}$. Since in a turbulent system the memory of the initial state decays rapidly, we can expect ω and $\mathbf{B}$ to be very similar, except for their magnitudes, $\mathbf{B} \ll \mathbf{v}$ being assumed (Batchelor, 1950). For $Pr_m \neq 1$ the strict analogy with the vorticity no longer holds, but $\mathbf{B}$ and ω should still be similar for scales exceeding the dissipation scales. For the inertial range spectrum E_k^M the analogy implies, assuming a Kolmogorov kinetic energy spectrum (Moffat, 1961),

$$E_k^M \propto \Omega_k = k^2 E_k^V \simeq \varepsilon^{2/3} k^{1/3} \, , \tag{7.183}$$

possibly valid up to the resistive dissipation range $k_\eta \simeq l_\eta^{-1} = (\varepsilon/\eta^3)^{1/4}$. This spectrum is rather different from the $k^{-3/2}$ law for the equipartitioned MHD spectrum $E_k^M \simeq E_k^V$, where the magnetic energy is contained in the large eddies. Since vorticity in Navier-Stokes turbulence is found to be concentrated in long, thin tubes (Siggia, 1981), the magnetic field is expected to exhibit the same behavior, which is in fact observed in simulations (Meneguzzi et al., 1981; Yanase et al., 1991; Nordlund et al., 1992). Note that the Alfvén effect is not present in this case.

There is, however, a caveat in the argument leading to the spectrum $E_k^M \propto k^{1/3}$. The enstrophy spectrum $\Omega_k \propto k^{1/3}$ is intimately connected with the cascade of kinetic energy (an ideal invariant in the limit $E_k^M \ll E_k^V$) described by the constant parameter ε. The induction equation is linear in $\mathbf{B}$; hence E_k^M should be proportional to the magnetic energy dissipation rate. However, E^M is not an ideal invariant, i.e. E_k^M is not a cascading quantity, so that the $k^{1/3}$ law for E_k^M cannot be inferred from a cascade argument but is only based on the formal analogy with the vorticity.

In addition, this analogy tells us nothing about the *magnitude* of E^M. Will a weak seed field decay or grow in time? We expect that for a smooth field distribution the magnetic energy will always grow initially owing to field line stretching, twisting and folding by the $\mathbf{B} \cdot \nabla \mathbf{v}$ term much as with vortex line stretching in Navier-Stokes turbulence. During this process the magnetic field is rapidly concentrated in rope-like structures of decreasing diameter d. If d becomes as small as l_η, Batchelor argues that the diffusion term $\eta \nabla^2 \mathbf{B}$ will dominate over the convective stretching term $\mathbf{B} \cdot \nabla \mathbf{v}$, leading to subsequent decay of E^M. Only if the magnetic Prandtl number exceeds some threshold which is of order unity, such that the viscous dissipation scale, the Kolmogorov micro-scale, exceeds the resistive scale, $l_K = (v^3/\varepsilon)^{1/4} > l_\eta$, will there be no small-scale vorticity structures to generate magnetic scales of order l_η. Hence the convective term remains dominant and growth of E^M continues until the Lorentz

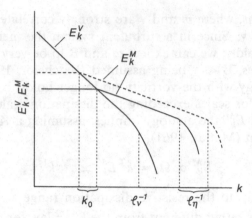

Fig. 7.12. Qualitative behavior of the spectra E_k^M, E_k^V in convectively driven 3-D MHD turbulence with $Pr_m = v/\eta > 1$. Decreasing η leads to a roughly self-similar increase of E_k^M (dashed line).

force becomes important leading to dynamic saturation. For conditions far above threshold, $Pr_m \gg 1$, the growth rate is determined only by the convection and no longer depends on η, which is called "fast dynamo" in the terminology of dynamo theory. There is some uncertainty about the saturated state. Does the field grow until equipartition $E_k^M \simeq E_k^V$ is reached in the largest scales, or is the growth terminated as soon as $E_k^M \simeq E_k^V$ at the smallest scales $kl_K \sim 1$? Three-dimensional simulations for $Pr_m \gg 1$ indicate some intermediate behavior, as illustrated in Fig. 7.12. For small k, where $E_k^M < E_k^V$, the magnetic spectrum is essentially flat, $E_k^M \simeq$ const., up to the point k_0, where the spectra cross. For larger k, E_k^M decays roughly as a shallow power law, $E_k^M \propto k^{-\alpha}$, $\alpha \sim 1$, until resistive dissipation sets in at $kl_\eta \sim 1$, where E_k^M has become much larger than E_k^V. The mean magnetic energy E^M may reach E^V for sufficiently large Pr_m. In this regime the energy dissipation is mainly resistive in spite of the fact that $\eta \ll v$.

The discussion of magnetoconvection given so far was concerned mainly with small-scale field excitation. Dynamo theory on the other hand considers the build-up of large-scale fields. As we have seen in section 7.3.2 these are driven by a *mean* kinetic helicity H^V, while excitation of small-scale fields may be associated with *local* fluctuations of kinetic helicity density (Kraichnan & Nagarajan, 1976). Simulations by Meneguzzi et al. (1981) with finite H^V show a linear growth of the large-scale field, a process related to the inverse cascade of the *magnetic* helicity, see eq. (7.52). If this large-scale field becomes strong enough the Alfvén effect wipes out the disparity of E_k^M and E_k^V at the smaller scales, as discussed in section 7.3.2.

7.8.2 Convection of magnetic flux in 2-D

It is well known (Cowling, 1934) that in two-dimensional systems there is no dynamo effect in the strict sense, meaning unlimited growth of an infinitesimal seed field by finite flow velocities or unlimited sustainment of a field against resistive decay by a weak flow $v = O(\eta)$. (Here two-dimensional means plane or axisymmetric. The antidynamo theorem does not apply to the more general case of helical symmetry, as will be discussed in section 9.3.2). It is in fact easy to see that the magnetic field will eventually decay independently of the type of two-dimensional motion. From eq. (7.29),

$$\frac{dA}{dt} = -\eta\, E^M \, ,$$

it follows that the mean square potential A can only decrease, such that the time integral of E^M is determined by the initial magnetic field,

$$\int_0^\infty E^M\, dt = \frac{1}{\eta}\, A(t = 0) \, . \tag{7.184}$$

In spite of its ultimate fate the magnetic field evolving from an initially smooth distribution due to convection by a two-dimensional turbulent flow exhibits interesting features which resemble the field line stretching in three-dimensional magnetoconvection. For small η or large R_m the magnetic energy E^M and even the average field magnitude $\langle B \rangle$ may be amplified enormously before finally decaying.

Neglecting the Lorentz force in the 2-D MHD equations,

$$\begin{aligned} \partial_t\psi + \mathbf{v}\cdot\nabla\psi &= \eta\nabla^2\psi \, , \\ \partial_t\omega + \mathbf{v}\cdot\nabla\omega &= \nu\nabla^2\omega \, , \end{aligned} \tag{7.185}$$

we expect that ψ and ω behave in a similar way. Hence the magnetic energy spectrum should relax to the form

$$E_k^M = k^2\psi_k^2 \propto k^2\omega_k^2 \simeq \eta_\Omega^{2/3} k \, , \tag{7.186}$$

since for 2-D Navier-Stokes turbulence the energy spectrum is determined by the enstrophy cascade, $E_k^V \simeq \eta_\Omega^{2/3} k^{-3}$, eq. (7.79). The induction equation is linear in ψ, hence $A_k = \psi_k^2$ should be proportional to the dissipation rate of mean square potential η_A, such that we find, using eq. (7.186),

$$A_k \simeq \eta_A \eta_\Omega^{-1/3} k^{-1} \, . \tag{7.187}$$

Since A_k is a cascading quantity in 2-D MHD the spectrum (7.187) is based on more solid ground than the corresponding magnetic energy spectrum in 3-D, eq. (7.183). The characteristic small-scale structures of 2-D Navier-Stokes turbulence are vorticity gradient sheets, hence the analogy between ω and ψ predicts the formation of flux gradient sheets, i.e.

the magnetic field should be concentrated in sheets. Figure 7.13 illustrates the transition from 2-D MHD turbulence with $E^M \sim E^V$, where the magnetic field is organized in eddies of roughly circular shape, to the regime of magnetoconvection $E^M \ll E^V$, where B is contained in thin sheets.

Let us investigate the process of field amplification more in detail. Two phases can be distinguished, a first phase of exponential growth, where B-sheets are formed, and a subsequent phase of linear growth, where these sheets are stretched. The first phase is due to an exponential shrinking of the sheet thickness δ, which is similar to the process of current sheet formation discussed in section 7.6.1 (corresponding to the exponential growth of $\varepsilon(t)$ in Fig. 7.9). The magnetic energy is

$$E^M \simeq N \left(\frac{\widetilde{\psi}}{\delta} \right)^2 L_0 \delta \propto \delta^{-1} , \qquad (7.188)$$

where $\widetilde{\psi} = k_0^{-1} |\nabla \psi|$ is a typical ψ-fluctuation amplitude, $L_0 = k_0^{-1}$ is the wavelength of a velocity eddy which determines the width of the sheet, $N \sim k_0^2$ is the number of sheets (roughly proportional to that of velocity eddies), and δ/k_0 is the area of a sheet. Exponential sheet thinning continues until resistive diffusion prevents further increase of $\nabla \psi$, the thickness δ being determined by the balance across the sheet,

$$[v_n] \frac{\widetilde{\psi}}{\delta} \simeq \eta \frac{\widetilde{\psi}}{\delta^2} . \qquad (7.189)$$

Here $[v_n]$ is the variation of the normal velocity across the sheet,

$$[v_n] \equiv (\partial_n v_n) \, \delta \simeq k_0 V \delta , \qquad (7.190)$$

and V is a typical velocity fluctuation amplitude. Equation (7.189) yields

$$\delta \simeq (\eta/k_0 V)^{1/2} . \qquad (7.191)$$

Note that during this sheet formation phase the average field $\langle B \rangle$ is essentially unchanged but $E^M = \langle B^2 \rangle$ increases due to a reduction of the filling factor. In the subsequent phase of sheet stretching, illustrated in Fig. 7.14, an area $F \simeq k_0^{-2}$ bounded by a sheet is stretched into an elongated shape with $F = L\Delta = $ const. because of incompressibility. Since $L \sim Vt$ and $\delta \simeq$ const., both the magnetic energy and the magnitude B increase linearly in time, now because of an increase of the filling factor.

As mentioned before, E_k^M does not grow indefinitely, but will eventually saturate and decay, which for sufficiently low initial level happens before dynamic interaction with the velocity field can occur. Strong resistive diffusion will certainly set in when the sheet separation Δ in Fig. 7.14 approaches the sheet thickness δ. We can use this to calculate an upper limit of the amplification factor of E^M. From eq. (7.188) with $L_0 \to L \sim$

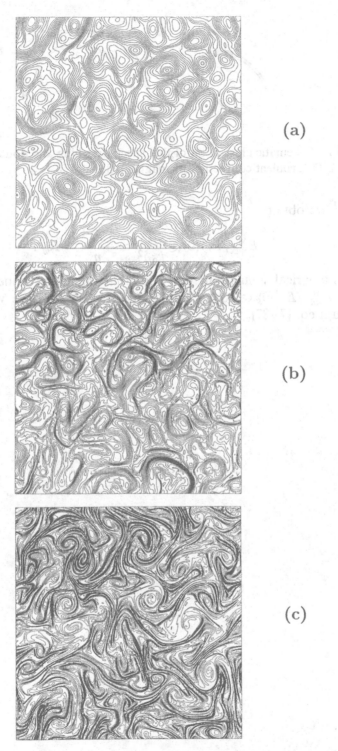

Fig. 7.13. ψ contours, showing the transition from circular flux tubes for $E^M \sim E^V$ to flux sheets for $E^M \ll E^V$. (a) $E^M = E^V$; (b) $E^M = 0.1E^V$; (c) $E^M = 0.01E^V$.

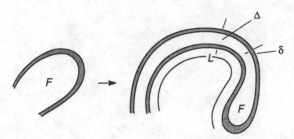

Fig. 7.14. Schematic illustration of flux sheet stretching and double sheet formation in 2-D turbulent convection of magnetic flux.

$(k_0^2 \delta)^{-1}$ we obtain

$$E_{max}^M / E^M(0) \lesssim \frac{1}{k_0^2 \delta^2} \simeq \frac{V k_0^{-1}}{\eta} = R_m \,. \qquad (7.192)$$

From numerical simulations one finds that the actual value is somewhat lower, $E_{max}^M / E^M(0) \propto R_m^{0.8}$. Simulations also confirm the validity of the spectrum eq. (7.187).

8
Disruptive processes in tokamak plasmas

Tokamaks constitute the best plasma physics laboratory available today. The largest devices (e.g. JET and DIII-D) confine plasmas of considerable volume (many m^3), high densities ($n_e \sim 10^{20}\,m^{-3}$) and high temperatures ($T_e \sim 10\,keV$) under quasi-stationary conditions (for an introduction to the general physics of tokamaks see Wesson, 1987). Tokamak plasmas exhibit a rich variety of MHD phenomena, being investigated by numerous diagnostic tools with high spatial and temporal resolution, which make theoretical interpretation a challenging task.

Particularly conspicuous MHD effects are the different kinds of disruptive events which affect global plasma confinement more or less severely. In this chapter we consider the three most important disruptive processes. Section 8.1 deals with the sawtooth oscillation, a quasi-periodic internal relaxation process, which is observed in most tokamak discharges. Their main effect is to limit the central temperature increase, generating a more uniform average temperature distribution. They also have the beneficial effect of preventing the central accumulation of impurity ions.

Section 8.2 considers major disruptions, which constitute the most violent processes in a tokamak plasma. Disruptions occur when certain limits in the plasma parameters are exceeded, causing loss of a large fraction of the plasma energy, which often leads to the termination of the discharge.

Section 8.3 is devoted to a further quasi-periodic relaxation phenomenon, called edge-localized mode (ELM). It appears in tokamak plasmas which are radially limited by a magnetic separatrix instead of a material limiter, allowing high temperatures and densities near the plasma edge. The resulting steep gradients are regulated by the ELM process, which is also found to control the impurity content of the plasma.

Though the basic MHD mechanisms of these disruptive phenomena seem to be understood, many details of the observations are still unexplained, while certain theoretically crucial quantities have not yet been measured accurately. A common feature of these phenomena is that they seem to be two-stage processes consisting of a coherent precursor which can be associated with some MHD instability and a usually faster turbulent phase, during which rapid transport occurs. The characteristic properties are summarized at the end of this chapter in Table 8.1.

A fourth type of disruptive process in tokamak plasmas, the fishbone oscillation, will not be considered, since it is basically a microscopic effect, a $(1, 1)$ oscillation driven by fast trapped particles. The interested reader is refered to the presentation by White (1989).

8.1   Sawtooth oscillations

In the presentation of this subject it appears to be most instructive to follow roughly the historical path from the first experimental observations and their theoretical interpretation to the present state characterized by an embarrassing wealth of observational results and several different theoretical models trying to put the pieces of the empirical puzzle together. (See also a recent review of the topic by Kuvshinov & Savrukhin, 1990.)

8.1.1   Early experimental observations

Sawtooth oscillations were first observed by Goeler et al. (1974) in the Princeton ST-tokamak. Analysis of the soft-X-ray (SX) emission from the hot plasma core showed the signal along a chord across the central region to be modulated by a periodic relaxation oscillation with an amplitude of 5–10 per cent consisting of a slowly rising part (the rise phase), followed by a rapid drop (the collapse or crash phase), which has the shape of a sawtooth, whence the name of the phenomenon. Shifting the observation chord somewhat out of the center showed the signal to be inverted with a sudden rise, coinciding with the drop of the central signal, followed by a more gradual decay. Moving the chord further out the oscillation amplitude becomes invisibly small, hence the phenomenon is localized in the central part of the plasma column. The experimental set-up and the observed signals are given schematically in Fig. 8.1.

Sawtooth oscillations have subsequently been found to occur in practically all tokamaks under various different operational conditions corresponding to a broad range of plasma parameters. They are primarily observed in the SX signals, where one now uses large arrays of diodes, each corresponding to a different observation chord, from which the spa-

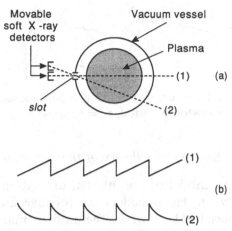

Fig. 8.1. Observation of sawtooth oscillations by soft-X-ray diagnostics. Schematic drawings of (a) the experimental set-up, (b) soft-X-ray signals along chords (1) and (2) .

tial distribution of the radiation from the plasma can be reconstructed. This diagnostic technique allows high time resolution, but it has the disadvantage that the radiation is predominantly due to impurity ions and hence is a complicated function of the electron temperature and density and the impurity ion concentration. Fortunately the analysis of the electron cyclotron emission (ECE), which represents a direct measure of the local electron temperature, gives amplitudes $\delta T_e/T_e$ and even distributions $T_e(r,\theta,t)$ in good agreement with those inferred from SX measurements. In addition, sawtooth oscillations are also observed in the electron density using microwave interferometry, which implies that density transport is also involved, though $\delta n_e/n_e$ is smaller than $\delta T_e/T_e$.

A first indication of the origin of the sawtooth relaxation phenomenon comes from the sinusoidal precursor oscillation, which is sometimes superimposed on the rise phase of the sawtooth as shown in Fig. 8.2. While the main sawtooth signal is symmetric ($m = 0, n = 0$), the precursor oscillation corresponds to a growing helical ($m = 1, n = 1$) distortion of the central plasma region, which gives rise to an oscillating signal owing to diamagnetic effects and plasma rotation. This leads to the interpretation of the sawtooth collapse in terms of the internal kink instability, giving rise to a flattening of the central temperature and density profiles. The collapse is therefore also called internal disruption. Since the internal kink mode is confined to a region inside the $q = 1$ surface, the ejected plasma energy appears just outside this surface, from where it is transported further outward diffusively, which explains the inverted signal along the eccentric chord (2) in Fig. 8.1.

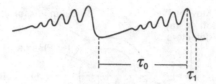

Fig. 8.2. $m = 1$, $n = 1$ precursor oscillation superimposed on the $m = 0$, $n = 0$ sawtooth signal. τ_0 = sawtooth period, τ_1 = collapse time.

8.1.2   *Kadomtsev's theory of the sawtooth collapse*

The first theoretical model of the internal disruption has been proposed by Kadomtsev (1975). He considers the resistive kink instability in the framework of reduced MHD, where the mode is marginally stable in the ideal limit. As discussed in section 4.7.2, this is a rather good approximation for not too small value of η, a slender tokamak column $a/R \ll 1$, and a large value of the safety factor $q_a \gg 1$, such that the resistive growth rate $\gamma = O(\eta^{1/3})$ is much larger than λ_H, the (normalized) ideal growth rate, $\lambda_H^2 \sim W_t = O((r_1/R)^4)$, see eqs (4.117) and (4.79). The nonlinear evolution of the resistive kink mode (section 6.6.2) is characterized by the nonlinear time scale $\tau = O(\eta^{-1/2})$, which is much shorter than the global resistive time. Hence the helical flux function ψ_* is not dissipated but only convected along with the plasma. From the equation

$$\frac{d\psi_*}{dt} \equiv \partial_t \psi_* + \mathbf{v} \cdot \nabla \psi_* = \eta j \qquad (8.1)$$

one obtains that the change of the value of ψ_* of a fluid element during the collapse time τ_1 vanishes in the limit $\eta \to 0$,

$$\delta \psi_* = \int_0^{\tau_1} \eta j \, dt = O\left(\eta^{1/2}\right) \to 0 . \qquad (8.2)$$

It is true that the current density in the sheet is high, $j = O(\eta^{-1/2})$, but the time a fluid element spends in the current sheet region is of the order of τ_A and hence this contribution does not change the order of magnitude of the time integral. Note that conservation of the local value of ψ_* is much more restrictive than conservation of the total magnetic helicity H, the latter giving rise to selective turbulence decay (section 7.3.1), which plays a central role in the dynamics of the reversed-field pinch (chapter 9).

The resistive kink mode evolves through a sequence of helical states characterized by a magnetic island which grows at the expense of the central plasma column until the latter has shrunk to zero and the configuration is again symmetric. This process, called complete reconnection (of the helical flux inside the $q = 1$ surface), is given schematically in

Fig. 8.3, which shows the evolution of the helical field $\mathbf{B}_* = \mathbf{e}_z \times \nabla\psi_* = \mathbf{B}_\perp - (rB_0/R)\mathbf{e}_\theta$. This behavior has been verified by numerical simulation studies (e.g. Sykes & Wesson, 1976; Waddell et al., 1976). The conservation property of ψ_* allows the calculation of the final helical flux distribution $\psi_\infty(r)$ from a given initial one $\psi_0(r)$. As indicated in Fig. 8.4, we use the two relations, area conservation

$$r_i |dr_i| + r_e |dr_e| = r |dr| \tag{8.3}$$

and helical flux conservation $d\psi_* = \text{const.}$,

$$d\psi_* = \left.\frac{d\psi_0}{dr}\right|_i dr_i = \left.\frac{d\psi_0}{dr}\right|_e dr_e = \frac{d\psi_\infty}{dr} dr \ . \tag{8.4}$$

Note that for $dr > 0$ one has $dr_e > 0$, $dr_i < 0$. As an example consider the initial helical flux function

$$\psi_0(r) = \frac{B_0}{R} \left(\frac{1}{q_0(0)} - 1\right) \frac{r^2}{2} \left(1 - \frac{r^2}{2r_1^2}\right) \ , \tag{8.5}$$

corresponding to the parabolic current profile

$$\begin{aligned} j_0(r) &= \nabla^2\psi_0 + 2B_0/R \\ &= \frac{2B_0}{Rq_0(0)} \left(1 - 2\left(1 - q_0(0)\right) \frac{r^2}{r_1^2}\right) \end{aligned} \tag{8.6}$$

and the safety factor profile

$$\begin{aligned} q_0(r) &= \frac{rB_0}{R} \left(\frac{d\psi_0}{dr} + \frac{rB_0}{R}\right)^{-1} \\ &= \left[1 + \left(\frac{1}{q_0(0)} - 1\right)\left(1 - \frac{r^2}{r_1^2}\right)\right]^{-1} \ , \end{aligned} \tag{8.7}$$

with $q_0(r_1) = 1$. Equation (8.3) can be solved immediately to yield

$$r_e^2 - r_i^2 = r^2 \ .$$

The first part of eq. (8.4) gives an equation for dr_i^2/dr

$$\begin{aligned} \left(1 - \frac{r_i^2}{r_1^2}\right) \frac{dr_i^2}{dr} &= \left(1 - \frac{r_e^2}{r_1^2}\right) \frac{dr_e^2}{dr} \\ &= \left(1 - \frac{r^2 + r_i^2}{r_1^2}\right) \left(\frac{dr_i^2}{dr} + 2r\right) \ , \end{aligned}$$

$$r\frac{dr_i^2}{dr} + 2\left(r_i^2 + r^2 - r_1^2\right) = 0 \ ,$$

which has the solution

$$r_i^2 = r_1^2 - r^2/2 \ . \tag{8.8}$$

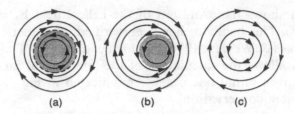

(a)   (b)   (c)

Fig. 8.3.   Schematic drawing of the evolution of the helical magnetic field **B.** in the nonlinear evolution of the resistive kink mode. The dashed line in (a) indicates the resonant surface, the shaded area the central plasma region which is removed by the convection associated with the kink mode.

Using the result in the second part of eq. (8.4) one obtains

$$\psi_\infty(r) = \frac{B_0}{R}\left(\frac{1}{q_0(0)}-1\right)\frac{r_1^2}{4}\left(1-\frac{r^4}{4r_1^4}\right), \tag{8.9}$$

$$j_\infty(r) = \frac{2B_0}{R}\left[1-\frac{1}{2}\left(\frac{1}{q_0(0)}-1\right)\frac{r^2}{r_1^2}\right], \tag{8.10}$$

$$q_\infty(r) = \left[1-\left(\frac{1}{q_0(0)}-1\right)\frac{r^2}{4r_1^2}\right]^{-1} \tag{8.11}$$

for $r < r_2 = \sqrt{2}\,r_1$, while $\psi_\infty(r) = \psi_0(r)$ for $r > r_2$. Note that $q_\infty(r) \geq 1$ with $q_\infty(0) = 1$. Hence in the final state the $q = 1$ surface is removed from the system and the current density is nearly (but not exactly) flat for $r < r_2$. Since the field is not changed for $r > r_2$, the total current is conserved, which gives rise to a negative sheet current at $r = r_2$ corresponding to the jump in $d\psi_\infty/dr$ (Fig. 8.5). But since in the final state the sheet current is no longer sustained by the convective flow associated with the kink mode, it is expected to decay rapidly due to resistive diffusion.[*]

The poloidal field energy $\int_0^{r_2} B_\theta^2 r\,dr = W_\theta$ has decreased. In fact the difference $\Delta W_\theta = W_{\theta 0} - W_{\theta\infty}$ is the free energy released by the instability. ΔW_θ can easily be calculated from the initial and final states, eqs (8.5) and (8.9),

$$\frac{\Delta W_\theta}{W_{\theta 0}} = \frac{5}{8}\left(\frac{1}{q_0(0)}-1\right)^2 \Bigg/ \left(\frac{1}{q_0^2(0)}-\frac{4}{q_0(0)}+6\right). \tag{8.12}$$

---

[*] Kadomtsev's model assumes that $\psi_\infty(r)$ has its maximum in the center $r = 0$. In principle more general final states are possible (Kolesnichenko et al., 1992), where the region of maximum ψ_* is not moved down to the center but inside only somewhat preserving a finite region with $q < 1$, while the adjacent regions of smaller values of ψ_*, those in the center and outside the original $q = 1$ radius, exchange positions along with their plasmas. In contrast to Kadomtsev's model there is, however, no simple dynamical process leading to such states.

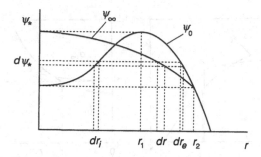

Fig. 8.4. Calculation of the final state $\psi_\infty(r)$ from the initial state $\psi_0(r)$ of the resistive kink mode.

Numerically this fraction is found to be rather small, $\Delta W_\theta / W_{\theta 0} \simeq 0.003$ for $q_0(0) = 0.9$ and $\simeq 0.015$ for $q_0(0) = 0.8$. (If ΔW_θ were to be referred to the total poloidal field energy $\int_0^a B_\theta^2 r dr$, the energy fraction available as free energy would even be smaller.) Usually ΔW_θ is much smaller than the loss of thermal plasma energy caused by the sawtooth collapse.

If prior to the internal disruption the electron temperature and the density are peaked on the axis, the profiles immediately after the disruption are expected to be hollow, since the hot and dense central plasma is now distributed in the region $r_1 \lesssim r < r_2$, while cooler, less dense plasma located in the region $r \sim r_1$ prior to the collapse now occupies the central region. The process is, however, more complicated because of parallel heat conduction, which is fast enough to compete with convective heat transport. Hence the final temperature indentation in the center will be rather shallow. Considering the temperature change $\delta T_e(r)$, one has $\delta T_e < 0$ for $r < r_1$, and $\delta T_e > 0$ for $r > r_1$; hence $r = r_1$, the original position of the $q = 1$ surface, is the experimentally observed inversion radius of the sawtooth signal.

Let us now discuss the time scale τ_1 of the sawtooth collapse predicted by Kadomtsev's model. From eq. (6.75) follows that

$$\tau_1 \simeq (\tau_A \tau_\eta)^{1/2} \propto \eta^{-1/2}, \tag{8.13}$$

where $\tau_A^{-1} = v_{A\theta}/r_1 = |\psi_0''|_{r=r_1} = (B_\theta/r_1) \, d \ln q / d \ln r|_{r=r_1}$ and $\tau_\eta = r_1^2/\eta$. Hence the collapse time is of the order $\tau_1 \sim (r_1/v_{A\theta}) S^{1/2}$, where $S = r_1 v_{A\theta}/\eta$ is the Lundquist number in terms of the poloidal Alfvén velocity. This result is roughly consistent with the sawtooth collapse times observed in smaller tokamaks with $r_1 \lesssim 10$ cm, $T_e \lesssim 1$ keV, $S \lesssim 10^6$, where $\tau_1 \sim 10$–$30\,\mu s$. The dynamical process implied in Kadomtsev's model of the sawtooth collapse has been verified by numerical simulations, notably by Sykes & Wesson (1976) and Waddell et al. (1976).

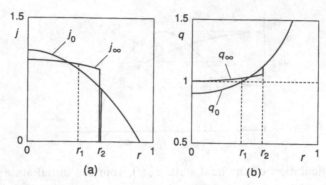

Fig. 8.5. Current profiles $j_0(r)$, eq. (8.6), and $j_\infty(r)$, eq. (8.10), (a); safety factor profiles $q_0(r)$, eq. (8.7), and $q_\infty(r)$, eq. (8.11), (b), for $q(0) = 0.9$, $r_1 = 0.4$.

Kadomtsev's theory leads to the following picture of the sawtooth oscillation. Owing to the good energy confinement in the central plasma region T_e increases, hence $\eta \propto T_e^{-3/2}$ decreases and the ohmic heat deposition E_0^2/η increases, E_0 = applied toroidal electrical field, which leads to further peaking of T_e. The process is called thermal instability, which also occurs for strong additional heating with central energy deposition. Since the current density tends to satisfy the equilibrium relation $\eta(r)j(r) \simeq E_0 \simeq$ const., j becomes more and more peaked, $q(0)$ drops below unity, such that the resistive kink mode starts to grow leading to ejection of the excess thermal energy, flattening the current profile and thus bringing the safety factor back to $q(r) \geq 1$, from where the sawtooth cycle restarts.

Closer inspection of this model, however, reveals a number of open questions and even inconsistencies.

(a) The problem of the time scale of the sawtooth disruption. The predicted η-dependence, eq. (8.13), leads to collapse times exceeding the times observed in large-diameter, high-temperature tokamak plasmas by more than an order of magnitude. Several modifications of Kadomtsev's theory have been proposed which are considered in section 8.1.4.

(b) The problem of fast disruption onset. As already outlined in the introduction of chapter 5 it is generally difficult to explain the sudden onset of a fast process in terms of a linear instability. Since the configuration slowly evolves across the point of marginal stability, the growth rate is small for an extended time, which should give rise to an equilibrium bifurcation rather than an explosive event. The problem is addressed in section 8.1.5.

(c) Direct observations of the poloidal magnetic field distribution indicate that $q(0) < 1$, typically $q(0) \sim 0.7$–0.8, and does not vary substantially during a sawtooth period. In particular $q(0)$ seems to remain below unity during a sawtooth collapse. This obviously contradicts the idea of complete reconnection and poses a serious difficulty in our understanding of the sawtooth phenomenon, as is discussed in section 8.1.7.

8.1.3 Sawtooth behavior in large-diameter, high-temperature tokamak plasmas

During the last two decades tokamak devices have grown considerably in size with linear dimensions increasing by about an order of magnitude from plasma radius $a \sim 0.1$ m in the ST-tokamak to $a \sim 1$ m in the JET tokamak. Plasma temperatures were raised by a similar factor from about 1 keV to 10 keV, while densities and magnetic field intensities remained essentially unchanged. Let us consider the observational results on sawtooth oscillations in large, hot tokamak plasmas in more detail.

Sawtooth periods have increased by about a factor of 10^2 from $\tau_0 \sim$ 1 ms to $\tau_0 \sim 10^2$ ms. This is mainly due to an enormous improvement of the energy confinement. The confinement time τ_{E_1} of the plasma energy W_1 inside the $q = 1$ surface is defined by the relation

$$\frac{W_1}{\tau_{E_1}} \simeq P ,\qquad (8.14)$$

where P is the heating power deposited inside this surface. The sawtooth collapse ejects a certain energy ΔW_1, thus generating an imbalance between heating power and energy transport from which the plasma (partially) recovers during the rise phase

$$\frac{dW_1}{dt} \simeq P - \frac{W_1 - \Delta W}{\tau_{E_1}} \simeq \frac{\Delta W}{\tau_{E_1}} .\qquad (8.15)$$

Since W_1 increases roughly linearly (Fig. 8.6a), one finds that the rise time, the sawtooth period τ_0, is given approximately by

$$\tau_0 \simeq \tau_{E_1} \propto r_1^2 \qquad (8.16)$$

assuming a diffusive heat loss. While this is approximately true for ohmic heating or moderate additional heating, sawtooth periods may become considerably longer than indicated by eq. (8.16) for very intense heating power, where the sawtooth collapse appears to be retarded, such that W_1 reaches considerably higher values ("giant sawteeth", Fig. 8.6b). The collapse mechanism may even be switched off for periods comprising many confinement times τ_{E_1}, such that the system reaches an equilibrium state with a stationary value of W_1 ("monster sawtooth", Fig. 8.6c).

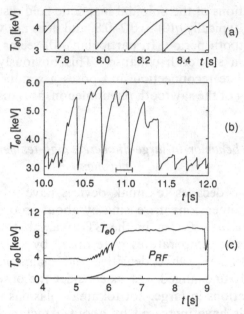

Fig. 8.6.   ECE measurements of different types of sawtooth oscillations observed in the JET-tokamak. (a) normal sawteeth $\tau_0 \sim 200$ ms (Edwards et al., 1986); (b) giant sawteeth $\tau_0 \sim 500$ ms (Campbell et al., 1987); (c) monster sawtooth $\tau_0 \sim 5$ s (Campbell et al., 1989). P_{RF} is the injected radiofrequency heating power.

By contrast the sawtooth collapse time τ_1 is found to be rather independent of the sawtooth period. For the sawteeth shown in Fig. 8.6, which are representative for large tokamak devices, the collapse times are $\tau_1 \sim 100$–$200\,\mu$s. Compared with previous small tokamaks, where $\tau_1 \sim 10$–$30\,\mu$s, τ_1 has increased at most linearly with the radius

$$\tau_1 \propto r_1 , \tag{8.17}$$

which excludes a strong dependence on the resistivity. The dynamics of the collapse process is revealed by tomographic reconstruction of the temperature distribution from SX or ECE signals with good temporal and satisfactory spatial resolution. An example is shown in Fig. 8.7. The collapse can be divided into two phases: (a) The shift of the hot plasma core. Measurements at different toroidal positions indicate that the process has a dominant $m = 1, n = 1$ structure. During this phase the core temperature does not change (frames A–D). (b) The collapse, properly speaking, where the core temperature decays and an (approximately) axisymmetric state is restored with a slightly hollow temperature distribution (frames E–H). Times are given in microseconds. It is seen that each phase takes about $100\,\mu$s.

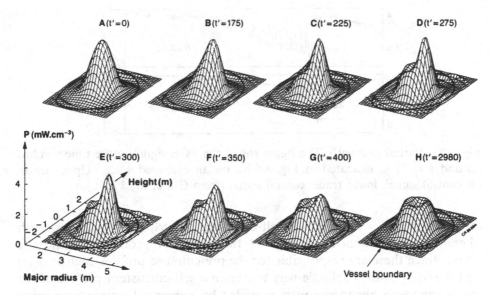

Fig. 8.7. Sawtooth collapse. Tomographically reconstructed 2-D X-ray emission profiles, times t' in μs (from Edwards et al., 1986).

The collapse may be preceded by a distinct precursor of relatively long growth time, ~ 5 ms, but often sets in abruptly without visible precursor. As a rule of thumb, discharges with dominant ion-cyclotron-radio-frequency heating (ICRH) are precursorless, while neutral-beam-heated discharges tend to exhibit a more or less pronounced precursor signal. But even in the presence of an extended precursor the time scales of the final helical shift and the temperature collapse are not significantly longer than in the case of a precursorless collapse.

In addition to the main collapse so-called subordinate or partial collapse events frequently occur in sawteeth of long period and large amplitude. They give rise to a compound sawtooth behavior with the subordinate collapse located midway between two main ones, clearly visible in the giant sawteeth in Fig. 8.6b. Figure 8.8 shows the ECE signals taken along the central chord (lower trace) and an off-central chord (upper trace). While the central temperature is only slightly affected, the off-center one exhibits a temperature drop comparable to that at the main collapse and is usually followed by a large amplitude weakly damped successor oscillation.

8.1.4 Numerical simulations of sawtooth oscillations

Kadomtsev's theory only deals with the collapse process. The initial state is not specified apart from some qualitative features, mainly the

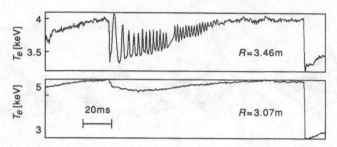

Fig. 8.8.   Partial sawtooth. The figure shows the ECE signal in the time window around $t = 11$ s, indicated in Fig. 8.6(b), on an extended scale. Upper trace: off-central signal, lower trace: central signal (from Gill et al., 1986).

existence of a $q = 1$ surface. A self-consistent description of the sawtooth phenomenon requires knowledge of the transport processes in the rise phase, since these are responsible for the pre-collapse profiles such as $q_0(r)$ and $T_{e0}(r)$. The only reliable way to obtain a self-consistent picture of the entire sawtooth phenomenon is provided by numerical simulations using some MHD model supplemented by a transport equation for the electron temperature,

$$\partial_t T_e + \mathbf{v} \cdot \nabla T_e = \kappa_\parallel \nabla_\parallel^2 T_e + \nabla_\perp \cdot \kappa_\perp \nabla_\perp T_e + P , \qquad (8.18)$$

and the classical relationship $\eta \propto T_e^{-3/2}$. Certain assumptions are made regarding the heat diffusivities $\kappa_\parallel$, $\kappa_\perp$, and the source term P is usually restricted to ohmic heating, $P = \eta j^2$. As expected, the sawtooth behavior is found to depend sensitively on the choice of $\kappa_\parallel$, $\kappa_\perp$.

Systematic simulation studies have been performed in the framework of reduced MHD (Denton et al., 1986, 1987; Vlad & Bondeson, 1989). Since $\kappa_\perp T_e/a^2 \sim P \sim \eta B_\theta^2/a^2$, $\kappa_\perp$ is related to η, $\kappa_\perp \sim \eta$ for $T_e \sim B_\theta^2$, i.e. $\beta_p \sim 1$.

It might appear that $\kappa_\parallel$ is essentially a free parameter. However, for $\kappa_\parallel$ below some threshold value no sawtooth oscillations occur. Instead the system relaxes to a stationary state, called a convection cell, where the temperature assumes a helical distribution sustained by a stationary convection flow, while the magnetic configuration is symmetric with $q \simeq$ const. slightly exceeding unity. Periodic sawtooth oscillations are only generated for sufficiently high values of $\kappa_\parallel$ as are typical for tokamak plasmas, which reduces $\nabla_\parallel T_e$ and thus prevents formation of a convection cell.

In the simulations the physical mechanism driving the sawtooth process is found to agree roughly with the picture envisioned by Kadomtsev. Ohmic heating increases the central temperature and hence reduces η, which makes $q(0) \propto j(0)^{-1}$ fall below unity, until the $m = 1$ mode is suffi-

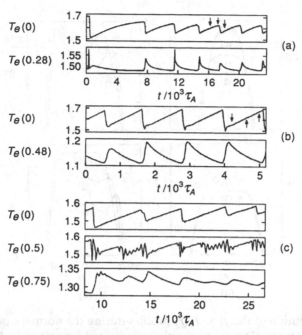

Fig. 8.9. Numerical simulations of electron temperature evolution for normal sawteeth with $\kappa_\perp = $ const. (a), $d\kappa_\perp/dr > 0$ (b), and compound sawteeth (c). Plotted are traces corresponding to observation chords across the center (upper curve) and outside the inversion radius (lower curve). The middle curve added in case (c) shows the behavior at the inversion radius. (From Denton et al., 1986.)

ciently unstable. Within this general picture, however, different sawtooth shapes, including compound sawteeth, may be obtained by varying the radial dependence of the perpendicular heat diffusivity, $\kappa_\perp(r)$ increasing from the center toward the edge. Denton et al. (1986) choose an ad hoc model $\kappa_\perp = \kappa_{\perp 0}(1 + \alpha|\nabla T|^2)$. Results are illustrated in Figs 8.9 and 8.10. For small α corresponding to nearly constant $\kappa_\perp$, a rounded temperature profile $T_e(r)$ is generated (Fig. 8.10a), which leads to saturation of the central temperature before q has fallen sufficiently to initiate the collapse. As a result sawteeth are rounded and the sawtooth amplitude δT_e is relatively small (Fig. 8.9a). For larger α the central heat loss is comparatively weak such that $T_e(0)$ grows linearly up to the collapse, which gives rise to a larger amplitude δT_e (Fig. 8.9b). The central T_e-profile remains flat (Fig. 8.10b). The sharp corners in this profile generate a skin current distribution instead of a peaked one, as is seen from the equation for the mean current distribution $j(r,t)$

$$\partial_t j = \nabla^2 \eta j \simeq \eta'' j , \tag{8.19}$$

since $\eta'' j > \eta j''$, $\eta''(r) \propto T_e''(r)$, the prime denoting the radial derivative.

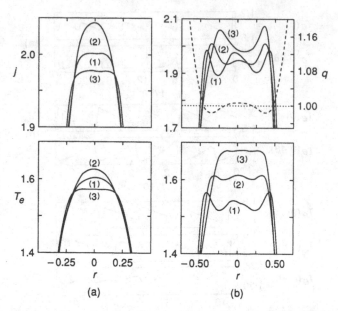

Fig. 8.10. Evolution of the $j(r)$, $T_e(r)$ profiles during the normal sawteeth plotted in Fig. 8.9a,b. (1), (2), (3) refer to the times indicated by arrows in Fig. 8.9.

For small values of the resistivity this leads to the development of a q-profile (inserted in the j-profile chart in Fig. 8.10b), which exhibits two $q = 1$ surfaces. Such a configuration gives rise to a double-tearing mode (section 8.2.4) localized between these surfaces, which flattens the q- and T_e-profiles locally but does not change the plasma core. This process can be associated with the subordinate collapse. Subsequently the central q-value falls below unity leading to the full sawtooth collapse. The $m = 1$ temperature oscillation at the intermediate radius $r = 0.5$ in Fig. 8.9c sustained during the period between partial and full collapse is consistent with experimental observations (Fig. 8.8).

These results do not depend sensitively on the choice of the reduced MHD model. Simulations using the full MHD equations in toroidal geometry (Aydemir et al., 1989) reveal essentially the same behavior. Here particular attention was given to the collapse time τ_1. Aydemir finds the approximate scaling $\tau_1 \propto S^{1/2}$, valid in the range $10^4 \lesssim S \lesssim 10^6$, consistent with the nonlinear time scale of the resistive kink mode, eq. (8.13). Extrapolating to S-values $\sim 10^8$, typical for large hot tokamak plasmas such as in the JET device, one obtains $\tau_1 \sim 5$ ms, which is much longer than the observed collapse times of $\sim 100\,\mu$s. Hence it appears to be necessary to consider other mechanisms to explain the fast collapse time scale.

8.1.5 *Alternative concepts of fast sawtooth reconnection*

Calculating the change $\delta q(0)$ during one sawtooth period due to resistive diffusion for the experimentally observed sawtooth period of a typical ohmic discharge in the JET tokamak, $\delta q(0)$ is found to be quite small, $\delta q(0) \sim 0.01$ instead of ~ 0.1 observed in the simulations, which is due to the much larger S-values, $S \sim 10^8$ in JET compared with $S \lesssim 10^6$ feasible in present-day simulations. Since the sawtooth collapse flattens the q-profile inside the inversion radius, at least according to Kadomtsev's complete reconnection model, $q(r)$ would remain flat and close to unity all the time. Wesson (1986) therefore suggested that the $m = 1$ displacement observed during the collapse (Fig. 8.7) should be associated with an ideal interchange mode driven by the pressure gradient rather than a resistive kink mode driven by the current density gradient. He shows that the stabilizing term in the mode energy W_t, eq. (4.79), derived for finite shear $s \propto 1 - q(0)$, vanishes for $q(0) \to 1$, i.e. $s \to 0$ and hence the mode can be unstable at arbitrarily low β_p. The instability is triggered by a slight change in the q-profile and evolves nonlinearly on a fast time scale, since practically no reconnection is involved. The process leads to the formation of a cool plasma bubble, as in the case of vacuum bubbles discussed in section 5.2.2. This behavior gives a seemingly natural explanation of the results of SX tomography in Fig. 8.7.

Unfortunately there is now strong evidence that q is not close to but distinctly below unity, $q(0) \lesssim 0.8$ (for more details see section 8.1.7), which essentially eliminates the interchange model. Hence one has to return to the kink mode model, which seems to require substantial reconnection. To avoid the relatively long time scales associated with resistive reconnection for small values of η, two different mechanisms have been proposed, which may allow shorter reconnection times. These are electron viscosity μ_e and electron inertia m_e, which enter the generalized Ohm's law in the following form,

$$\mathbf{E} + \mathbf{v} \times \mathbf{B} = \eta \, \mathbf{j} - \eta_2 \nabla^2 \mathbf{j} + \alpha \frac{d\mathbf{j}}{dt} \,, \tag{8.20}$$

$$\eta_2 = (c/\omega_{pe})^2 \, \mu_e \,, \quad \alpha = (c/\omega_{pe})^2 \propto m_e \,,$$

c/ω_{pe} is the electron inertial skin depth. Assuming electron viscosity to be sufficiently anomalous for the η_2-term to become the dominant dissipative effect in Ohm's law, the modified Sweet-Parker theory (section 6.1.2) yields a reconnection rate $M \sim \eta_2^{1/4}$, eq. (6.9), which is rather insensitive to the actual value of η_2. Modifying the analysis in section 6.6.2, we replace eq. (6.72) by

$$u_0 B_* \simeq \eta_2 B_* / \delta^3 \,, \tag{8.21}$$

neglecting the θ-dependence for simplicity. This leads to an island evolution equation, replacing eq. (6.74),

$$\dot{w} \simeq (\eta_2 |\psi_0''| w^3 / r_1^3)^{1/4} , \qquad (8.22)$$

with the similarity solution $w \propto t^4$. An estimate of the collapse time is obtained by setting $w \simeq r_1$, which yields

$$\tau_1' \simeq r_1 (\eta_2 |\psi_0''|^3)^{-1/4} . \qquad (8.23)$$

The anomalous electron viscosity μ_e has been associated with the radial momentum transport caused by small-scale field line stochasticity (Kaw et al., 1979; Aydemir, 1990). Experimental observation of this quantity is very difficult and hence neither the magnitude of μ_e nor its dependence on temperature and density are known directly. If we make the assumptions that (a) the observed anomalous electron heat transport is (at least partly) due to field line stochasticity and (b) the electron viscosity, i.e. the electron momentum diffusivity, is of the same order as the average heat diffusivity, $\mu_e \sim \kappa_e \sim 1\,\mathrm{m^2/s}$ typically, we find that the reconnection time due to the μ_e-effect eq. (8.23) becomes shorter than the resistive Kadomtsev time, eq. (8.13), $\tau_1' < \tau_1$, for $S \gtrsim 10^8$. However, in order to explain the observed collapse time $\tau_1 \sim 100\,\mu s$, μ_e has to be further enhanced locally by at least a factor of 10^2. Since reconnection takes place in the current sheet, the original X-point, where field line stochastization is expected to be particularly pronounced, such enhancement over the average value is not impossible and hence electron viscosity could lead to sufficiently fast reconnection rates.

Electron inertia represented by the last term on the r.h.s. of eq. (8.20) becomes important when the current density, which is mainly carried by parallel electron flow, rises abruptly (in time or space) to large values. For quasi-stationary conditions one has $dj/dt = \mathbf{v} \cdot \nabla \mathbf{j}$. From eq. (8.20) we obtain the relation, modifying eq. (6.2),

$$u B_* \simeq \alpha u B_* / \delta^2 ,$$

hence δ is the electron skin depth,

$$\delta = \alpha^{1/2} = c/\omega_{pe} . \qquad (8.24)$$

Insertion into the continuity equation $u r_1 = v_A \delta$ gives directly $u \simeq v_A c / \omega_{pe} r_1$, and the collapse time τ_1'' due to electron inertia becomes

$$\tau_1'' \simeq \tau_A \frac{r_1}{c/\omega_{pe}} . \qquad (8.25)$$

For typical plasma conditions in the JET tokamak this quantity is significantly shorter than the Kadomtsev time $\tau_1'' \sim 300\,\mu s \sim 0.1\tau_1$, which is not too far from the observed collapse time of $\sim 100\,\mu s$ (Wesson, 1990). An even more efficient reconnection process is due to the electron pressure term in the generalized Ohm's law not included in eq. (8.20) (Aydemir, 1992; Kleva et al., 1995).

It therefore appears that the fast collapse time is not incompatible with complete reconnection as involved in Kadomtsev's model, several mechanisms being available in principle. However, the experimental observation of $q(0) < 1$ during the entire sawtooth cycle indicates that complete reconnection does in general *not* occur, such that the problem of the reconnection time scales is actually of minor importance. Unfortunately the observation of $q(0) < 1$ raises severe questions of a different kind. Before dealing with these in more detail, we consider the problem of sudden collapse.

8.1.6 *Kink-mode stabilization and the problem of sudden collapse onset*

It is generally accepted that the sawtooth collapse is caused by an $m = 1$ instability. Natural as this might appear in view of the tomographic results of the temperature evolution during the collapse, the concept leads to the problem of explaining the abrupt onset of the collapse of the typical precursorless sawteeth. Since the equilibrium evolves very slowly on the time scale of the sawtooth period τ_0, the growth rate of the $m = 1$ mode, vanishing at the threshold, is small for a considerable time until the configuration has developed deeper into the unstable regime. As a result a soft process is expected, in the sense of an equilibrium bifurcation to a slowly evolving helical state.

Two basically different approaches to the onset problem are conceivable. The first explores the fact that the onset of instability is not identical with the appearance of a macroscopic dynamic process. Starting at very low amplitude, the unstable perturbation may grow for many exponential growth times until it reaches an observable level. Such behavior has in fact been found in numerical sawtooth simulations (Aydemir et al., 1989). As can be seen in Fig. 8.11, where the time evolution of the growth rate of the $m = 1$ mode and the central plasma temperature are plotted, the kink mode is unstable long before the collapse occurs. Since the $m = 1$ mode is damped for a certain period of time after the collapse, the mode amplitude decays to a very low level, from where it restarts to grow when the configuration becomes unstable. During the growth period comprising many growth times, $\int^t \gamma dt' \gg 1$, $q(0)$ decreases from unity to about 0.9. When the mode amplitude finally reaches the nonlinear regime, the growth rate has increased to a substantial value, leading to rapid reconnection and temperature drop in roughly one growth time.

However, such an explanation seems somewhat improbable for sawtooth oscillations in plasmas like in JET, where the ratio of sawtooth period to collapse time is large, $\tau_0/\tau_1 \sim 10^3$, since it would imply 10^3 growth times.

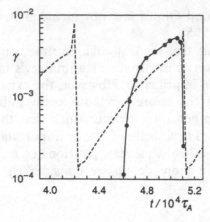

Fig. 8.11. Time evolution of the kink mode growth rate and the central temperature (dashed line, plotted on a linear scale) in a sawtooth simulation (from Aydemir et al., 1989).

Real systems have a finite "thermal" fluctuation level, below which mode amplitudes cannot decay. If a mode becomes unstable, it will hence reach a nonlinear level within a limited number of growth times. In addition, an exponentially growing mode amplitude should give rise to a precursor oscillation, which is often not observed, and even if a precursor exists, its amplitude is still small at the time of the collapse and the precursor growth is much slower than the collapse process, such that the latter cannot be interpreted as the result of the precursor oscillation reaching nonlinear amplitudes (Dubois et al., 1983). Thus one has to conclude that the $m = 1$ mode remains stable within a certain range of $\delta q = 1 - q(0) > 0$.

It has in fact been calculated that the resistive kink mode may be stable in a toroidal configuration with $q(0) < 1$, if the q-profile has a plateau of a certain minimum size with $q'(r_s) \simeq 0$ at the $q = 1$ surface (Soltwisch et al., 1987; Holmes et al., 1989). The presence of such plateaus has in fact been observed experimentally, as is described in section 8.1.7. Tight aspect ratio $A < 3$ (Hastie et al., 1987) and triangularity of the plasma cross-section (Strauss et al., 1989) yield additional stabilizing effects. Also a local flattening of the pressure profile enhances stability, which is due to a reduction of the free energy $-\delta W_t$ of the ideal kink mode, eq. (4.79), since $\beta_{p\,eff}$ depends sensitively on the pressure gradient in the vicinity of the $q = 1$ surface. The stabilizing effect of a local pressure profile modification has clearly been observed in tokamak experiments with electron cyclotron resonance heating (ECRH) where the radial heat deposition can be localized rather accurately. The importance of the normalized temperature gradient $d \ln T_e /\, d \ln r|_{r=r_1}$ ($\simeq d \ln p_e / d \ln r$ since dn_e/dr is small) as an apparent sawtooth trigger mechanism has been stressed by Savrukhin et al. (1991) by comparing various tokamak experiments.

For a monster sawtooth (Fig. 8.6c), however, which corresponds to a collapse-free period of several seconds exceeding the global resistive diffusion time, and which does not exhibit any MHD activity in the vicinity of the $q = 1$ surface, a plateau in the q-profile could not be sustained. Hence stabilization must be caused by a different effect. Since monster sawteeth are observed only during high-power neutral beam injection or ICRH, where a considerable population of high-energy ions is generated, it is natural to associate sawtooth stabilization with the presence of energetic ions. Detailed calculations (White et al., 1988; Pegoraro et al., 1989; White et al., 1989) show that energetic trapped ions have a strong stabilizing effect on the resistive kink mode, which leads to plasma parameter regimes where the resistive kink mode is completely stable for $q(0) < 1$ and a smooth monotonically decreasing current profile.

Since in the stability analysis the energetic particle contribution is important only in a narrow layer around the resonant surface, it will probably be switched off at small amplitude as soon as the island size exceeds this layer width. Hence the growth rate should suddenly increase during the nonlinear evolution, rapidly reaching that of the (resistive) MHD kink mode, which would lead to an abrupt onset of the macroscopic dynamics and explain the sudden appearance of the collapse in a monster sawtooth after the termination of high-power heating. In normal sawteeth, however, such as in ohmic or ECRH plasmas, energetic ion effects are negligible. In this case thermal processes, e.g. diamagnetic flows or ion Larmor radius effects, must be invoked to lead to temporary stabilization of the kink mode, though the actual mechanism is not yet well understood.

8.1.7 *Experimental observation of partial reconnection in the sawtooth collapse*

As has already been mentioned in the preceding sections, experimental observations indicate that contrary to Kadomtsev's model, the sawtooth collapse does not seem to flatten the central q-profile, $q(0)$ remaining distinctly below unity. Because of its fundamental significance let us consider the experimental results in somewhat more detail. The radial distribution of the poloidal magnetic field B_p has first been determined by Soltwisch (1986) in the TEXTOR-tokamak by measuring the Faraday rotation angle θ of the polarisation plane of a far-infrared laser beam after passage through the plasma column, $\theta \propto \int n_e B_{p\parallel} ds$, where $B_{p\parallel}$ is the component of the poloidal field parallel to the beam. These experiments, which have since been further refined, constitute the most reliable and convincing current profile measurements available to date. A typical measured current profile is shown in Fig. 8.12 for a low-q discharge, $q_a = 2.1$, exhibiting a broad shoulder around the $q = 1$ surface. (Note

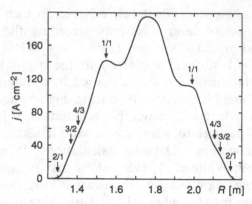

Fig. 8.12. Current density profile measured in TEXTOR (from Soltwisch et al., 1987).

that in a toroidal configuration j is not a flux function, whence the inside outside asymmetry of the shoulder.) Stability analysis of this experimental profile reveals that the resistive kink mode is (weakly) stable (Soltwisch et al., 1987). Increasing q_a, $q(0)$ is found to increase, too, where the experimental values follow approximately the law

$$q(0) \simeq \frac{q_a}{1 + q_a} \, . \tag{8.26}$$

The change of $q(0)$ in a sawtooth collapse, $\delta q(0)$, can be determined quite accurately by making use of the very regular sawtooth oscillations in ohmic discharges in TEXTOR. Superposition of many equivalent saw-teeth yields a substantial increase of the experimental signal-to-noise ratio (Soltwisch, 1988). For a discharge with $q_a = 3.7$ the average value central q-value is $q(0) = 0.77 \pm 0.1$. At the sawtooth collapse $q(0)$ increases abruptly and subsequently slowly decays until the next collapse occurs, but the jump $\delta q(0)$ is small, $\delta q(0)/q(0) \simeq 0.08$, so that $q(0)$ remains clearly below unity.

Since TEXTOR is a typical tokamak device of medium size and circular cross-section, the results should be considered representative of tokamaks rather than exceptional. In fact similar, though less reliable, results concerning the current distribution have been obtained in several other tokamak devices including JET (O'Rourke, 1991).

In the latter device an interesting phenomenon has been observed, which corroborates the persistence of a peaked current profile with shoulders, as shown in Fig. 8.12. After injection of a pellet of frozen deuterium into an ohmic plasma, but also spontaneously after the onset of the sawtooth oscillation, a large pressure perturbation localized both radially and azimuthally is sometimes generated (Gill et al., 1992). Since it exhibits

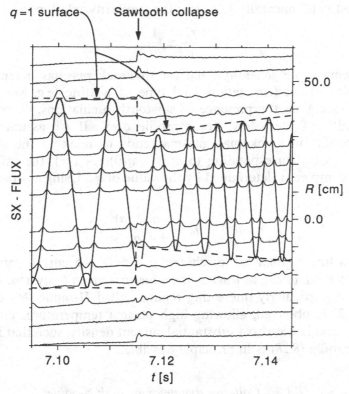

Fig. 8.13. X-ray flux (vertical-camera signals) showing the inward shift of the snake during a sawtooth collapse. The solid line follows the point of maximum emission and the dashed line shows the inferred radius of the $q = 1$ surface (from Weller et al., 1987).

an $m = 1, n = 1$ helical structure, its location can be identified with the $q = 1$ surface. Because of its appearance as a rotating region of enhanced SX emission the phenomenon has been termed "snake" (Fig. 8.13). The snake can persist for times which are long compared with the local density diffusion time and hence constitutes a helical equilibrium state on the transport time scale. The most interesting feature in the present context is that the snake survives several sawtooth collapses and thus monitors the time behavior of the $q = 1$ surface during sawtooth activity. Figure 8.13 shows the effect of a sawtooth collapse on the snake radius, which shrinks by about 30 per cent but does not vanish. Hence the $q = 1$ surface is merely displaced inward by a certain distance Δr, which can be associated with a flat region in the safety factor profile, the shoulder in the current distribution.

The position of the $q = 1$ surface given by the sawtooth inversion radius

r_1 is found experimentally to follow the similarity relation

$$\frac{r_1}{a} \simeq \frac{1}{q_a} \tag{8.27}$$

which seems to be generally valid for tokamak plasmas independent of plasma size, shape (elongation) and type of heating (e.g. Snider, 1990). Relation (8.27), i.e. the presence of sawtooth oscillations, determines the radial width of the safety factor profile $q(r)$. If we assume that the current profile in a tokamak plasma tends to relax to the self-similar shape eq. (3.55) corresponding to $q(r) = q(0)(1 + cr^2)$, the safety factor profile is completely determined by the value of q_a. Using eqs (8.26), (8.27) one obtains

$$q(r) = q_a \frac{1 + q_a(r/a)^2}{1 + q_a} . \tag{8.28}$$

Since in a limiter tokamak discharge $j(a) \simeq 0$, the current profile must deviate from eq. (8.28), at least in the edge region. In a divertor discharge, however, a particularly interesting regime called H-mode (for details see section 8.3) is observed, allowing high plasma temperatures close to the separatrix and hence also a substantial current density, such that relaxation to the q-profile (8.28) is in principle possible.

8.1.8   Collapse dynamics at high S-values

The observation of $q(0) < 1$ has severe consequences for our understanding of the sawtooth collapse. Since Kadomtsev's model of complete reconnection no longer applies, two new problems arise: What mechanism prevents complete reconnection? And, if only partial reconnection occurs, how can the energy escape so rapidly? Numerical simulations have failed so far to reproduce partial reconnection of the resistive kink mode. (Partial reconnection may, however, occur due to diamagnetic effects, see Biskamp, 1981; Rogers & Zakharov, 1995.) Combining the experimental results and our general knowledge of linear and nonlinear MHD processes suggests the following picture of typical high S-number sawtooth oscillations. During the sawtooth rise phase the kink mode is stabilized by a plateau in the q-profile, by thermal ion Larmor radius effects or by energetic ions. When dp/dr becomes large enough or the energetic ion source is switched off, the kink mode is abruptly destabilized; the details of this process still need more elucidation. The kink mode now grows at a fast rate gaining considerable momentum, which can lead to a deformation of the hot plasma core such that the temperature distribution looks crescent-like, as observed in JET. There are, however, also observations of the sawtooth collapse, e.g. in the TFTR tokamak

(Nagayama et al., 1991), showing a slower initial growth with an extended $m = 1$ precursor oscillation. In this case the hot plasma core remains circular.

The subsequent phase, the sawtooth collapse, properly speaking, is characterized by rapid heat transport, which seems to be caused by a destruction of magnetic surfaces. Field line stochastization can be generated by toroidally as well as nonlinearly induced coupling of the $(m, n) = (1, 1)$ mode to other resonances such as (3,4), (4,5), (5,6) as observed in numerical simulations (Aydemir et al., 1989; Holmes, 1991; Baty et al., 1992). The coupling is the more efficient the smaller is the mode number of these resonances, i.e. the lower is $q(0)$. For sufficiently low values, $q(0) \lesssim 0.8$, Baty et al. find a sudden onset of large-scale stochasticity at some helical amplitude, while for $q(0) \gtrsim 0.9$ stochasticity remains weak, such that full reconnection appears to be necessary to account for the heat loss. This suggests that there are two different types of sawtooth oscillations, those with $q(0)$ close to unity, showing complete reconnection, and those with $q(0)$ significantly below unity. The latter type seems to be the more typical one. After flattening of the temperature profile the kink mode is no longer driven, such that the collapse leaves the system in a helical state with mainly unchanged magnetic topology, since only partial reconnection has occurred. The remaining weak helical variations of the temperature and impurity density, which slowly decay as the system returns to axisymmetry, give rise to the successor oscillation often observed experimentally.

Numerical simulations have not yet been able to show such behavior, which may be due to the relatively low S-values, typically $S \sim 10^5$. Such behavior can also be expected from linear theory, section 4.7.2. Ideal MHD effects such as pressure and toroidicity only exert an appreciable stabilizing influence on the resistive kink mode if $\lambda_H < 0$, $|\lambda_H| > \hat{\eta}^{1/3}$, i.e. if the ideal damping rate $|\lambda_H| \sim (r_1/R)^2$ exceeds the growth rate of the resistive mode, which implies $S \gg (R/r_1)^6 \sim 10^6$. Unfortunately, simulations for such high S-values, say $S \gtrsim 10^7$, are at present not feasible. It also appears that for plasma conditions typical for present-day large tokamak devices collisionality is low enough that collisionless reconnection processes such as electron inertia or electron pressure in Ohm's law are more important than resistivity.

A different concept has recently been proposed, based on experimental studies of the sawtooth collapse in TFTR (Nagayama et al., 1996), which emphasizes the deviations from helical symmetry. Reconnection is claimed to occur only in a small toroidal section, where the X-point of the primarily helical $m = 1$ perturbation is located on the outer torus side, since here reconnection will be favored due to the compression of the flux surfaces. Such local reconnection can be faster than in the strictly helical case resulting in efficient destruction of magnetic surfaces and hence heat transport.

8.2   Major disruptions

Since the very first tokamak experiments major disruptions, also called external disruptions, current disruptions or just disruptions, have limited stable tokamak operation. An example is given in Fig. 8.14. Such events not only terminate tokamak discharges abruptly, but also spoil wall surface conditions, which leads to contamination of subsequent discharges by high contents of impurity ions. During two decades of tokamak research much experience has been gained, mostly empirically, to avoid disruptions or at least reduce their frequency, which is a main prerequisite for the useful operation of the present large devices such as the JET tokamak, since disruptions may cause, and actually have caused, major mechanical and electrical damage.

Conditions leading to disruptions, as well as processes occuring during them, involve complicated plasma transport and atomic processes, the detailed description of which is beyond the scope of this book. Instead consideration will be concentrated on the main MHD aspects, discussing transport effects only briefly to give a self-contained picture of the disruption phenomenon.

8.2.1   Disruption-imposed operational limits

Though disruptions may occur in tokamak plasmas under almost any conditions, the probability is rather low, if plasma parameters are maintained within certain limits. Such operational limits are primarily determined by critical values of the electron density n_e, the safety factor q_a, and the plasma pressure β. When these limits are approached the probability of a disruption rises to practically unity. The existence of pressure and current limits is not surprising, since MHD stability theory predicts pressure- and current-driven instabilities for sufficiently large gradients of plasma pressure and toroidal current density. The existence of an independent density limit, however, indicates that in addition to MHD effects transport processes play an important role.

For conditions well below the critical β the operational regime of a tokamak is conveniently described by the Hugill diagram (Fielding et al., 1977) showing the operational limits in the $n_e R/B_t$, q_a^{-1} plane, where $n_e R/B_t$ is called the Murakami parameter (Murakami et al., 1976), B_t being the toroidal field. A typical Hugill diagram, drawn schematically, is shown in Fig. 8.15. The upper boundary of the operational regime,

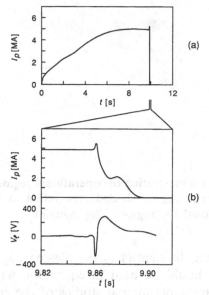

Fig. 8.14. Disruption observed in JET: (a) plasma current I_p; (b) expanded trace of plasma current and loop voltage V_l. (From Ward et al., 1988.)

the current limit, is sharp and rather universally valid for tokamaks (e.g. Wesson et al., 1989):

$$q_a = 2 , \qquad (8.29)$$

where $q_a = q(\psi_a)$ is the true value of the safety factor as defined in eq. (3.26) at the plasma boundary flux surface, and not some technical quantity, e.g. the "cylindrical q" often used in experimental work. The q-limit (8.29) has a simple MHD interpretation in terms of external $m = 2$ kink instability which sets in if q_a falls below 2.

The right-hand boundary, the upper density limit, depends on the experimental or operational conditions, in particular the way of density supply and the type and intensity of heating, and on the impurity ion content. In ohmically heated plasmas one finds a roughly linear relation for the critical mean electron density $\bar{n}_e$,

$$\bar{n}_e = \alpha_n B_t / q_a R , \qquad \alpha_n \simeq 10\text{–}30 , \qquad (8.30)$$

where n_e is measured in $10^{19} \, \text{m}^{-3}$, B_t in T, R in m. The variation of α_n is mainly a density profile effect; for flat profiles (as generated by gas puffing) $\alpha_n \sim 10$, while for peaked profiles (generated by pellet injection) $\alpha_n \sim 20$ (JET team, 1989). For additionally heated plasmas with beryllium limiter, giving rise to a particularly low impurity contamination, even higher values have been reached, $\alpha_n \simeq 33$ (JET team, 1991). This

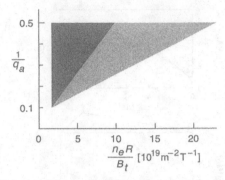

Fig. 8.15. Hugill-diagram illustrating the operational regime in a typical tokamak (schematic drawing). The darker shaded area refers to ohmic discharges, the lighter to strong additional heating or pellet injection.

behavior indicates that the density in the plasma edge region rather than the average density should be used in eq. (8.30), which would make α_n more universal. In fact experimental studies of the energy balance reveal that a disruption becomes very probable, if the radiative heat loss P_{rad} in the edge region becomes the sole loss channel. This allows a simple qualitative interpretation of the experimental result (8.30). Assuming that a disruption sets in, when the ohmic input power is completely balanced by radiation losses,

$$P_{in} = \eta \, \bar{j}^2 = P_{rad} \propto \bar{n}_e^2 \, , \tag{8.31}$$

one obtains the relation $\bar{j} \propto \bar{n}_e$. This is essentially identical with eq. (8.30), since

$$\frac{B_t}{q_a R} = \frac{B_{\theta a}}{a} = \frac{I}{2\pi a^2} = \bar{j} \, . \tag{8.32}$$

Here we have neglected the temperature effect introduced by $\eta \propto T_e^{-3/2}$, conforming with the observation that T_e does not vary strongly in ohmic discharges. Equation (8.31) also explains the general finding that the critical density increases with input power, ohmic or additional, $\bar{n}_e \sim P_{in}^{1/2}$ (though the beneficial effect can be compensated by an increased impurity content).

The observed β-limit is satisfactorily described by the simple relation (Troyon et al., 1984),

$$\beta = \alpha_p \, I / a \, B_t \, , \quad \alpha_p \simeq 3 \, , \tag{8.33}$$

with β in per cent, the plasma current I in MA, B_t in T, and a in m, where a is the plasma radius in the torus midplane. The latter definition is essential, since eq. (8.33) is also valid for elongated plasma shape with vertical radius $b > a$. Relation (8.33) has been obtained from MHD theory as the absolute stability limit of the $n = 1$ free boundary mode by optimizing pressure and current profiles. While the q-limit (8.29) is due to a single robust MHD effect, rather independent of internal field and pressure distributions, experimental observation of the β-limit (8.33) implies that the plasma actually assumes profiles close to the optimum ones, though often at the expense of deteriorated energy confinement. Since the maximum β conditions may be approached by various heating methods with different radial heat deposition distribution, disruptions at the β-limit appear in different forms, in contrast to the rather uniform dynamic behavior observed at the density and q-limits, to which the discussion is therefore mainly restricted.

In addition, disruptions occur when the current rise in the discharge is too rapid, in which case hollow current profiles may be generated. These give rise to double-tearing modes, which will be considered in section 8.2.4.

8.2.2 Disruption dynamics

The dynamics of the disruption at the density limit, the most commonly encountered disruption type, is illustrated schematically in Fig. 8.16. The extended pre-disruption phase τ_0 sets in when the radiated power P_{rad}, rising gradually during the discharge due to increasing density and/or impurity ion content, reaches the input power level. Since at this moment heat conduction to the limiter has ceased, the hot plasma column can detach itself from the limiter by shrinking in radius, surrounding itself by a cold, resistive, radiating plasma mantle. Hence also the current distribution shrinks. Since the total plasma current is maintained constant externally, the current profile becomes steeper, which destabilizes tearing modes, above all the (2,1) mode. Density limit disruptions usually show a slowly growing magnetic precursor oscillation, corresponding to rotating (2,1) magnetic islands. As the magnetic perturbation becomes larger, the interaction with the surrounding vessel slows down the frequency, finally locking the helical perturbation into a fixed position. Though the time-dependent signal $\delta \dot{B}_\theta$ measured by induction (Mirnov) coils has ceased, the mode amplitude δB_r continues to grow until the disruption sets in.

The disruption, properly speaking, is primarily characterized by a rapid redistribution and loss of thermal plasma energy. According to the extent of this loss disruptions are classified as *minor* or soft, or *major* or hard. While a minor disruption consists mainly of a flattening of the (electron) temperature profile around the $q = 2$ radius with only a small central drop in temperature, from which the plasma usually recovers, a major disruption leads to the loss of most of the plasma energy and the subsequent

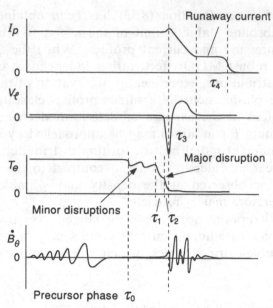

Fig. 8.16.   Different phases preceding and following a major disruption (schematic plot).

termination of the discharge. Often a major disruption is preceded by one or several minor disruptions. The temperature evolution observed in a typical major disruption is shown in Fig. 8.17. The temperature distribution exhibits a pronounced $m = 1$ structure, indicating a strong coupling of the large amplitude $m = 2$ mode to the weakly stable $m = 1$ mode. In spite of the similarity of the evolution of the temperature distribution the major disruption is dynamically rather different from the sawtooth collapse, the latter corresponding to a spontaneous $m = 1$ instability, proceeding on a much faster time scale.

The final drop in the temperature appears to be caused by intense interaction of the expanding hot plasma with the limiter or some other nearby metallic part (Karger et al., 1975; Wesson et al., 1989). This process occurs on a shorter time scale τ_2 than the internal temperature flattening τ_1, and leads to massive contamination of the entire plasma by impurity ions, which results in rapid radiative cooling. The fast radiation burst is clearly seen in Fig. 8.17.

Though the rapid loss of plasma energy is the most conspicuous process during the disruption, it seems to be only a secondary effect caused by large-scale destruction of magnetic surfaces, which is the consequence of an internal MHD process. The basic relaxation process is the rapid broadening of the current profile which has slowly contracted during the

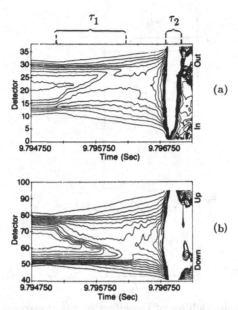

Fig. 8.17. X-ray emission in a major disruption in the JET tokamak. Radial emission profile versus time. (a) Vertical camera; (b) horizontal camera (from Wesson et al., 1989).

predisruption phase. This relaxation corresponds to a decrease of the poloidal field energy, as in Kadomtsev's model of the sawtooth collapse. While, however, in reality the sawtooth collapse is too rapid to allow full reconnection, i.e. full flattening of the central current profile, the major disruption proceeds on a slower time scale such that the current profile does in fact broaden, thus lowering the poloidal field energy. To illustrate this effect let us consider a simple model with constant current profiles as indicated in Fig. 8.18, with initial (just before the disruption) current density j_0 and width r_0, and final values j_1, r_1. The corresponding fields are

$$B_{\theta 0}(r) = \begin{cases} j_0\, r/2 & r < r_0 \\ j_0\, r_0^2/2r & r > r_0 \,, \end{cases} \tag{8.34}$$

$$B_{\theta 1}(r) = \begin{cases} j_1\, r/2 & r < r_1 \\ B_{\theta 0}(r) & r > r_1 \,. \end{cases} \tag{8.35}$$

Since the field is not changed for $r > r_1$, the total current is conserved. However, the current profile cannot simply relax to a pure square shape with $j_0 r_0^2 = j_1 r_1^2$, since in a fast process the dynamics is constrained by the conservation of total magnetic helicity H (section 7.3.1), which in the

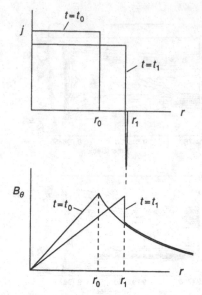

Fig. 8.18. Schematic modelling of the effect of rapid current profile expansion.

limit of large aspect ratio degenerates to the expression (2.55)

$$H = \int_V \psi d^3 x = (2\pi)^2 R \int_0^{r_1} \psi(r) r dr \;, \tag{8.36}$$
$$\psi = \int_{r_1}^r B_\theta(r) dr \;.$$

Note that conservation of H is much less restrictive than the conservation of the local helical flux $\psi_*(r)$, eq. (8.2), the latter arising from helical symmetry of the dynamic process, completely determining the final profiles. The disruption dynamics is not helically symmetric but turbulent with many different helicities excited, such that only the total helicity is conserved. Hence the current profile is not determined and the choice of a flat final profile is only the simplest assumption.

From eqs (8.34), (8.35) one obtains the corresponding flux distributions

$$\psi_0(r) = \begin{cases} \dfrac{j_0}{4}(r^2 - r_0^2) + \dfrac{j_0}{2}r_0^2 \ln \dfrac{r_0}{r_1} & r < r_0 \\[2ex] \dfrac{j_0}{2}r_0^2 \ln \dfrac{r}{r_1} & r > r_0 \;, \end{cases} \tag{8.37}$$

$$\psi_1(r) = \begin{cases} \dfrac{j_1}{4}(r^2 - r_1^2) & r < r_1 \\[1.5ex] \psi_0(r) & r > r_1 \;, \end{cases} \tag{8.38}$$

with the common normalization $\psi_0(r_1) = \psi_1(r_1) = 0$. Using these expressions one easily computes

$$H_0 = \int_0^a \psi_0(r)r\,dr = -\frac{j_0}{16}\,r_0^2(2r_1^2 - r_0^2) \tag{8.39}$$

$$H_1 = \int_0^a \psi_1(r)r\,dr = -\frac{j_1}{16}\,r_1^4 . \tag{8.40}$$

Conservation of magnetic helicity $H_0 = H_1$ yields the current density j_1 in the relaxed state,

$$j_1\,r_1^2 = j_0 r_0^2\left[1 + \left(1 - \frac{r_0^2}{r_1^2}\right)\right] > j_0\,r_0^2 . \tag{8.41}$$

Since the total current remains constant, $I = j_0\,r_0^2$, there is a negative skin current I_s at $r = r_1$ as in Kadomtsev's model,

$$I_s = -\left(1 - \frac{r_0^2}{r_1^2}\right)I . \tag{8.42}$$

The poloidal field energy decreases. Using eqs (8.34), (8.35) and (8.41) one obtains

$$\frac{\Delta W_\theta}{W_{\theta 0}} = 1 - \frac{(2 - (r_0/r_1)^2)^2}{1 + 4\ln(r_1/r_0)} > 0 \tag{8.43}$$

for $r_1 > r_0$. (The inequality can be checked by showing that it is true for small $1 - r_0^2/r_1^2$ and that the derivative of ΔW_θ with respect to r_0/r_1 does not vanish for $r_1 > r_0$.) As in the sawtooth case, eq. (8.12), the decrease of the poloidal field energy is relatively small. Even for rather extreme conditions, assuming $r_1/r_0 = 2$, one obtains only $\Delta W_\theta \simeq 0.2W_{\theta 0}$. ΔW_θ is usually small compared with the change of the plasma energy, the latter being almost entirely lost during the fast disruption phase.

While the change of the current distribution caused by the internal relaxation process is initially shielded by the highly conducting plasma region outside the $q = 2$ surface, the shielding becomes ineffective owing to the strong resistivity increase after contact with the limiter, such that the field outside the plasma starts to change. In the first rapid phase this change corresponds to the resistive decay of the negative skin current, eq. (8.42), which leads to an increase of the total current. The corresponding increase of the poloidal field induces a negative toroidal electric field, which is measured as a *negative voltage spike* in a toroidal loop placed somewhere outside the plasma (Wesson et al., 1990). This negative voltage spike was originally considered as *the* experimental signature of a major disruption. Subsequently the plasma current decays mainly resistively,

$$L \frac{dI}{dt} = -RI \,,$$

$$I(t) = I(0) \exp \left\{ - \int_0^t (R/L) dt' \right\} \,, \tag{8.44}$$

where L and R are the inductance and resistance, respectively, of the plasma column. In this phase the toroidal voltage has a large positive value, which may give rise to efficient runaway electron acceleration to high energies (Dreicer, 1959), which finally carry the entire remaining plasma current. Since high-energy electrons are practically collisionless, the corresponding resistivity is much lower than that of the ambient low-temperature plasma, such that the runaway current can be sustained for a substantial time τ_4 until it, too, disrupts, thus completely terminating the discharge.

The different phases during a disruptive event, as illustrated in Fig. 8.16, are qualitatively similar in most tokamaks, but in large devices such as JET these phases are particularly clearly separated, allowing good diagnostics. Let us discuss the relevant time scales $\tau_0, \ldots, \tau_4$. The pre-disruption phase τ_0 characterized by the growth of the $m = 2$ precursor oscillation is dominated by transport processes ($\tau_0 \sim 10^2$ ms in JET). The time scale τ_1 of the internal temperature relaxation appears to be associated with a fast resistive MHD process ($\tau_1 \sim 1$ ms in JET). The MHD approach to the disruption phenomenon deals mainly with these two phases, to which the subsequent sections of this chapter are hence restricted.

The fast phase of energy loss τ_2 is dominated by plasma–wall interaction effects and the time scale is expected to depend on limiter shape and material ($\tau_2 \sim 0.1$ ms in JET, $\tau_2 \ll \tau_1$). The mechanism by which impurity ions can penetrate into the central plasma region in such a short time is not yet well understood.

The current decay time $\tau_3 = I/(dI/dt)$ depends not only on the plasma resistance R, which may vary significantly, but also on the possibility of controlling the radial position of the plasma column by adjusting the external vertical field to the decreasing value of the plasma current. If the current decrease is too fast, the vertical field cannot be reduced sufficiently fast such that the plasma column is pushed against the inner wall, which further accelerates the current decay. As a result τ_3 tends to become shorter for higher plasma currents (Wesson et al., 1989). The magnitude and sustainment time τ_4 of the runaway current vary considerably and the cause of its abrupt termination is still unclear.

At the q-limit, $q_a \simeq 2$, the dynamics is more violent. No extended magnetic precursor phase is observed, but there is experimental evidence for a locked (2,1) mode growing rapidly, leading to disruption dynamics similar to a major disruption at the density limit.

8.2.3 Single-helicity models

The growth of a helical magnetic perturbation, usually of (2,1) signature, seems to be the most important effect leading to the disruption. Hence the theory starts with a single-helicity model of the $(m, n) = (2, 1)$ tearing mode. Because the growth of the magnetic precursor is in general slower than the saturation time of the tearing mode, one essentially observes the saturated island states growing in size owing to a change of the background profiles.

What are the conditions on current and resistivity profiles giving rise to large island size? As discussed in section 5.5.2 large islands appear if the current density gradient is large just inside the resonant surface. This can directly be seen from the destabilizing contribution in the tearing mode energy integral (4.140) for the (2,1) mode,

$$\delta W \propto \int_0^a \frac{dj/dr}{B_\theta(2-q)} |\psi_1|^2 r\, dr \; . \tag{8.45}$$

When the current profile steepens owing to the contraction of the current channel or when the $q = 2$ surface moves further out by an increase of the current, δW in eq. (8.45) becomes larger. This property has been used by Turner & Wesson (1982) to model the occurrence of disruptions, though the computations assumed that initially $q(0) \simeq 1$. For smaller values observed experimentally, $q(0) \lesssim 0.8$ typically, generation of large $m = 2$ islands by local steepening of the current density gradient close to the $q = 2$ surface becomes more difficult. Here an additional effect may be important, that of the resistivity profile inside the islands, which has a strong influence on the tearing mode island size (section 5.5.3). As illustrated in Figs 5.16, 5.17, an enhancement of η in the island center, $\eta_O > \eta_X$, leads to a significant increase of the island size. Using eq. (5.108), we can estimate the effect of the resisitivity profile on the saturated island size w,

$$\Delta'(w) + \alpha_\eta w = 0 \; , \tag{8.46}$$

$$\alpha_\eta \equiv j d \ln \eta / d\psi \simeq \frac{\eta_O - \eta_X}{\eta_X} \cdot \frac{1}{w^2} \; .$$

For $\alpha_\eta > 0$, corresponding to $\eta_O > \eta_X$, $\Delta'(w)$ is negative, leading to larger islands $w > w_0$, where w_0 is the island size for a flat η-profile in the islands, $\Delta'(w_0) = 0$. Expanding $\Delta'(w)$ about $w = w_0$ gives

$$w = w_0 \left(1 - \frac{\alpha_\eta}{\Delta''}\right)^{-1} \; , \tag{8.47}$$

where

$$\Delta'' = -\left.\frac{d}{dw}\Delta'(w)\right|_{w_0} > 0 .$$

The resistivity distribution is determined by the electron temperature and the impurity density

$$\eta \propto Z_{eff}^2 \, T_e^{-3/2} , \tag{8.48}$$
$$Z_{eff}^2 \equiv \sum_j Z_j^2 \, n_j/n_e \propto n_I ,$$

where n_j is the number density of ions of charge number Z_j and n_I is an average impurity ion density. The temperature distribution in the islands depends on the balance between thermal conduction from the separatrix into the islands and energy loss by impurity radiation,

$$\kappa_\perp \nabla_\perp^2 \, T_e = P_{rad} , \tag{8.49}$$

where

$$\nabla_\perp^2 \, T_e \simeq (T_{eX} - T_{eO})\,/w^2 ,$$
$$P_{rad} \propto n_I . \tag{8.50}$$

In contrast to the electron temperature the impurity density is not a flux function, since parallel diffusion is much slower for ions than for electrons and cross-field diffusion much faster. Hence we can assume $n_I \simeq n_I(r)$, such that Z_{eff}^2 is about equal at the X- and the O-points and

$$\frac{\eta_O - \eta_X}{\eta_X} \simeq \frac{3}{2}\frac{T_{eX} - T_{eO}}{T_{eX}} . \tag{8.51}$$

Combining eqs (8.49)–(8.51), equation (8.47) becomes

$$w = w_0 \left(1 - c\,n_I/\Delta''\right)^{-1} , \tag{8.52}$$

where the factor c depends only on T_e. Hence the growth of $m = 2$ islands in the predisruption phase can be attributed (at least partly) to an increase of impurity ion density. Note that no relative impurity enhancement in the island centers is required.

When the $m = 2$ island size reaches a threshold value, the disruption sets in, which is characterized by a fast change of the temperature. Attempts have also been made to describe this process in the single-helicity framework (Sykes & Wesson, 1980; Drake & Kleva, 1984). In these numerical studies the $m = 2$ islands are found to expand inward down to the center, which leads to sidewise ejection of the central plasma. However, the process requires $q(0)$ to become rather high, $q(0) > 1.5$, to reduce the repelling effect of the magnetic shear, a process for which there is no experimental evidence. Indeed both observations and theoretical considerations suggest that the disruption is caused by a three-dimensional dynamic process, which is most conveniently described in terms of a destabilization of tearing modes of different helicities by the large-amplitude (2,1) mode, notably the (3,2) and (1,1) modes. Before we turn to these multi-helicity models, we first consider the particular case of the double tearing mode, where two island chains of the same helicity interact strongly, which may be a model of disruptions occurring for a nonmonotonic current profile.

8.2.4 *Hollow current profile disruptions*

Hollow current profiles may arise either in the initial phase of the discharge, if the current rise is faster than the resistive time scale of the plasma, such that a skin current distribution is temporarily generated, or in quasi-stationary discharges, if accumulation of impurity ions and the resulting radiative cooling of the plasma center lead to an enhanced resistivity in this region. Associated with a hollow j-profile is a nonmonotonic q-profile with a local maximum in the center. As can easily be seen the radius r_j, where j reaches its maximum, is smaller than r_q, where q has its minimum. Typical q and j profiles are plotted in Fig. 8.19. In such a configuration two resonant surfaces r_1, r_2 may exist for a mode (m, n), $q(r_1) = q(r_2) = m/n$. If this double tearing mode is unstable, islands grow at both surfaces with a poloidal phase shift π between the two island chains. In the nonlinear evolution two different types of behavior are possible (Carreras et al., 1979b):

(a) Island saturation. For weak instability and sufficiently wide separation of the two resonant surfaces, the perturbations on each surface effectively decouple and islands saturate at some finite size, as in the case of a single resonant surface, which leaves the system in a helical state (Fig. 8.20a).

(b) Complete reconnection. For a smaller distance between the resonant surfaces the nonlinear perturbations on each surface interact strongly. The bulging of the islands of one chain drives helical flux toward the X-points of the other island chain, which gives rise to current sheet formation and fast reconnection (Fig. 8.20b). The process continues until the entire helical flux between the resonant surfaces is reconnected and the system has returned to a symmetric state. In this case the double tearing mode evolves in a way similar to the $m = 1$ resistive kink mode, flattening the current and safety factor profiles.

As in Kadomtsev's model of the $m = 1$ mode (section 8.1.2), the final state in the case of complete reconnection can be calculated from the initial

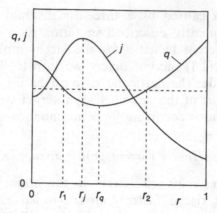

Fig. 8.19. Hollow current profile and corresponding q-profile.

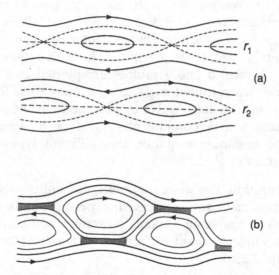

Fig. 8.20. Nonlinear evolution of the double tearing mode. (a) Saturation for weak island interaction; (b) fast reconnection for strong island interaction. (Schematic drawings.)

helical flux distribution, as is illustrated in Fig. 8.21. The radial extent of the region affected by the reconnection is either localized in a radial belt if $\psi_*(0) > \psi_*(r_2)$ or includes the center if $\psi_*(0) < \psi_*(r_2)$. However, complete reconnection is found to occur only if $\psi_*(0)$ is sufficiently close to $\psi_*(r_2)$.

Practically speaking, however, the double tearing mode is of minor importance, since three-dimensional multi-helical dynamics seems to become important before conditions for double tearing instability are reached.

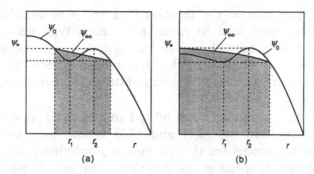

Fig. 8.21. Helical flux $\psi_*(r)$ of a configuration with two resonant surfaces r_1, r_2. The shaded regions give the extent of the reconnection, which is either radially localized (a) or reaches down to the center (b) (schematic drawing) (from Carreras et al., 1979b).

8.2.5   *Multi-helicity dynamics*

For large amplitudes of the (2,1) mode the plasma column tends to become unstable with respect to further modes. To investigate this question either one may adopt the quasi-linear point of view, where the flattening of the average current density in the island region gives rise to a steepening of the current profile in the adjacent region destabilizing tearing modes with $q < 2$, notably the (3,2) mode, or one may consider explicitly the properties of the helical configuration produced by the large-amplitude (2,1) mode which may become unstable, even ideally. In either case the secondary instability is likely to introduce a fast dynamic time scale which is independent of the slow growth of the (2,1) mode, the precursor oscillation. More importantly, the interaction of resonant modes of different helicity leads to destruction of flux surfaces and stochastization of magnetic field lines, which ergodically fill a region of large radial extent (Finn, 1975). Because of the high parallel electron mobility the electron temperature distribution will rapidly be flattened over this region, which provides a natural mechanism for the fast temperature relaxation in the disruption.

Most 3-D simulation studies of the disruption dynamics are restricted to the framework of the reduced MHD equations, written here once more for convenience:

$$\partial_t \psi - \mathbf{B} \cdot \nabla \phi = \eta\, j - E \,, \tag{8.53}$$

$$\partial_t \omega + \mathbf{v} \cdot \nabla \omega - \mathbf{B} \cdot \nabla j = \nu \nabla^2 \omega \,, \tag{8.54}$$

$$j = \nabla_\perp^2 \psi \,, \quad \omega = \nabla_\perp^2 \phi \,,$$

$$\mathbf{B} = \mathbf{e}_z \times \nabla \psi + \mathbf{e}_z B_z \,, \quad \mathbf{v} = \mathbf{e}_z \times \nabla \phi \,.$$

Since in this approximation the resistive kink mode is unstable for any $q(0) < 1$, which would lead to rapid Kadomtsev-type relaxation, the initial state has usually been chosen to have $q(0) > 1$. This eliminates the possibility of efficient coupling to the (1,1) mode. Hence the dominant secondary mode stimulated by a large-amplitude (2,1) perturbation is the (3,2) tearing mode.

Assume that the (3,2) mode is destabilized and grows to finite amplitude. The interaction between the (2,1) and (3,2) modes then leads to the excitation of further modes (m, n). Applying the ordering $\psi_{00} \sim 1$, $\psi_{21} \sim \varepsilon$, $\psi_{32} \sim \varepsilon^2$, the nonlinearities in the dynamic equations, in particular the Lorentz force $\mathbf{B} \cdot \nabla j$, excite modes by the coupling process

$$\partial_t (m_1 \pm m_2, n_1 \pm n_2) = (m_1, n_1) * (m_2, n_2) ,$$

with the following ordering hierarchy (Hicks et al., 1981):

$$
\begin{array}{ll}
\varepsilon & (2,1) \\
\varepsilon^2 & (3,2) , (4,2) \\
\varepsilon^3 & (1,1) , (5,3) , (6,3) \\
\varepsilon^4 & (1,0) , (6,4) , (7,4) , (8,4) \\
\varepsilon^5 & (3,1) , (4,3) , (8,5) , (9,5) , (10,5)
\end{array}
\tag{8.55}
$$

Figure 8.22 indicates all modes up to order ε^{30}, the radius of a disc being proportional to the ε-order of the mode. We see that the modes fall essentially into the cone $3/2 \lesssim m/n \lesssim 2$, spreading slowly beyond these boundaries with increasing order. This ordering scheme serves as a guideline for the selection of modes to be included in pertaining numerical simulations.

The main dynamical effect caused by the interaction of tearing modes of different helicity appears to be an acceleration of mode growth (Waddell et al., 1979; Carreras et al., 1981), which sets in as soon as the islands of the individual modes, computed separately with their individual helical flux functions, start to overlap, which implies the generation of regions of stochastic field line behavior. These first appear at the separatrices of the individual islands, in particular around the X-points, broadening as the mode amplitudes continue to grow. The size of the stochastic regions is determined by the amplitude of the main perturbing helicity, ψ_{21} at the $q = 3/2$ radius r_{32}, and ψ_{32} at the $q = 2$ radius r_{21}. Since the tearing mode eigenfunction ψ_{mn} is located mainly inside its resonant radius r_{mn} as indicated in Fig. 4.5b, the effect of the (2,1) mode on the (3,2) mode is significantly stronger than the inverse effect. Hence the (3,2) islands are destroyed first, as seen in Fig. 8.23, which shows the global field line stochastization obtained from a typical simulation run.

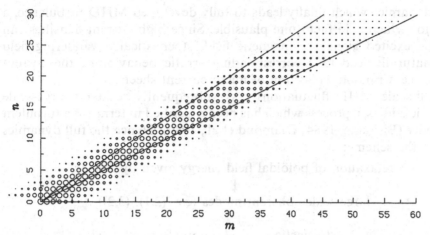

Fig. 8.22. Distribution of interacting modes (m, n) up to order ε^{30}. The two lines indicate the helicities $m/n = 3/2$ and $m/n = 2$.

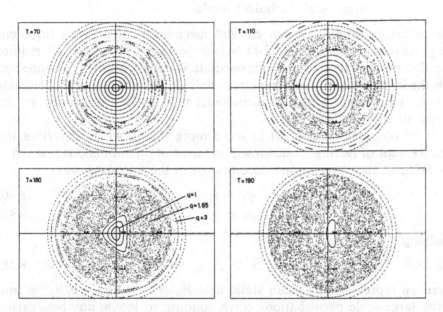

Fig. 8.23. Rapid field line stochastization obtained from a disruption simulation. Times are in Alfvén times.

Contrary to the visual impression stochastic field lines are generated primarily by smooth large-scale modes and do not require small-scale magnetic perturbations. Keeping only the (2,1) and (3,2) modes in the field line computations in Fig. 8.23 results in stochastic regions of similar extent. Stochastic field lines do, however, stimulate the *generation* of small-scale dynamics which finally leads to fully developed MHD turbulence, a behavior which is in fact quite plausible. Since high current densities can only be excited along the magnetic field, stochastically wandering field lines naturally lead to a corresponding erratic behavior of the current density, i.e a random collection of micro-current sheets.

Small-scale MHD fluctuations may subsequently react on large-scale magnetic eddies, a process which has been described in terms of a turbulent resistivity (Biskamp, 1984; Diamond et al., 1984). Hence the full dynamics follows the scheme:

relaxation of poloidal field energy given by $d\,j_0/dr$

$\downarrow$

large-scale coherent modes (e.g. (2,1), (3,2))

$\Big\downarrow$ stochastic field lines (a) $\Big\uparrow$ anomalous resistivity (b)

small-scale turbulent modes.

Process (a) corresponds to an efficient mechanism of a direct turbulent energy cascade, which unlike 2-D MHD (section 7.6.1) does not require extended current sheets as an intermediate state. Process (b) is connected with the inverse cascade of magnetic potential. The energy for this circular process is provided by the mean poloidal field, i.e. the relaxation of the current density gradient $d\,j_0/dr$.

Let us consider process (b) in some more detail, which describes the enhancement of tearing mode growth in terms of an anomalous resistivity. The fields ψ and ϕ (or ω) are decomposed in the following way:

$$\psi = \psi_0 + \psi_l + \psi_s , \tag{8.56}$$

$$\phi = \phi_l + \phi_s , \tag{8.57}$$

satisfying

$$\psi_0 \gg \psi_l \gg \psi_s , \quad \phi_l \gg \phi_s , \tag{8.58}$$

where ψ_0 represents the mean static field $\mathbf{B}_0 = \mathbf{e}_z \times \nabla\psi_0 + \mathbf{e}_z B_z$, ψ_l and ϕ_l the large-scale perturbations corresponding to low-m-number tearing modes, ψ_s and ϕ_s the small-scale, high-m-number perturbations. From the reduced equations (8.53), (8.54) one obtains the approximate equations

$$\partial_t \psi_l - \mathbf{B}_0 \cdot \nabla\phi_l = \langle \mathbf{B}_s \cdot \nabla\phi_s \rangle + \eta\,\nabla_\perp^2 \psi_l , \tag{8.59}$$

$$\partial_t \omega_l - \mathbf{B}_0 \cdot \nabla j_l - \mathbf{B}_l \cdot \nabla j_0 = \tag{8.60}$$

$$\qquad - \langle \mathbf{v}_s \cdot \nabla\omega_s \rangle + \langle \mathbf{B}_s \cdot \nabla j_s \rangle + \nu\nabla_\perp^2 \omega ,$$

$$\partial_t \psi_s - \mathbf{B}_0 \cdot \nabla\phi_s = \mathbf{B}_l \cdot \nabla\phi_s + \mathbf{B}_s \cdot \nabla\phi_l + \eta\,\nabla_\perp^2 \psi_s , \tag{8.61}$$

$$\partial_t \omega_s - \mathbf{B}_0 \cdot \nabla j_s = -\mathbf{v}_l \cdot \nabla\omega_s + \mathbf{B}_l \cdot \nabla j_s + \nu\nabla_\perp^2 \omega_s . \tag{8.62}$$

Here self-interaction terms such as $\mathbf{B}_l \cdot \nabla \phi_l$ or $\mathbf{B}_s \cdot \nabla \phi_s$ have been omitted, the former constituting an additive effect in eqs (8.59), (8.60) not considered here, the latter being small according to the ordering (8.58). We have also used relations such as $\mathbf{B}_s \cdot \nabla j_l \ll \mathbf{B}_l \cdot \nabla j_s$. The small-scale averages, $\langle \mathbf{B}_s \cdot \nabla \phi_s \rangle$ in eq. (8.59) and $-\langle \mathbf{v}_s \cdot \nabla \omega_s \rangle + \langle \mathbf{B}_s \cdot \nabla j_s \rangle$ in eq. (8.60), give rise to anomalous resistivity and viscosity contributions, respectively. To evaluate these terms we expand the small-scale quantities ψ_s, ω_s in the amplitude of the large-scale perturbations, $\varepsilon \sim B_l / B_0$,

$$\psi_s = \psi_s^{(0)}(x_s) + \varepsilon \psi_s^{(1)}(x_s, x_l) + \cdots, \qquad (8.63)$$

and similarly for ϕ_s. Here $\psi_s^{(0)}, \psi_s^{(1)}, \ldots$, are periodic functions in the small-scale variable x_s. To lowest order we write

$$\partial_t \psi_s^{(0)} - \mathbf{B}_0 \cdot \nabla \phi_s^{(0)} = f_s, \qquad (8.64)$$

$$\partial_t \omega_s^{(0)} - \mathbf{B}_0 \cdot \nabla j_s^{(0)} = g_s. \qquad (8.65)$$

Here f_s, g_s are introduced as phenomenological random stirring forces representing process (a), the excitation of small-scale modes by large-scale ones. f_s, g_s determine primarily the fluctuation levels $\langle v_s^2 \rangle, \langle B_s^2 \rangle$. The first-order quantities obey the equations (neglecting dissipation terms)

$$\partial_t \psi_s^{(1)} - \mathbf{B}_0 \cdot \nabla \phi_s^{(1)} = \mathbf{B}_l \cdot \nabla \phi_s^{(0)} + \mathbf{B}_s^{(0)} \cdot \nabla \phi_l, \qquad (8.66)$$

$$\partial_t \omega_s^{(1)} - \mathbf{B}_0 \cdot \nabla j_s^{(1)} = -\mathbf{v}_l \cdot \nabla \omega_s^{(0)} + \mathbf{B}_l \cdot \nabla j_s^{(0)}. \qquad (8.67)$$

These equations are formally integrated along the mean field $\mathbf{B}_0$ and the result is substituted into the average terms in eqs. (8.59), (8.60) to lowest order, e.g. $\langle \mathbf{B}_s^{(0)} \cdot \nabla \phi_s^{(1)} \rangle$. Assuming isotropy of the small-scale fluctuations in the poloidal plane one obtains the turbulent resistivity and viscosity terms in eqs (8.59), (8.60) (Biskamp, 1984),

$$\langle \mathbf{B}_s \cdot \nabla \phi_s \rangle = \eta_a \nabla^2 \psi_l,$$

$$\langle \mathbf{B}_s \cdot \nabla j_s \rangle - \langle \mathbf{v}_s \cdot \nabla \omega_s \rangle = v_a \nabla^2 \omega_l,$$

where η_a and v_a are given approximately by

$$\eta_a = \frac{\tau}{2} (\langle v_s^2 \rangle - \langle B_s^2 \rangle), \qquad (8.68)$$

$$v_a = \frac{\tau}{2} \langle B_s^2 \rangle, \qquad (8.69)$$

and τ is a correlation time resulting from expressions such as $\int^t dt' \mathbf{B}_s(t) \cdot \mathbf{B}_s(t') \simeq \tau B_s^2(t)$. These expressions are essentially identical with the results obtained in 2-D MHD turbulence (Pouquet, 1978), the 3-D structure of the turbulence appearing only in the time integration.[†]

[†] A somewhat different result has been obtained by Montgomery and Hatori (1984), who stress the difference between 2-D MHD and 3-D reduced MHD.

The two-scale formalism leading to expressions (8.68), (8.69) is, strictly speaking, based on a k-spectrum characterized by a gap between large-scale and small-scale fluctuations. In real systems fluctuation spectra are in general continuously decreasing. To apply the results we therefore assume that all modes with $k > k_l$ contribute to the turbulent resistivity experienced by the mode k_l. Hence the change of ψ_l due to small-scale fluctuations becomes

$$
\begin{aligned}
\partial_t \, \psi_l|_s &= \langle \mathbf{B}_s \cdot \nabla \phi_s \rangle \\
&\simeq -k_l^2 \psi_l \int_{k_l}^{\infty} dk \frac{\tau_k}{2} (|v_k|^2 - |B_k|^2) \\
&\simeq k_l^2 \psi_l \int_{k_l}^{\infty} dk \frac{1}{2kv_A} \frac{\varepsilon}{v_A k^2} \\
&\simeq \frac{1}{4} \frac{\varepsilon}{v_A^2} \psi_l \, ,
\end{aligned}
\tag{8.70}
$$

using $\tau_k \simeq (kv_A)^{-1}$ and the reduced spectrum $E_k^R \simeq \varepsilon/v_A k^2$, eq. (7.115), together with $E_k^R < 0$. Equation (8.70) gives the amplification of ψ_l by the turbulent eddies of smaller scale, which is consistent with the inverse cascade of the magnetic potential.

The main assumption in the derivation of expressions (8.68), (8.69) is that of isotropy of the small-scale turbulence, which may appear doubtful in the sheared magnetic field $\mathbf{B}_0(r)$, where modes behave rather differently in radial and azimuthal directions. For not too strong magnetic perturbations as implied in the ordering eq. (8.58), current sheets are essentially aligned azimuthally, $\partial_r \gg \partial_\theta$. Hence an alternative approximation is to assume small-scale modes to be (driven) linear eigenmodes, i.e. high-m-number tearing modes ψ_{mn}, ϕ_{mn} (Diamond et al., 1984). These have a definite parity with respect to their resonant surfaces, ψ_{mn} being even and ϕ_{mn} odd. Assuming that the driving terms, the r.h.s. of eqs (8.66), (8.67), are smooth, i.e. have even parity with respect to the resonances of the driven mode, only the r.h.s. of the $\psi_s^{(1)}$ equation (8.66) has the correct parity to contribute to the anomalous transport coefficients. As a result the magnetic contribution in the anomalous resistivity vanishes, hence

$$
\eta_a' = \frac{\tau}{2} \langle v_s^2 \rangle \, ,
\tag{8.71}
$$

which is reminiscent of the β-term, eq. (7.50), in kinematic dynamo theory. The anomalous viscosity v_a' is essentially identical with that in the isotropic case, eq. (8.69).

Expressions η_a, η_a' are *not* at variance with the result of section 7.3.2, eq. (7.56), that for finite $\mathbf{B}_0$ the effect of small-scale turbulence is described by a hyperresistivity instead of a resistivity. In the framework of reduced MHD, appropriate to the case of strongly magnetized systems such as a tokamak plasma (in contrast to an RFP; see chapter 9) the magnetic helicity degenerates to the linear expression $\int \psi d^3 x$, which is in fact conserved in the presence of η_a. It should be noted that expression (8.71) does not account for the Alfvén effect, which makes the contribution from the smallest scales weak, $\langle \mathbf{B}_s \times \mathbf{v}_s \rangle \rightarrow 0$, and hence does not apply to fully developed MHD turbulence, in contrast to expression (8.68). We should also note that in spite of its popularity in nonlinear MHD the 3-D reduced MHD model has to be considered with some caution, in particular in turbulence studies. As discussed in section 2.4, these equations have only two quadratic invariants, in contrast to three for both 2-D and 3-D MHD. Whether there is an inverse cascade in 3-D reduced MHD is still an open question.

Previous high-S-number simulations, $S \sim 10^5$–10^6, such as those described in the work of Diamond et al. do not reproduce the experimentally observed strong coupling to the $m = 1$ mode (see Fig. 8.17), mainly because of the choice of the initial q-profile, $q(0) > 1$. This coupling has been found in numerical simulations of density limit disruptions (Bondeson et al., 1991), which while remaining in the framework of reduced MHD include an electron temperature transport equation (as in the sawtooth simulations by Denton et al., 1986). Recently, 3-D full MHD toroidal simulations of high-β disruptions (Park et al., 1995) have revealed the following sequence of dynamic events, which is in agreement with experimental observations. The first phase consists of a global $m = 1$ mode with a dominant $m = 1$ component. This leads to a pressure profile steepening on the outer torus region, which destabilizes higher-n ballooning modes. The latter grow rapidly without saturation, self-accelerating by further increasing the pressure gradient in a narrow toroidally and poloidally confined region. This dynamics provides a rather convincing model of the thermal quench τ_2 in Fig. 8.16.

8.3 Edge-localized modes

The third type of disruptive MHD process in a tokamak plasma is the edge-localized mode (ELM), a relaxation oscillation which affects primarily the plasma edge region. In contrast to sawteeth and major disruptions, which are caused by low-mode-number, current-driven instabilities, the edge-localized mode is driven by a steep pressure gradient and is characterized by high mode numbers. Since this process occurs only in a particular plasma state, the so-called H-mode, we start this section by a brief introduction to the physics of this plasma state.

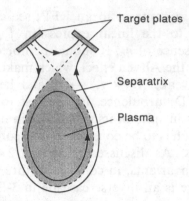

Fig. 8.24.   Schematic drawing of a divertor configuration.

8.3.1   The L → H transition

In a tokamak, plasma and energy transport across the magnetic field are mainly caused by small-scale turbulent convection, i.e. turbulent density and temperature fluctuations with wavenumbers of the order of the ion Larmor radius ρ_i, $k_\perp \rho_i \lesssim 1$, which are observed in practically all tokamak discharges. These fluctuations are usually associated with drift waves (see e.g. Liewer, 1985), and simple estimates of their amplitudes yield transport coefficients of the observed order of magnitude, though a more detailed theory is very difficult and constitutes one of the most actively pursued fields in tokamak research.

Special transport effects arise at the plasma edge. In the usual case, where the plasma is in contact with a material limiter and the plasma temperature is low, the relative density, temperature and electric potential fluctuation amplitudes are particularly large, as are transport coefficients. Such edge behavior is also observed in plasmas limited by a magnetic separatrix at low heating power. Plasma and heat flows across the separatrix are usually diverted onto special target plates where the heat load is deposited and the plasma is neutralized. Since the neutral gas may stream back into the discharge column and become ionized again, the process is called recycling. A divertor configuration is shown schematically in Fig. 8.24. The intensity of the line radiation from the neutralized gas in front of the target plates (mainly Hα or Dα) is a measure of the plasma flow leaving the discharge across the separatrix.

In a separatrix-limited plasma high edge temperatures are in principle allowed. If the input power exceeds a threshold level then the plasma suddenly switches to a completely different transport behavior, which is characterized by an almost perfect transport barrier in a narrow zone just inside the separatrix. As a result the plasma flow across the separatrix is drastically reduced and particles and energy diffusing from the central plasma region toward the edge are piling up in front of the separatrix, giving rise to steep density and temperature gradients. The evolution of the density profile is illustrated in Fig. 8.25. Since the plasma confinement is improved, the new plasma state is called the *H*-mode ("*H*" for "high confinement"), as compared with the previous state of poorer confinement, the *L*-mode ("*L*" for "low confinement"). Here the term "mode" means "regime" or "phase" and has no relation to a dynamic (eigen-)mode such as the tearing mode. Simultaneously with the sudden decrease of the plasma flow across the separatrix at the $L \rightarrow H$ transition a switch-off of the density fluctuations in the edge zone is observed. The radial extent of this zone coincides with the narrow region of the transport barrier. This provides strong evidence that the reduction of the transport coefficients is *caused* by the reduction of the turbulence level.

A quantitative theory of the $L \rightarrow H$ transition has not yet been developed. Here we briefly outline a possible mechanism, which has recently attracted considerable attention (see e.g. Shaing et al., 1989; Doyle et al., 1991). In turbulent convection both electrons and ions are moved around essentially in the same way. Hence their cross-field transport rates are equal, and the transport is said to be ambipolar. In the absence of fluctuations transport is reduced to the much lower collisional level, such that particles move essentially on free orbits determined by the magnetic geometry. Since cross-field excursions from the average or drift orbits along field lines are very different for electrons and ions because of the largely different gyro-radii, cross-field transport is no longer intrinsically ambipolar in the strongly inhomogeneous edge zone with gradient scale length of the order of typical ion excursions. Hence a negative radial electric field is built up, which slows down the ion loss and speeds up the electron cross-field motion until both fluxes are equal. Experimentally, an increase of the electric field intensity during the $L \rightarrow H$ transition is in fact observed, restricted to the narrow transport barrier region. Hence $E_r(r)$ is strongly inhomogeneous, which corresponds to a sheared poloidal plasma rotation $v_\theta = E_r(r)/B_t$. A poloidal shear flow tends to suppress the convective turbulence by distorting and tearing apart turbulent eddies, pictorially speaking. Hence one could have the following self-enforcing process:

$$\begin{array}{c} \longrightarrow \text{reduction of convective turbulence} \\ \downarrow \\ \text{reduction of transport} \\ \downarrow \\ \text{increase of electric field in the edge zone} \\ \downarrow \\ \longleftarrow \text{increase of shear of the poloidal plasma flow.} \end{array}$$

This scheme appears to be a promising starting point for a self-consistent theoretical model.

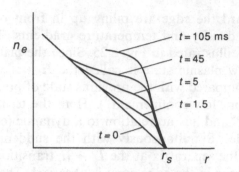

Fig. 8.25. Evolution of the density profile following the $L \to H$ transition (at $t = 0$) (after Wagner et al., 1991).

It is interesting to note from Fig. 8.25 that the transport barrier zone, given by the region of strong density gradients, expands in such a way that the gradient remains approximately constant. Since the plasma flux Γ from the interior does not vary significantly, this implies that the transport coefficient $D \simeq \Gamma/|\nabla n_e|$ is constant across the barrier zone. Hence one would also expect the electric field to be roughly constant. Consequently the shear of the poloidal velocity dv_θ/dr will decrease as the barrier region expands, which would lead finally to the resurrection of the convective turbulence, causing the plasma to fall back to the L-mode.

8.3.2   The ELM phenomenon

It is observed that some time after the $L \to H$ transition a sudden disruptive event occurs, called edge-localized mode or ELM, by which the plasma, accumulated near the edge owing to the presence of the transport barrier, is ejected, thus reducing density and temperature gradients. The process does not, however, eliminate the barrier altogether, so that the steep gradients can reform until the next event occurs. ELMs usually occur rather erratically but may also appear in a periodic way like the sawtooth oscillation in the plasma center. ELMs enhance the average energy loss from the plasma, but they have also the beneficial effect of removing impurity ions, which in the absence of ELMs would accumulate in the plasma center and terminate the discharge by excessive radiative cooling.

ELMs appear in a variety of forms. One mainly distinguishes between two classes, a fast process called type-III ELM, which occurs at heating power levels not far above the threshold power needed for the $L \to H$ transition, and a slower process called type-I or giant ELM, which dominates at higher input power levels.

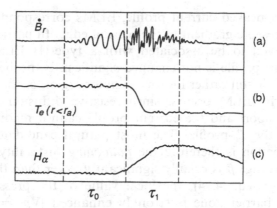

Fig. 8.26. Time evolution at an ELM event of the magnetic oscillation signal (a), the electron temperature inside the barrier zone (b), and the $H\alpha$-signal far outside the plasma column (c) (schematic drawing).

Type-III ELMs are characteristic for H-mode plasmas in the ASDEX tokamak (Zohm et al., 1991), the device where the $L \rightarrow H$ transition and the ELM phenomenon have first been observed. The time evolution during an ELM event is shown schematically in Fig. 8.26. The magnetic signal starts with a coherent precursor oscillation growing over a period τ_0 ($\tau_0 \sim$ 100–500 μs in ASDEX, but can be much longer, up to 10 ms in DIII-D), which suddenly turns turbulent, exhibiting a broad frequency spectrum. The turbulence lasts for a short time, $\tau_1 < 500\,\mu s$, and subsequently decays rapidly. The remaining traces in Fig. 8.26 illustrate the plasma behavior, viz. the temperature inside the transport barrier zone (b), and the plasma outflow measured by the $H\alpha$-signal at a target plate (c). While the plasma is not affected during the coherent precursor phase, plasma conditions start to change as soon as turbulence sets in. This leads to a rapid temperature and density decrease in the plasma edge region, reaching inside to nearly half the plasma radius, and a strong increase of the recycled plasma at the target plates.

The sequence of events in an ELM is reminiscent of a major disruption, where magnetic turbulence is suddenly generated by a large-amplitude, coherent (2,1) mode, giving rise to the destruction of magnetic surfaces and the rapid outflow of plasma energy. This similarity is, however, only superficial, since the basic MHD ingredients are quite different. In contrast to sawtooth collapse and major disruption, which are characterized by low-m modes, the ELM precursor has high mode numbers, $m \sim 15$, corresponding to $n \sim 5$ for $q_a \sim 3$. Since for such modes rational surfaces are rather densely located in radius, the occurrence of ELMs is found to be insensitive to the value of q_a. While disruptions correspond primarily to an expansion of a peaked current profile, ELMs correspond to a relaxation of the steep pressure gradient at the plasma edge. In particular no voltage spike is observed to be associated with a type-III ELM. This implies that the current profile is not changed significantly, probably because the current density is still rather low in the outer region of the plasma column by the time the ELM occurs, since resistive diffusion is much slower than heat transport and the current profile cannot readily adjust to the steepening of the T_e-profile. The most natural candidate to explain the ELM phenomenon is therefore the *ballooning* instability (section 4.6.1). Though the average β is usually significantly lower than the critical value, $\beta \ll \beta_c \sim \varepsilon/q^2$, eq. (4.74), the local value of the pressure gradient in the transport barrier zone is strongly enhanced $|\nabla p| = p/l_p, l_p \sim 0.1a$, such that the effective ballooning mode threshold value of β is given by eq. (4.77),

$$\beta_c \sim l_p/R q^2 .$$

Quantitative evaluations of the ballooning stability criterion show that plasma conditions at the onset of a type-III ELM are usually somewhat below the ideal ballooning threshold β_c, $\beta \lesssim 0.5\beta_c$. A possible explanation is resistive effects, which may lower the instability threshold. In any case the picture of linear localized ballooning modes has to be modified in several respects for a more quantitative dynamic theory of the ELM:

(a) Since the effective ion Larmor radius is not much smaller than the pressure gradient scale length in the barrier zone, kinetic effects are not negligible. They provide an upper limit of the poloidal mode number m for which the MHD approximation is valid,

$$k_\theta \, \rho_{eff} < 1 ,$$

where $k_\theta = m/a$ and ρ_{eff} is of the order of the poloidal Larmor radius, $\rho_{eff} \sim \sqrt{T_i m_i}/eB_\theta$. In ASDEX $a = 40$ cm, $\rho_{eff} \sim 2$ cm, hence $m < 20$, consistent with the observations.

(b) ELMs can be suppressed by metallic parts located in the vicinity of the separatrix. Thus the unstable modes seem to have a finite radial width, reaching beyond the separatrix.

(c) Like sawtooth and major disruptions the ELM is a two-stage process as seen in Fig. 8.26. The onset of the turbulent disruptive phase τ_1 is a complicated nonlinear phenomenon, which cannot be described in terms of a linear instability.

As the power supplied to the plasma is increased, the frequency of type-III ELMs decreases. This may be due to the increase of the edge temperature which reduces resistive MHD effects. At high input power the edge pressure gradient reaches the limit for ideal ballooning modes, $\beta \simeq \beta_c$. Here a different type, the type-I or giant ELM, occurs, which is the standard ELM in the DIII-D tokamak, where particularly high plasma heating powers are used. The fraction of the plasma energy lost in a type-I ELM is significantly larger than in a type-III one, causing a strong perturbation of the entire discharge. A type-I ELM is usually preceded by a gradual increase of incoherent density fluctuations, which can be interpreted as a resurrection of the drift wave turbulence characteristic of the anomalous transport in the L-mode, such that the type-I ELM appears to be a short excursion back into the L-mode. It should, however, be noted that these precursor density fluctuations do not give rise to an increased plasma flow across the separatrix, since the Hα signal from the divertor does not increase. Hence the transport barrier appears to be eroded from the inside, finally collapsing at the ELM. The onset of a type-I ELM is a distinct rapid MHD process, which, in contrast to the type-III ELM, is accompanied by a negative voltage spike. Since the type-I ELM occurs for edge profiles close to the ideal ballooning limit, the MHD process is naturally associated with an ideal ballooning mode, though the details of the sudden onset are not understood. The total duration of a type-I ELM characterized by strong fluctuations and high transport, $\sim 10\,\mathrm{ms}$, is much longer than in a type-III ELM, and, in contrast to the latter, its frequency increases with increasing heating power.

Table 8.1 summarizes the main features of the three important disruptive processes which may occur in a tokamak plasma. All seem to be two-stage processes exhibiting a coherent precursor oscillation and a more turbulent phase where rapid plasma relaxation takes place. The table should, however, be considered with some caution. Many features given are not based on solid theories, but only on qualitative, often speculative and sometimes still controversial arguments drawn from experimental observations, linear stability theory and numerical modelling.

Table 8.1. MHD properties of disruptive processes in tokamak plasmas

	sawtooth collapse	major disruption	ELM (type III)
Free energy source	∇j and ∇p	∇j	∇p
Primary (precursor) phase	(1,1) kink mode	(2,1) tearing mode	Single $m,n \gg 1$ resistive ballooning mode
Precursor time scale	$\tau_A S^v, v \lesssim 1/2$	$\tau_A S$	τ_A
Main mode coupling	$(1,1) \leftrightarrow (3,4),(4,5)$	$(2,1) \leftrightarrow (1,1),(3,2)$	Broad spectrum of ballooning modes
Disruption time scale	τ_A	$\tau_A S^v, v \lesssim 1/2$	τ_A
Main relaxation process	T_e transport along stochastic field lines	T_e transport along stochastic field lines	Convective transport of T_e, T_i, n
Reconnection of B_θ	Partial	Yes	Probably no
Voltage spike	No	Yes	No

9

Dynamics of the reversed-field pinch

While in the preceding chapters the emphasis was on systems embedded in a strong axial field, $B_z \gg B_p$, we now turn attention to pinch configurations, where both magnetic field components are comparable, $B_z \sim B_p$. The most important of such systems regarding practical applications is the reversed-field pinch (RFP), which exhibits a number of very interesting magnetohydrodynamic phenomena.

Let us compare the ordering of the important RFP parameters in terms of the inverse aspect ratio $\varepsilon = a/R$ with the corresponding tokamak ordering:

	RFP	Tokamak
B_p/B_z	1	ε
$q = \dfrac{rB_z}{RB_p}$	ε	1
$\beta_p = \dfrac{2p}{B_p^2}$	1	1
$\beta_t = \dfrac{2p}{B_z^2}$	1	ε^2

$$(9.1)$$

Hence the RFP is a low-q, high-β plasma. Since in real RFP experiments β_p is considerably below unity, the toroidal outward shift of the magnetic surfaces, the Shafranov shift, is small, such that equilibrium and stability are well described in the cylindrical approximation. Since $q \ll 1$, Mercier's criterion (4.73) reduces to Suydam's criterion (4.65). The physical reason is that the cylindrical curvature of a field line $\kappa_c = -B_\theta^2/rB_z^2 \sim 1/r$ is much larger than the toroidal one, the field lines winding more frequently around the minor torus axis, the magnetic axis, than around the major one.

There is a class of stable equilibrium configurations, derived by Taylor from a variational principle. We discuss Taylor's theory in section 9.1. Since in real experiments transport processes drive the plasma away from the stable state, instability will eventually set in. Section 9.2 gives an overview of the ideal and resistive stability properties of general RFP configurations, which differ considerably from those in the tokamak case. The nonlinear evolution of an individual mode, the single-helicity case, is considered in section 9.3. However, in contrast to typical tokamak instabilities, where the single helicity behavior dominates, instability in the RFP plasma usually gives rise to multi-helicity mode excitation, leading to fully developed MHD turbulence. This dynamic behavior is considered in section 9.4.

9.1 Minimum-energy states

9.1.1 *Taylor's theory*

In a typical RFP experiment the discharge is generated by inducing a strong toroidal current in a preionized plasma with a rather weak, initially homogeneous toroidal magnetic field B_0, corresponding to the toroidal flux $\psi_t = \pi a^2 B_0$, where a is the wall radius. In the initial phase the discharge is usually rather turbulent, since the corresponding pinch configurations are highly unstable. Such turbulent states are expected to show self-organization, in particular selective decay as discussed in section 7.3.1, viz. the decay of the energy* W under the constraint of constant total magnetic helicity H. If the relaxation process is rapid compared with the change of the external parameters then the relaxed state is determined by the variational principle (7.39),

$$\delta \left[W - \mu H \right] = \delta \left[\tfrac{1}{2} \int_V B^2 d^3x - \frac{\mu}{2} \int_V \mathbf{A} \cdot \mathbf{B} d^3x \right] = 0 \,, \qquad (9.2)$$

neglecting kinetic and thermal energy contributions which are usually small compared with the magnetic one. While in a strictly ideal MHD system not only the total helicity H but also the infinite set of helicities H_ε corresponding to individual flux tubes V_ε would be conserved (section 2.3), small-scale reconnection rapidly destroys the individuality of these flux tubes and hence the meaning of H_ε, preserving only the total helicity. This is the basis of Taylor's theory (Taylor, 1974, 1986). While in a turbulent system the energy decay rate is independent of η,

$$\varepsilon = \left| \frac{dW}{dt} \right| \simeq \eta \int j^2 d^3x = O(\eta^0) \,, \qquad (9.3)$$

---

* Following conventional notation the energy is denoted by W instead of E.

the helicity decays much more slowly:

$$\left| \frac{dH}{dt} \right| = \eta \int \mathbf{j} \cdot \mathbf{B} d^3 x \sim \varepsilon \, l_d = O(\eta^{2/3}) \,, \tag{9.4}$$

where l_d is the turbulent dissipation scale defined in eq. (7.88).

Before deriving explicitly the minimum energy solution, let us first return to the problem of gauge invariance of the helicity, briefly considered in section 2.2. Performing a gauge transformation $\mathbf{A}' = \mathbf{A} + \nabla \chi$, eq. (2.26)

$$H' - H = \oint \chi \mathbf{B} \cdot d\mathbf{F}$$

seems to indicate that $B_n = 0$ at the boundary is sufficient for gauge invariance. Since, however, the volume of interest here is a torus, which is multiply connected, the gauge function need not be single-valued. Cutting the torus in a poloidal plane to enforce single valuedness, the surface term becomes

$$\oint \chi \mathbf{B} \cdot d\mathbf{F} = \psi_t \Delta \chi \,, \tag{9.5}$$

where $\Delta \chi$ is the increment of χ once around the torus and ψ_t is the toroidal flux. Ignoring this effect may lead to erroneous results as regards the minimum energy state, as has been pointed out by Reiman (1980). An obviously gauge-invariant generalization of the usual expression of the helicity has been given by Taylor (1986),

$$\mathbf{H} = \int \mathbf{A} \cdot \mathbf{B} d^3 x - \oint \mathbf{A} \cdot d\mathbf{l} \oint \mathbf{A} \cdot d\mathbf{s} \,, \tag{9.6}$$

where $d\mathbf{l}$ and $d\mathbf{s}$ denote the line elements the long and the short way around the torus surface. (A somewhat different gauge-invariant expression of helicity has been given in chapter 6, eq. (6.78).) Note that $\oint \mathbf{A} \cdot d\mathbf{s} = \psi_t$. In a compact (i.e. unslit) conducting chamber both loop integrals are constant, and the variational principle (9.2) remains unchanged if H is replaced by expression (9.6). Variation with respect to $\mathbf{A}$ with $\delta \mathbf{A} = 0$ at the wall (vanishing of the tangential component is actually sufficient) yields eq. (7.40),

$$\nabla \times \mathbf{B} = \mu \mathbf{B} \,, \tag{9.7}$$

where μ is a constant parameter.

The general solution of eq. (9.7) in cylindrical coordinates is given in eq. (3.48). The analysis of Taylor (1974) and Reiman (1980) shows that the minimum energy state can be only either the $m=0$ symmetric solution,

$$\begin{aligned} B_z &= B_0 J_0(\mu r) \,, \\ B_\theta &= B_0 J_1(\mu r) \,, \end{aligned} \tag{9.8}$$

or a helical state consisting of a superposition of the $m=0$ and the $m=1$ Bessel function solutions,

$$B_z = B_0 J_0(\mu r) + B_1 J_1(\alpha r) \cos(\theta + kz) ,$$

$$B_\theta = B_0 J_1(\mu r) - B_1 \left[\frac{\mu}{\alpha} J_1'(\alpha r) + \frac{k}{r\alpha^2} J_1(\alpha r) \right] \cos(\theta + kz) ,$$

$$B_r = -B_1 \left[\frac{k}{\alpha} J_1'(\alpha r) + \frac{\mu}{r\alpha^2} J_1(\alpha r) \right] \sin(\theta + kz) , \qquad (9.9)$$

$$\alpha^2 = \mu^2 - k^2 .$$

The parameters μ and B_0 as well as the helical amplitude B_1 are determined by the given values of H and ψ_t and the boundary condition $B_r(a) = 0$. Only the symmetric part contributes to the toroidal flux, which can easily be computed,

$$\psi_t = 2\pi B_0 \int_0^a J_0(\mu r) r\, dr = 2\pi a B_0 J_1(\mu a)/\mu , \qquad (9.10)$$

using the relation $J_0''(x) + J_0'(x)/x + J_0(x) = 0$. In order to decide which solution has the lower energy, we explicitly calculate the energy and the helicity. Since H, eq. (9.6), is gauge-invariant we can use the special gauge

$$\mathbf{A} = \frac{1}{\mu} (\mathbf{B} - B_0 J_0(\mu a)\mathbf{e}_z) , \qquad (9.11)$$

which has the property $\oint \mathbf{A} \cdot d\mathbf{l} = 0$ such that the loop integral term in expression (9.6) vanishes.

First consider the symmetric solution, eq. (9.8). Inserting (9.11) into the helicity we obtain

$$H = \frac{1}{\mu} \int \mathbf{B} \cdot (\mathbf{B} - B_0 J_0(\mu a)\mathbf{e}_z) d^3x$$

$$= \frac{2}{\mu} W - \frac{B_0}{\mu} J_0(\mu a) 2\pi R \psi_t , \qquad (9.12)$$

where $2\pi R$ is the length of the cylinder. The energy can easily be calculated,

$$W = \frac{2\pi^2 R B_0^2}{\mu^2} [(\mu a)^2 (J_0^2(\mu a) + J_1^2(\mu a)) - \mu a J_0(\mu a) J_1(\mu a)] . \qquad (9.13)$$

Eliminating B_0 by use of eq. (9.10) and introducing the normalized quantities

$$\widehat{W} = 2a^2 W/R\psi_t^2 , \quad \widehat{H} = aH/R\psi_t^2 , \quad \hat{\mu} = \mu a ,$$

eqs (9.12), (9.13) yield explicit expressions for $\widehat{W}$ and $\widehat{H}$ in terms of $\hat{\mu}$:

$$\widehat{W} = \hat{\mu}^2 \left[1 + \frac{J_0^2(\hat{\mu})}{J_1^2(\hat{\mu})} \right] - \hat{\mu} \frac{J_0(\hat{\mu})}{J_1(\hat{\mu})} ,$$

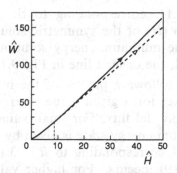

Fig. 9.1. Energy of the force-free constant-μ solution as a function of the helicity. The full line represents the cylindrical solution, the dashed line the helical one, which has lower energy above the bifurcation point $\widehat{H} \simeq 8.21$. The arrows indicate the direction of increasing $\hat{\mu}$ and increasing B_1/B_0 (from Reiman, 1980).

$$\widehat{H} = \frac{\widehat{W}}{\hat{\mu}} - \frac{J_0(\hat{\mu})}{J_1(\hat{\mu})} \tag{9.14}$$

$$= \hat{\mu} \left[1 + \frac{J_0^2(\hat{\mu})}{J_1^2(\hat{\mu})} \right] - 2 \frac{J_0(\hat{\mu})}{J_1(\hat{\mu})} \,.$$

This is a parametric representation of the function $\widehat{W}(\widehat{H})$ for the symmetric solution, the lowest branch of which is given by the full line in Fig. 9.1. Note that $\widehat{W}$ and $\widehat{H}$ are monotonically increasing with $\hat{\mu}$.

Now consider the helical solution. Since the normal field component must vanish at the boundary, $B_r(a) = 0$,

$$\hat{k}\,\hat{\alpha}\,J_1'(\hat{\alpha}) + \hat{\mu}\,J_1(\hat{\alpha}) = 0 \,, \tag{9.15}$$

with $\hat{k} = ka$, $\hat{\alpha}^2 = \hat{\mu}^2 - \hat{k}^2$. Since $\hat{k} = na/R$ has only discrete values, the mixed solution is possible only for discrete values of μ. The smallest value of $\hat{k}$ for which eq. (9.15) has a solution is $\hat{k} = 1.25$, corresponding to $\hat{\mu} = 3.11$. Since $a/R \ll 1$, the mode number n is large for $\hat{k} > 1$ such that the allowed values of $\hat{\mu}$ are quasi-continuous.

Inserting solution (9.9) into H would give a rather complicated expression $\widehat{H}(B_1/B_0, \hat{\mu})$, which, however, we need not compute explicitly, since we are interested only in the relation $\widehat{W}(\widehat{H})$. This can be obtained from eq. (9.12),

$$\widehat{W} = \hat{\mu} \left[\widehat{H} + \frac{J_0(\hat{\mu})}{J_1(\hat{\mu})} \right] \,, \tag{9.16}$$

which is formally identical with the expression for the symmetric solution (9.14). The difference, however, is that $\widehat{H}$ is now a monotonically increasing function of B_1/B_0 for fixed $\hat{\mu}$. This means that $\widehat{W}$ is a linear function of $\hat{H}$ for any $\hat{\mu} \geq 3.11$, corresponding to the right-hand ray of the tangent to the curve $\widehat{W}(\hat{H})$ of the symmetric solution (9.14) at the point $\hat{H}(B_1/B_0 = 0, \hat{\mu})$. The minimum-energy solution is the tangent at the smallest value $\hat{\mu} = 3.11$, the dashed line in Fig. 9.1.

Thus we obtain the following picture of the minimum-energy state to which the system relaxes for a given value of $\hat{H}$, the helicity normalized to the square of the toroidal flux. For small values of $\hat{H}$, corresponding to $\hat{\mu} < 3.11$, the minimum-energy state is given by the symmetric solution, eq. (9.8). At $\hat{H} = 8.21$ corresponding to $\hat{\mu} = 3.11$, a bifurcation to the helical solution, eq. (9.9), occurs. For higher values of $\hat{H}$, $\hat{\mu}$ no longer changes. Instead the helical amplitude B_1/B_0 increases, following the equation (Martin and Taylor, 1974)

$$\hat{H} = 8.21 + 4.49 \left(B_1/B_0\right)^2 . \tag{9.17}$$

9.1.2   Experimental results

The predictions of Taylor's theory are in surprisingly good agreement with experimental observations obtained from various RFP devices. An experimental torus is not compact but slit, in both poloidal and toroidal directions. While the latter slit is necessary to introduce the initial toroidal field but is often bridged during the discharge to preserve $\psi_t = $ const., the voltage V_t applied to the poloidal slit is used to induce the toroidal current I, which simultaneously feeds helicity into the plasma,

$$\frac{dH}{dt} = 2V_t\psi_t , \tag{9.18}$$

as can be derived from eq. (9.6); see eq. (6.77).

Under most discharge conditions the toroidal current rise time is long compared with the time scale of the internal relaxation processes, so that the plasma follows a sequence of minimum energy states. To compare theory and experiment the following easily measurable quantities are very convenient: (a) the normalized toroidal field at the boundary,

$$F = \frac{B_{z0}(a)}{\langle B_z \rangle} , \tag{9.19}$$

called the *field reversal parameter*, and (b) the normalized toroidal current I,

$$\theta = \frac{B_{\theta 0}(a)}{\langle B_z \rangle} = \frac{aI}{2\psi_t} , \tag{9.20}$$

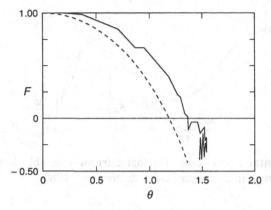

Fig. 9.2. Experimental F–θ curve (full line, from DiMarco, 1983) and theoretical curve eqs (9.21), (9.22) (dashed line).

called the *pinch parameter*. Here the subscript zero denotes the poloidal average. Taylor's theory predicts:

$$F = \frac{\hat{\mu}J_0(\hat{\mu})}{2J_1(\hat{\mu})},\qquad (9.21)$$

$$\theta = \frac{\hat{\mu}}{2}.\qquad (9.22)$$

In Fig. 9.2 a typical discharge history is plotted in the F, θ plane, which agrees rather well with the theoretical predictions (9.21), (9.22). As the current increases, $B_z(a)$ decreases and finally becomes negative, corresponding to field reversal. The current does not increase beyond $\theta \sim 1.5$, or $\hat{\mu} \simeq 3$ according to (9.20), which agrees well with the theoretical bifurcation point $\hat{\mu} = 3.11$. Further increase of the voltage V_t does not lead to a higher current. Instead the theory predicts that for higher values of the helicity the plasma column should become increasingly helically deformed, as is in fact observed experimentally. Figure 9.3 illustrates two RFP discharges characterized by (a) a rapid current rise, where the current overshoots before relaxing to a value $\theta \simeq 1.5$, and (b) a slow current rise, where the system remains on the minimum-energy curve $F(\theta)$.

Though agreement of experimental observations with Taylor's theory is qualitatively very good, there is a systematic deviation from the theoretical $F(\theta)$ curve as seen for instance in Fig. 9.2, F being somewhat larger, or less strongly reversed, than predicted. This is mainly caused by the behavior of the current density j_z in the region close to the boundary. While the theory predicts $j_z = \mu B_z$, which implies reversal of j_z and a finite value at the wall, the low plasma temperature in the edge zone makes resistivity

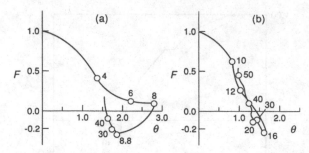

Fig. 9.3.   Experimental F–θ curves: (a) fast current rise, (b) slow current rise. Numbers indicate times in μs (from Bodin & Newton, 1980).

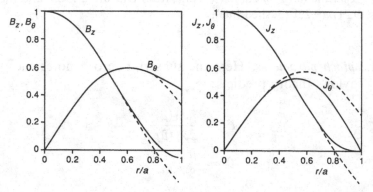

Fig. 9.4.   Modified Bessel function profiles: $\mu(r) = \mu_0(1 - (r/a)^4)$, $\beta_0 = 0.1$, $\theta_0 = 1.6$. The dashed line indicates the Bessel function model eq. (9.8).

effects dominant, such that j_z becomes very small in this region. Hence μ is constant only in the central part of the plasma column, typically falling off to zero in the outer third of the radius. In addition, finite β slightly modifies the current profile. A modified Bessel function model accounting for both effects is plotted in Fig. 9.4. The toroidal current j_z is no longer reversed. The modifications shift the $F(\theta)$ curve to the right in agreement with the experimental observations in Fig. 9.2.

On a longer time scale resistive diffusion tends to also wipe out the B_z-reversal, while ohmic heating leads to an enhanced peaking of the toroidal current density in the center. These transport processes carry the system into an unstable regime where relaxation restarts. The dynamic behavior resulting from the interplay between resistive destabilization and the MHD processes trying to restore a minimum energy state will be discussed in sections 9.3 and 9.4.

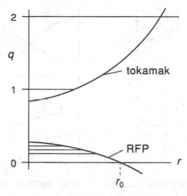

Fig. 9.5. Schematic drawing of q-profiles and possible resonances in the tokamak and the RFP.

9.2 Stability properties of RFP equilibria

Since the stability properties of high-β pinch configurations such as the RFP differ significantly from those of tokamaks, to which chapter 4 was essentially restricted, a brief overview of RFP stability theory is given here. It is instructive to compare q-profiles and possible resonances in the tokamak and the RFP (Fig. 9.5). While in the former q is monotonically increasing from a value slightly below 1, such that the main resonances $q = m/n$ are $m = 1, 2, 3, \ldots$ with $n = 1$, in the latter these are $m = 1$, $n = n_0, n_0 + 1, \ldots$, with $n_0^{-1} \simeq q_{max} \sim \varepsilon$. Hence there are many $m=1$ resonances in the RFP. The radial density of resonances is much higher than in a tokamak, with an accumulation point at the field reversal radius r_0.

9.2.1   Ideal instabilities

We first note that $q(r) = rB_z/RB_\theta$ cannot have a minimum but has to fall off monotonically reversing sign together with B_z. The situation is illustrated in Fig. 9.6. If there is a minimum of q, then for any finite pressure gradient Suydam's criterion (4.65) indicates interchange instability in a finite radial region, which implies the presence of a nonlocalized $m=1$ mode (Robinson, 1971; Goedbloed & Sakanaka, 1974).

In the following we restrict consideration to the $m=1$ modes which are in general the most unstable ones. Since owing to the low average q the configuration is very unstable to external modes, the RFP has to be surrounded by a conducting wall close to the plasma boundary. In addition, interest is mainly in configurations of relatively low β close to a force-free state. Hence we will only discuss internal current-driven instabilities. Ideal instabilities of a cylindrical pinch are conveniently

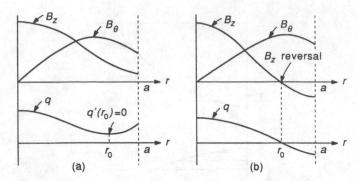

Fig. 9.6.   (a) Unstable profiles exhibiting a minimum in the q-profile, (b) stable profiles due to B_z-reversal (schematic drawing).

studied using the energy principle, eq. (4.57),

$$\delta W = \int_0^a \left[f \left(\frac{d\xi}{dr} \right)^2 + g\xi^2 \right] r dr \, ,$$

where for current-driven modes, i.e. $p_0' = 0$, and $m = 1$, $k_z = -n/R$,

$$f = B_\theta^2 \frac{(nq - 1)^2}{1 + k_z^2 r^2} \, ,$$

$$g = B_\theta^2 k_z^2 \frac{(nq - 1) \left[nq \left(3 + k_z^2 r^2 \right) + 1 - k_z^2 r^2 \right]}{\left(1 + k_z^2 r^2 \right)^2} \, . \tag{9.23}$$

While the first term in δW representing the effect of field line bending is positive definite, the second may become negative. If one can find a trial function ξ such that $\delta W < 0$, there is an unstable mode (in general not identical with the trial function) with mode numbers $(1, n)$. Consider a resonance at $r = r_s$, $q(r_s) = 1/n$, i.e. $f(r_s) = 0$. In the tokamak case, where q is increasing with radius, g is negative inside the resonant radius and positive outside. By choosing ξ to be a step function, $\xi = 1$ for $r < r_s$, $\xi = 0$ for $r > r_s$, the first term in δW vanishes and one obtains $\delta W = \int_0^{r_s} gr dr < 0$, which implies instability. (Note that in the tokamak case the form of $\hat{g}$ given in eq. (9.23) applies only for $n > 1$, since for $n = 1$ toroidal corrections are important; see section 4.7.)

In the RFP, by contrast, q decreases with radius. We now have to distinguish between the two cases $r_s < r_0$ and $r_s > r_0$, where r_0 is the field reversal point. If $r_s < r_0$, where $q(r_s) > 0$, we have $g > 0$ for $r < r_s$ and $g < 0$ for $r > r_s$, which is just the opposite of the behavior in a tokamak. Hence for instability ξ should now be located in the outer region $r > r_s$. However, ξ can be constant only if there is no conducting wall at the

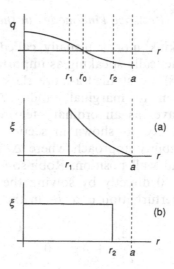

Fig. 9.7. Unstable trial functions in the RFP. (a) Resonant radius $r_1 < r_0$, (b) $r_2 > r_0$ (schematic drawing).

plasma boundary (in this case the configuration is usually unstable), while in the presence of such a wall ξ must decrease to zero (Fig. 9.7a). Here the f-term in δW does not vanish and minimization of δW has to be carried out numerically. Note that there may be ideal instabilities even in the absence of a resonant radius, i.e. $m/n > q(r)$ for all r. Such modes exhibit special nonlinear properties (section 9.3).

If the resonance is outside the reversal point, $r_s > r_0$, such that $q(r_s) < 0$, $n < 0$, the regions inside and outside r_s just change roles, g now being negative on the inside. In this case the minimizing displacement ξ behaves as in the tokamak case, Fig. 9.7b, and the wall has no stabilizing effect. If $\int_0^{r_s} g r \, dr < 0$ then the mode is unstable, which occurs for a sufficiently extended reversal region $r_0 < r < a$. The symmetric Bessel function model eq. (9.8), for instance, becomes ideally unstable for $\mu a > 3.176$, or $\theta > 1.59$, which corresponds to $a/r_0 = 1.32$ (Voslamber & Callebaut, 1962). A necessary condition for stability is $\psi_t = 2\pi \int_0^a B_z r \, dr > 0$. Since in an RFP device the field reversal is less pronounced than in the Bessel function model corresponding to the same value of θ, the negative $q(r_s)$ instabilities are not likely to occur. Typically the lower-n modes with $n \sim q_{max}^{-1}$, which have resonances close to the axis or are slightly nonresonant from above ($1/n > q_{max}$), are most unstable, while higher-n modes with resonances close to the field reversal radius are stable. Ideally stable RFP configurations with finite β allowing even a narrow vacuum region between the plasma boundary and the conducting wall have been obtained by Robinson (1971).

9.2.2   Resistive kink modes in the RFP

The $m = 1$ resistive kink modes (usually called tearing modes in the RFP literature) are practically speaking as important as the ideal internal modes. Robinson (1978) has studied a large class of pinch equilibria and determined the conditions for marginal stability, $\Delta' = 0$. (Remember that the $m = 1$ mode behaves as an ordinary tearing mode for conditions close to marginal stability, as shown in section 4.7.2.) In contrast to the usual tearing instability approach where Δ' is determined for fixed equilibrium profiles and wall position, Robinson considers the marginal stability problem $\Delta' = 0$ directly by solving the ideal equation for the radial magnetic field perturbation $\psi \propto \tilde{B}_r$ in the marginal stability limit $\omega = 0$:

$$\frac{d^2\psi}{dr^2} + A\psi = 0 \, , \tag{9.24}$$

$$A = \frac{d\sigma}{dr} \frac{B_z - k_z r B_\theta}{rF} + A_1(\sigma, n, r) \, ,$$

$$\sigma = \frac{\mathbf{j} \cdot \mathbf{B}}{B^2} \, , \quad F = \frac{B_\theta}{r} - \frac{n}{R} B_z \, ,$$

and $A_1(\sigma, n, r)$ is a nonsingular expression. The tearing mode is mainly driven by the first term in A, which becomes singular at the resonance. Choosing equilibrium functions $B_\theta(r)$, $B_z(r)$ and a mode number n, eq. (9.24) is integrated from the center smoothly across the singular radius r_s up to the point where $\psi = 0$, which gives the marginally stable position of the conducting wall.

The condition for stability on the axis, $r_s \to 0$, can be obtained analytically:

$$-\frac{4}{5} < \gamma < -\frac{4}{9} \, . \tag{9.25}$$

Here γ is a dimensionless measure of the curvature of the q-profile in the center:

$$\gamma = \frac{P}{2} \frac{d^2 P}{dr^2}\bigg|_{r=0} \, , \quad P = r B_z / B_\theta \, , \tag{9.26}$$

where $P = qR$ is called the magnetic pitch. Condition (9.25) implies $\gamma < 0$, i.e. q must decrease from the center. The lower limit is imposed by ideal stability, the upper by tearing stability. The condition requires the current density j_z to be peaked on the axis (a flat current profile corresponds to $\gamma = -1$), but not too strongly.

The Bessel function model eq. (9.8) represents a special case, since $d\sigma/dr = 0$. Here the stability condition can be obtained analytically,

$\mu a < 3.104$ or $\theta < 1.55$ (Gibson & Whiteman, 1968), which is not much different from the ideal stability limit. This is not surprising, since the main driving effect of the tearing instability, $d\sigma/dr$, is absent.

Tearing-mode-stable, zero-β RFP configurations have been given by Robinson (1978). These consist of the Taylor profiles, eq. (9.8), in the central part followed by an outer region carrying very little current, which corresponds to a modified Bessel function model. These configurations may have large values of the pinch parameter, $\theta \lesssim 3.7$, in contrast to the $\mu = $ const. case, where tearing instability limits θ to $\theta < 1.55$.

9.3 Single-helicity behavior

Kink modes in an RFP discharge are destabilized by central current density peaking caused by the thermal instability in the same way as in a tokamak discharge, discussed in section 8.1.2, which leads to a continuous reduction of the safety factor, in particular of $q(0)$. However, because of the radially decreasing q-profile the characteristics of the kink instability in the RFP differ substantially from the resistive kink mode in the tokamak. This is evident from the shape of the trial function $\xi(r)$, Fig. 9.7a, typical for the most unstable $q(r_s) > 0$ modes. While in the tokamak the convective flow generated by the instability is confined to the central region $r \lesssim r_s$, in the RFP it is primarily located in the outside region $r > r_s$. Hence the mode will also influence the reversal region, acting to preserve or restore toroidal field reversal, which is called the RFP dynamo effect.

In this section consideration is restricted to the case of single helicity $m/n = $ const. While in tokamaks this constitutes in general a good approximation of the nonlinear behavior because of the wide radial separation of possibly unstable resonances, it is less well justified in the RFP, where $m=1$ resonances with different mode number n are rather densely packed so that a single-helicity behavior can only be expected for conditions close to marginal stability. Single-helicity calculations can nevertheless provide important information about the basic nonlinear dynamics in an RFP plasma. Nearly single-helicity configurations may, however, occur, as is discussed in section 9.4.

9.3.1 *Nonlinear evolution of the $m = 1$ instability*

Since $m=1$ instabilities, either nonresonant ideal kink modes or resonant resistive kink modes, evolve on a fast time scale, the helical flux function is conserved. Integrating Faraday's law (2.58) one obtains

$$\partial_t \mathbf{A} = \mathbf{v} \times \mathbf{B} - \nabla\phi - \eta\mathbf{j} \,. \tag{9.27}$$

In a helical configuration, where physical quantities depend only on r and $u = \theta + kz$, scalar multiplication of eq. (9.27) by $\mathbf{h}$, defined in eq. (3.12), gives an equation for the helical flux function $\psi = krA_\theta - A_z$, eq. (3.15),

$$\partial_t \psi + \mathbf{v} \cdot \nabla \psi = \eta \left(j_z - kr j_\theta \right)$$

$$= \eta \left(\mathscr{L}\psi - \frac{2k}{1 + k^2 r^2} f \right) , \qquad (9.28)$$

using eq. (3.17); $f = \mathbf{B} \cdot \mathbf{h}/h^2$ is the magnetic field component in the direction of the helix, defined in eq. (3.16). For a rapidly evolving kink mode, however, the resistive term is negligible except at the location of current singularities, i.e. ψ is conserved practically everywhere,

$$\frac{d\psi}{dt} = 0 . \qquad (9.29)$$

(Note that this local condition, valid only in helical symmetry, is much more restrictive than Taylor's conservation of the total helicity H, which seems to be the only invariant in general nonsymmetric fast processes.)

In the tokamak case this behavior allows us to predict the final symmetric state of the resistive kink mode (section 8.1.2) or the vacuum bubble state resulting from an ideal kink instability (section 5.2.2). In the present case of an unstable RFP configuration, however, the final state is in general not cylindrically symmetric but remains helical such that eq. (9.29) is of no direct use. Instead the most practical way is to solve the dynamic MHD equations numerically (Caramana et al., 1983; Holmes et al., 1985; Schnack et al., 1985). As the initial state in these studies the following force-free equilibrium is chosen, characterized by the q-profile:

$$q(r) = q(0)(1 - 1.87(r/a)^2 + 0.83(r/a)^4) , \qquad (9.30)$$

which is obtained by resistive diffusion from the stable configuration given by Robinson (1978). Depending on the choice of $q(0)$, various $m=1$ modes with different values of n, either resonant, $q(0) > 1/n$, or nonresonant, $q(0) < 1/n$, can be unstable. For sufficiently low $q(0)$ many n are simultaneously unstable. The nonlinear evolution of a typical resonant mode consists of two stages illustrated in the ψ-contours in Fig. 9.8:

(a) A fast reconnection phase similar to the resistive $m=1$ mode in a tokamak configuration. The flux inside the resonant radius is reconnected until the original magnetic axis vanishes ($t \lesssim 100$). Contrary to the tokamak behavior the configuration does not relax into a stable cylindrical state, since removal of the resonance has lowered $q(0)$ thus making the configuration even more unstable.

(b) In the second phase ($t > 100$) the helical distortion becomes more pronounced and a second magnetic axis may finally be reintroduced

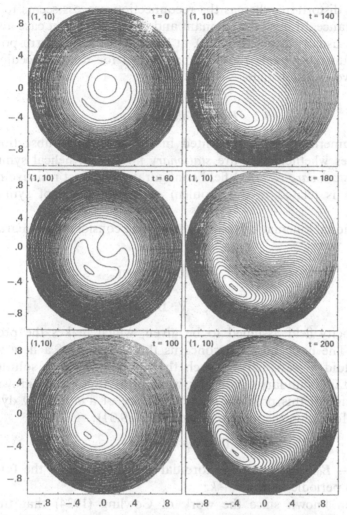

Fig. 9.8. Contour lines of the helical flux function ψ for the single helicity instability evolution with $m/n = 0.1$ for an initial $q(0) = 0.125$ (from Schnack et al., 1985).

in the central region. The latter reconnection process, however, occurs on the global resistive time scale and is hence affected by the slow equilibrium evolution.

For a nonresonant $m=1$ instability the nonlinear behavior is quite similar, except that the initial reconnection phase (a) is missing. From the ψ-contours of the time sequence in Fig. 9.8 for the resonant mode it appears, however, that the distinction between two different phases should not be overemphasized, since the main feature is the continuous growth of the helical distortion, whereas the reconnection process seems to be of minor importance. In both the resonant and the nonresonant case average B_z-reversal is usually attained, $F < 0$, at least for helicities corresponding to the most unstable modes, and is maintained for times exceeding the global resistive diffusion time.

9.3.2   *Helical ohmic states*

From the numerical results presented in the preceding subsection the question arises whether there are *stationary* flows with helical symmetry sustaining a *stationary* reversed-field configuration against resistive decay. The question is related to the problem of the existence of symmetric dynamos.

In conventional kinematic dynamo theory one considers only Faraday's law,

$$\partial_t \mathbf{B} = \nabla \times (\mathbf{v} \times \mathbf{B}) - \nabla \times \eta \mathbf{j} \,, \tag{9.31}$$

$$\mathbf{j} = \nabla \times \mathbf{B} \,, \quad \nabla \cdot \mathbf{B} = 0 \,,$$

with the velocity field assumed to be given, which makes the problem linear in $\mathbf{B}$. The basic question concerns the existence of a flow $\mathbf{v}$ in a conducting fluid of resistivity η, such that eq. (9.31) has a solution $\mathbf{B}$, satisfying certain boundary conditions, which is asymptotically growing in time. A variant of this problem is the existence of a *stationary* dynamo solution $\mathbf{B}, \mathbf{v}$ following the integrated equation (9.31):

$$\mathbf{v} \times \mathbf{B} - \nabla \phi = \eta \mathbf{j} \,. \tag{9.32}$$

Here $\phi = \varphi - E_0 z$, $E_0 =$ applied toroidal electric field and the function $\varphi = \varphi(r, u)$ is periodic in $u = \theta + kz$.

It has been known since the work of Cowling (1934) that the dynamo problem (9.31) or (9.32) has no solution in plane or axisymmetry. (Cowling's theorem is also valid if η is a function of position with the same symmetry, but is no longer true if the resistivity is allowed to be anisotropic, $\eta_\parallel \neq \eta_\perp$; see Lortz, 1989.)

The nonexistence of dynamo solutions in plane and axisymmetry may lead to the impression that no symmetric dynamos exist at all. As discussed in section 3.2, the most general continuous symmetry is the helical one, of which plane and axisymmetry are special cases. It turns out that Cowling's theorem is *not* valid for general helical symmetry. In fact an explicit solution of eq. (9.32) has been given by Lortz (1968). In the present context of RFP configurations it is, however, not sufficient to consider the kinematic problem, i.e. to find some arbitrary helical fields $\mathbf{B}, \mathbf{v}, \phi$ satisfying eq. (9.32). Since flows should only sustain the configuration against resistive decay, they are expected to be slow,

$$\mathbf{v} = O(\eta) \,,$$

such that inertia terms in the equation of motion are negligible, i.e. **B** should satisfy the equilibrium equation. Restricting consideration to force-free states, the equation

$$\mathbf{j} = \mu \mathbf{B} \tag{9.33}$$

has to be added to eq. (9.32). In helical symmetry the condition $\mathbf{B} \cdot \nabla \mu = 0$, eq. (3.46), implies that μ depends only on the helical flux function ψ, $\mu = \mu(\psi)$. Using the components of **B** and **j** parallel to **h** in eqs (3.14), (3.17), eq. (9.33) yields the relation

$$\mathcal{L}\psi - \frac{2kl}{l^2 + k^2 r^2} f = \mu f \,, \tag{9.34}$$

while the terms perpendicular to **h** give

$$\mu = -\frac{df}{d\psi} \,. \tag{9.35}$$

Hence the equilibrium equation (9.33) assumes the form, eq. (3.21),

$$\mathcal{L}\psi = \frac{2kl}{l^2 + k^2 r^2} f - f\frac{df}{d\psi} \,. \tag{9.36}$$

The coefficient

$$\frac{2kl}{l^2 + k^2 r^2} \propto \mathbf{h} \cdot \nabla \times \mathbf{h} \,, \tag{9.37}$$

which vanishes in the cases of plane symmetry ($k = 0$) and axisymmetry ($l = 0$), describes the helical twist of the configuration. It allows closed flux surfaces ($\mathcal{L}\psi \neq 0$) also for vacuum fields $\mu = 0$, i.e. a helical stellarator.

Now consider Ohm's law (9.32). Multiplying by **B**,

$$\mathbf{E} \cdot \mathbf{B} = \eta \mathbf{j} \cdot \mathbf{B} = \eta \mu B^2 \,, \tag{9.38}$$

using $\mathbf{E} = E_0 \mathbf{e}_z - \nabla\varphi$ and averaging over flux surfaces one obtains the relation

$$E_0 \langle B_z \rangle_\psi = \mu(\psi) \langle \eta B^2 \rangle_\psi \,, \tag{9.39}$$

since $\langle \mathbf{B} \cdot \nabla\varphi \rangle_\psi = 0$. From the definition, eq. (3.16),

$$f = \mathbf{h} \cdot \mathbf{B}/h^2 = lB_z - krB_\theta \tag{9.40}$$

follows

$$\mu(\psi) = -\frac{df}{d\psi} = k\frac{d}{d\psi}\langle rB_\theta \rangle_\psi - l\frac{d}{d\psi}\langle B_z \rangle_\psi \,,$$

which, inserted into eq. (9.39), gives the equation for $\langle B_z \rangle_\psi$ (Finn et al., 1992):

$$\frac{d}{d\psi}\langle B_z \rangle_\psi = -a(\psi)\langle B_z \rangle_\psi + b(\psi) , \qquad (9.41)$$

$$a(\psi) = \frac{E_0}{l\langle \eta B^2 \rangle_\psi} , \quad b(\psi) = \frac{k}{l}\frac{d}{d\psi}\langle rB_\theta \rangle_\psi .$$

In the case of cylindrical symmetry $k = 0, l = 1, \psi = -A_z$, $b(\psi)$ vanishes, and eq. (9.41) implies that $\langle B_z \rangle_\psi$ cannot change sign, $\langle B_z \rangle_\psi \propto \exp\{-\int^\psi a\,d\psi'\}$, though it may become small at the plasma boundary owing to the paramagnetic pinch effect considered below. In the general case of helical symmetry $b(\psi) \neq 0$ provides a coupling between B_z and B_θ. Since in a steady RFP discharge B_θ is continuously supplied in order to keep the toroidal current I at a prescribed level, this coupling in principle allows B_z-reversal to be sustained against resistive decay. In order to determine quantitatively the helical ohmic state of pitch k for given applied field E_0, resistivity profile $\eta(r)$ and boundary values, one derives an equation for $f(\psi)$ from eq. (9.39). Using eqs (3.12), (3.14) one obtains

$$B^2 = \frac{1}{l^2 + k^2r^2}[f^2 + (\nabla\psi)^2] ,$$

while eq. (9.40) and the relation $\partial_r\psi = lB_\theta + krB_z$ allow us to express B_z in terms of ψ and f. Insertion into eq. (9.39) gives

$$\left[\left\langle \frac{\eta}{l^2 + k^2r^2}(\nabla\psi)^2 \right\rangle_\psi + f^2 \left\langle \frac{\eta}{l^2 + k^2r^2} \right\rangle_\psi\right]\frac{df}{d\psi}$$

$$+ E_0\left[l\left\langle \frac{1}{l^2 + k^2r^2} \right\rangle_\psi f + k\left\langle \frac{r}{l^2 + k^2r^2}\partial_r\psi \right\rangle_\psi\right] = 0 . \qquad (9.42)$$

The set of equations (9.36), (9.42) consisting of a partial and an ordinary differential equation, is called a generalized differential equation by Grad et al. (1975). Such equations are treated iteratively, the first step being to solve the equilibrium equation for $\psi^{(n+1)}(r, u)$ with $f^{(n)}(\psi)$ given from the n-th iteration step using numerical techniques as described in section 3.4. Subsequently $f^{(n)}$ is updated, $f^{(n)}(\psi) \rightarrow f^{(n+1)}(\psi)$, by solving the ordinary differential equation (9.42) using the flux-surface averages computed with the updated flux function $\psi^{n+1}(r, u)$. Solutions $\psi(r, u)$ and $f(\psi)$ obtained in this way are found to be essentially identical with the time-asymptotic states in dynamic simulations using the same E_0, η-profile and boundary conditions (Aydemir & Barnes, 1984; Dobrott et al., 1985). These numerical results, which are discussed in more detail in section 9.3.3, demonstrate that field-reversed states are in fact generated and sustained for periods exceeding the global resistive diffusion time.

Knowledge of the magnetic configuration satisfying eqs (9.36), (9.42) allows us to calculate the flow from the perpendicular component of Ohm's law (9.32),

$$\mathbf{v}_\perp = \mathbf{E} \times \mathbf{B}/B^2 , \tag{9.43}$$

by solving eq. (9.38) for φ on each flux surface

$$\mathbf{B} \cdot \nabla\varphi = E_0 B_z - \eta\mu(\psi)B^2 . \tag{9.44}$$

Since $\langle \mathbf{B} \cdot \nabla\varphi \rangle_\psi = 0$, the solvability condition for eq. (9.44) is $\langle E_0 B_z - \eta\mu B^2 \rangle_\psi = 0$, which is guaranteed by eq. (9.39) or (9.42). The RFP dynamo effect sustains the poloidal current j_θ against resistive decay, which is expressed by the equation

$$E_{\theta 0} = v_{r0}B_{z0} - \langle (\tilde{\mathbf{v}} \times \tilde{\mathbf{B}})_\theta \rangle + \eta j_{\theta 0} , \tag{9.45}$$

the brackets denoting the average over θ and z and $\tilde{\mathbf{v}} = \mathbf{v} - \mathbf{v}_0$ etc. Since in steady state $E_{\theta 0} = 0$, $j_{\theta 0}$ is maintained by two terms. The term $-\langle (\tilde{\mathbf{v}} \times \tilde{\mathbf{B}})_\theta \rangle$ is usually called the dynamo term (section 7.3.2). The term proportional to v_{r0} is also found to contribute to a negative dB_z/dr, viz. a weakening of B_z toward the boundary and a corresponding increase in the central region, though a cylindrically symmetric flow alone cannot lead to field reversal, as is clear from eq. (9.41) for $k = 0$. The predominantly inward velocity v_{r0} is therefore called *paramagnetic pinch velocity*. The paramagnetic pinch effect, which was first discussed by Bickerton (1958), indicates the general tendency of a pinch discharge toward a force-free state.

9.3.3 Numerical simulation of helical states

Dynamic numerical simulation is the most convenient approach to obtain helical RFP configurations (or three-dimensional turbulent states as discussed in section 9.4). As in section 9.3.1 computations start from an unstable cylindrical configuration, but interest is now in the asymptotic steady helical state, which is primarily determined by the stationary boundary conditions and the resistivity profile and is independent of the initial state. The boundary conditions usually chosen correspond to constant toroidal current, $B_\theta(a) = $ const., and constant toroidal flux $\psi_t = $ const. The resistivity profile is taken to be constant over the main central part of the plasma, increasing sharply toward the edge to simulate the effect of a highly resistive boundary layer. A particular helicity given by the parameter k is assumed and maintained, the choice of which is rather arbitrary, corresponding, for instance, to the most unstable $m=1$ mode of the initial state. The question which helicity corresponds to the most probable, i.e. lowest energy, state for the boundary conditions chosen, and whether such single helicity states exist at all, can be decided only by fully three-dimensional simulations. The value of single-helicity states is to show that the basic dynamo effect in the RFP may be explained qualitatively by quasi-static helical perturbations.

As an example we consider a recent numerical study by Finn et al. (1992), which uses the semi-implicit code developed by Schnack et al. (1986). Any dynamic variable $A(r, u)$ is Fourier analysed in $u = \theta - n_1 z/R$,

$$A(r, u) = \sum_{-M}^{M} A_m(r)e^{imu} ,$$

with $n_1 = 10$, $R/a = 5$, and the initial $q(r)$ given in eq. (9.30) with $q(0) = 0.08$. Hence the mode (1,10) is nonresonant $q(0) < m/n = 0.1$, such that the fast reconnection process shown in Fig. 9.8 does not occur. The Lundquist number is $S = 10^4$ and the pinch parameter eq. (9.20) is $\theta = 1.7$. The plasma pressure is assumed to be zero and the density to be homogeneous $\rho = 1$. Relatively modest spectral resolution, $M \simeq 5$, is sufficient to achieve good numerical accuracy.

The asymptotic stationary profiles $B_{z0}(r)$, $B_{\theta 0}(r)$, $\mu(r) = \mathbf{B} \cdot \mathbf{j}/B^2$ obtained are similar to those of the modified Bessel function model, Fig. 9.4. $B_{z0}(r)$ exhibits field reversal, $B_{z0} < 0$ for $r/a \gtrsim 0.8$. The helical perturbation is finite at the boundary $\tilde{B}_z(a) \neq 0$, but its amplitude is small such that $B_z(a, u) < 0$ throughout.

Figure 9.9 gives the individual contributions of eq. (9.45). A negative dynamo term $-\langle (\tilde{\mathbf{v}} \times \tilde{\mathbf{B}})_\theta \rangle$, which sustains a positive poloidal current, i.e. a negative gradient of B_{z0} since $j_{\theta 0} = -dB_{z0}/dr$, is observed for $r/a > 0.35$. The negative paramagnetic pinch term $v_{r0}B_{z0}$, sustaining positive $j_{\theta 0}$, is found only inside the field reversal point at $r_0 \simeq 0.7r$, changing sign where B_{z0} does. Note that $v_{r0} < 0$ over the entire plasma cross-section. Hence only the dynamo term due to the helical perturbations $\tilde{\mathbf{v}}, \tilde{\mathbf{B}}$ can sustain field reversal. The poloidal electric field vanishes $E_{\theta 0} \simeq 0$ indicating steady state conditions.

The helical flux distribution is similar to that displayed in the last frame of Fig. 9.8. The appearence of a secondary magnetic axis in the central region, corresponding to a shallow minimum of ψ, is typical for states with a sufficiently strong helical amplitude.

Finally we should point out a formal inconsistency of the approximation used in these, and in fact in most, computations of RFP dynamics. Since plasma compressibility plays an important role (the presence of the radial flow $v_{r0} < 0$ requires $\nabla \cdot \mathbf{v} \neq 0$), the density following the continuity equation would not remain homogeneous but would become more and more concentrated in the center. The problem is circumvented by assuming an anomalous particle transport process which on the resistive time scale ($v_r = O(\eta)$) decouples particle and magnetic field behavior.

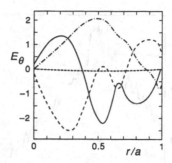

Fig. 9.9. Individual contribution to eq. (9.45): —— dynamo term $-\langle(\widetilde{\mathbf{v}} \times \widetilde{\mathbf{B}})_\theta\rangle$, - - - paramagnetic pinch term $v_{r0}B_{z0}$, – · – resistive term $\eta j_{\theta 0}$, ---- total poloidal electric field $E_{\theta 0}$ (from Finn et al., 1991).

9.4 Three-dimensional RFP dynamics

It has been shown in the preceding section that the dynamo effect in the RFP may be described by a stationary helical deformation of the plasma column, a helical ohmic state. Experimentally, however, such states are not observed. Instead RFP plasmas exhibit a strongly dynamic behavior characterized by magnetic fluctuations with a broad range of spatial scales. While these fluctuations are usually uncorrelated, giving rise to a quasi-stationary level of magnetic turbulence, more coherent oscillations reminiscent of sawtooth oscillations in tokamak plasmas may also occur.

The absence of steady helical states is a natural consequence of the close radial spacing of resonances. In a typical RFP device the aspect ratio is $R/a \sim 5$ and the central value of the safety factor is $q(0) \sim 0.1$. Hence there are many potentially unstable modes $(m,n) = (1,n)$ with resonant surfaces in the central part of the plasma, as indicated qualitatively in Fig. 9.5. The helicity of each of these modes may be singled out to produce a helical ohmic equilibrium, but this configuration will very probably be unstable with respect to neighboring helicities corresponding to somewhat smaller or larger n. As a result a broad spectrum of magnetic fluctuations should be excited, as is observed experimentally.

9.4.1 Turbulent RFP dynamo

Numerous simulation studies of 3-D quasi-stationary MHD turbulence in an RFP configuration have been performed (Sykes & Wesson, 1977; Aydemir & Barnes, 1984; Holmes et al., 1985; An et al., 1987; Nebel et al., 1989; Kusano & Sato, 1990; Finn et al., 1992). Boundary conditions in most of these computations are chosen as in the single-helicity case discussed in section 9.3.3, viz. $\theta = \text{const.}(\simeq 1.7)$ and $\psi_t = \text{const.}$

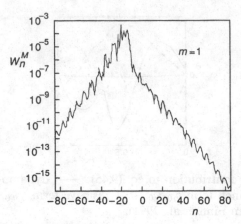

Fig. 9.10.   Spectrum of magnetic energies of modes $(1, n)$ (from Finn et al., 1991)

in most of these computations are chosen as in the single-helicity case discussed in section 9.3.3, viz. $\theta = $ const. $(\simeq 1.7)$ and $\psi_t = $ const.

While the average radial profiles, in particular $B_{z0}(r)$, are similar to those found for a single-helicity ohmic state, the mode spectrum is quite different, many values of n being excited for $m=1$, corresponding to a truly multi-helicity behavior; see Fig. 9.10 obtained in a typical simulation. Here n is defined such that resonance occurs for $q = -m/n$. Hence modes with $n \lesssim -10$ have their resonances inside the reversal radius r_0, those with sufficiently high positive n outside r_0, while the remaining modes are nonresonant. The nonlinear mode spectrum is, however, rather insensitive to these linear properties. The only conspicuous feature is the drop of the spectrum at $n \sim -10$, indicating that modes with q significantly above $q(0)$ are more difficult to excite. It should be noted that because of the logarithmic scale the width of the spectrum in Fig. 9.10 is exaggerated. In fact, the energy is contained in the band $-20 < n < -10$, while the wings of the spectrum displayed fall off exponentially. The magnetic Reynolds number is still quite low, $R_m = S v_0 / v_A \sim 10^2$, where $v_0 \sim 10^{-2} v_A$ is given by the average kinetic energy, $v_0 = (E^V)^{1/2}$. (Note that R_m and not the Lundquist number S characterizes a turbulent state.) Experimentally observed magnetic fluctuation spectra are significantly broader than those obtained in the simulations (Prager et al., 1990), presumably because of higher values of S, $S \sim 10^6$, and hence higher Reynolds numbers R_m.

For sufficiently high S small-scale MHD turbulence with the spectral properties discussed in chapter 7 is to be expected. Because of the Alfvén effect, however, small-scale fluctuations are aligned, such that their contribution to the dynamo term $\widetilde{\mathbf{v}} \times \widetilde{\mathbf{B}}$ is negligible. The main contribution comes from the large-scale fluctuations, which owing to the process of inverse magnetic helicity cascade (eq. (7.52)) are dominantly magnetic, $E_k^M \gg E_k^V$.

An important practical question concerns the S-scaling of the magnetic amplitude $\tilde{B}$, since multi-helical magnetic perturbations lead to destruction of magnetic surfaces and hence strongly deteriorated plasma confinement. Resistive diffusion of B_z decreases with increasing S, and hence one could expect the fluctuation amplitudes necessary to provide the dynamo effect to decrease, too. Consider Ohm's law (9.45) taken at the reversal point ($B_{z0} = 0$):

$$\eta j_{\theta 0} = \langle (\tilde{\mathbf{v}} \times \tilde{\mathbf{B}})_\theta \rangle . \tag{9.46}$$

If small-scale fluctuations with $\tilde{v} \sim \tilde{B}$ could provide the dominant contribution to the r.h.s., one would have the very favorable amplitude scaling $\tilde{B} = O(S^{-1/2})$. As mentioned above, however, these fluctuations, being Alfvén waves, do not contribute. A somewhat different scaling can be derived by starting from the expression (7.56), which describes the effect of MHD turbulence by a current diffusion term:

$$\langle (\tilde{\mathbf{v}} \times \tilde{\mathbf{B}})_\theta \rangle = (B_{0\theta}/B_0^2)\nabla \cdot D\nabla(\mathbf{j}_0 \cdot \mathbf{B}_0/B_0^2) . \tag{9.47}$$

Using a DIA-type approach (section 7.5.1), Strauss (1986) derived an expression for the hyperresistivity D generated by multi-helicity tearing mode turbulence, which scales as

$$D \sim \gamma \delta B^2 . \tag{9.48}$$

Here γ^{-1} is a typical nonlinear time. One now identifies γ with the linear tearing mode growth rate due to hyperresistivity $\eta_2 = D$ (eq. (4.90a)),

$$\gamma \sim D^{1/3} , \tag{9.49}$$

assuming that a tearing mode immersed in a turbulent bath does not exhibit the nonlinear growth rate reduction of an isolated tearing mode, the Rutherford regime (chapter 5.4.1). Combining eqs (9.48), (9.49) with eq. (9.47) at the reversal point $r \simeq r_0$, one obtains the amplitude scaling

$$\tilde{B} = O(S^{-1/3}) , \tag{9.50}$$

which is weaker than the naive estimate $\tilde{B} = O(S^{-1/2})$. An even weaker S-dependence is obtained by assuming that the dynamics of the field fluctuations, mainly different $m=1$ modes, is dominated by current sheet reconnection processes, $\tilde{v}/\tilde{B} = O(S^{-1/2})$, eq. (6.6). Insertion into eq. (9.46) gives the scaling law

$$\tilde{B} = O(S^{-1/4}) . \tag{9.51}$$

Neither RFP experiments nor numerical simulations have yet given a definite answer. Some simulation studies (An et al., 1987, Nebel et al.,

1989) find $\tilde{B}$ independent of S, while others (e.g. Merlin & Biskamp, 1988) obtain a scaling consistent with (9.51). The simulations are, however, limited to the rather narrow range $10^3 \lesssim S \lesssim 10^4$. In this range low-current experiments in the OHTE toroidal pinch (La Haye et al., 1984), where the radial profile of the magnetic fluctuations can be measured by probes, indicate the scaling $\tilde{B} \propto S^{-1/2}$ in the range $10^2 \le S \le 10^4$. In high current experiments, where only edge fluctuation are measured, no clear scaling law has been obtained to date.

For a small aspect ratio $A \sim 1$, where the radial spacing of $m=1$ resonances is more sparse, the dynamics is controlled by one or at most a few $m=1$ modes with different n. As a result the process of field reversal sustainment becomes more intermittent, giving rise to sawtooth-like oscillations (Kusano & Sato, 1990). Here phases of slow resistive diffusion leading to loss of field reversal, accompanied by an increase of magnetic energy, alternate with rapid relaxation caused by $m=1$ instability, where the field energy drops and field reversal is restored. Figure 9.11 shows the time history of $B_z(r)$ for (a) a small aspect ratio $A = 1.6$ exhibiting distinct relaxation oscillations, and (b) a larger aspect ratio $A = 4.8$, where the dynamics is more continuous. RFP sawtooth oscillations have also been observed experimentally (Prager et al., 1990).

9.4.2   *Quasi-linear theory of RFP dynamics*

In the discussion of dynamic MHD processes the question arises of how important nonlinear interactions (mode-coupling) are compared with the quasi-linear effect, the interaction of the fluctuations with the background configuration. In the case of RFP dynamics mode-coupling processes are represented by diagram A.

Diagram A: Typical mode-coupling processes.

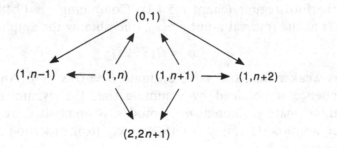

Consider two $m=1$ modes, $(1,n)$ and $(1, n + 1)$, with neighboring values of n, where mode interaction is expected to be strongest. Two types of coupling processes can be distinguished. Exciting the linearly stable $(2, 2n+1)$ mode implies the generation of smaller spatial scales, since such

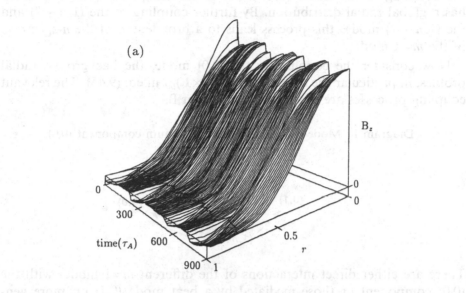

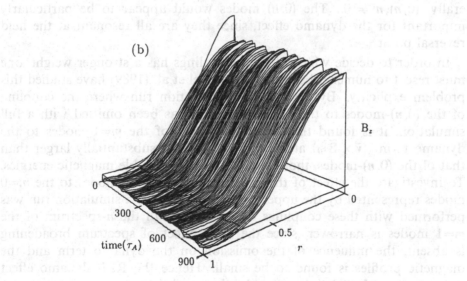

Fig. 9.11. Time history of the $B_z(r)$ profile for two 3-D simulation runs for different aspect ratio (a) $A = 1.6$, (b) $A = 4.8$ (from Kusano & Sato, 1990).

modes are rather strongly localized radially. This process corresponds to the direct turbulent energy cascade to small dissipative scales. By contrast the coupling to the stable $(0,1)$ mode is nondissipative since this mode has a global radial distribution. By further coupling to the $(1, n+2)$ and the $(1, n-1)$ modes this process leads to a broadening of the n-spectrum of the $m=1$ modes.

Now consider the coupling to the $(0,0)$ mode, the background radial profiles, in particular the dynamo term $\langle (\tilde{\mathbf{v}} \times \tilde{\mathbf{B}})_\theta \rangle$ in eq. (9.45). The relevant coupling processes are illustrated in diagram B.

Diagram B: Mode-coupling to the equilibrium component $(0,0)$.

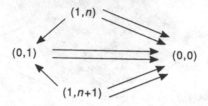

There are either direct interactions of the different $m=1$ modes with the $(0,0)$ component or those mediated by a beat mode $(0,1)$ or, more generally, $(0,n), n \neq 0$. The $(0,n)$ modes would appear to be particularly important for the dynamo effect, since they are all resonant at the field reversal point.

In order to decide which of these couplings has a stronger weight, one must resort to numerical simulations. Nebel et al. (1989) have studied this problem explicitly. By comparing a simulation run where the coupling of the $(0,n)$-modes to the $(0,0)$-component has been omitted with a full simulation, it is found that the contribution of the $m=1$ modes to the dynamo term $\langle (\tilde{\mathbf{v}} \times \tilde{\mathbf{B}})_\theta \rangle$ at the reversal point is substantially larger than that of the $(0,n)$-modes, though both have comparable magnetic energies. To investigate the effect of the couplings of the $m=1$ modes to the $m=0$ modes represented by the upper part in diagram A, a simulation run was performed with these couplings omitted. Though the n-spectrum of the $m=1$ modes is narrower, since the main effect of spectrum broadening is absent, the influence of the omission on the dynamo term and the magnetic profiles is found to be small. Hence the RFP dynamo effect seems to be rather well described by the quasi-linear approximation.

This has a direct consequence for the S-scaling of the fluctuating quantities. Consider the linearized form of Faraday's law for $\tilde{\mathbf{B}}$,

$$\partial_t \tilde{\mathbf{B}} - \nabla \times (\mathbf{v}_0 \times \tilde{\mathbf{B}})$$
$$= \nabla \times (\tilde{\mathbf{v}} \times \mathbf{B}_0) - \eta \nabla \times \nabla \times \tilde{\mathbf{B}} \simeq 0 , \tag{9.52}$$

since the dominant part of the perturbation $\widetilde{\mathbf{B}}$ is quasi-static such that the l.h.s. of eq. (9.52) is small. One thus obtains

$$\widetilde{v} \sim \eta \widetilde{B} \,. \tag{9.53}$$

From the $(0, 0)$-component of Ohm's law for steady state taken at the field reversal point, eq. (9.46),

$$\eta j_{\theta 0} = \langle (\widetilde{\mathbf{v}} \times \widetilde{\mathbf{B}})_{\theta} \rangle \,,$$

then follows

$$\widetilde{B} \propto \eta^0 \,, \quad \widetilde{v} \propto \eta \,, \tag{9.54}$$

i.e. the magnetic perturbation amplitude is independent of S, a result which agrees with numerical simulations in the range $10^2 < S \lesssim 10^4$ by Nebel (1989).

The scaling law, eq. (9.54), is the same as for a single-helicity ohmic equilibrium. Also the magnetic profiles, in particular the field reversal parameter F, are very similar to the single-helicity case for the same values of the current $I \propto B_{\theta}(a)$ and ψ_t. It thus appears that the RFP dynamo effect is determined mainly by the $m=1$ magnetic perturbation, either stationary single helicity $n=n_0$ or fluctuating multi-helicity with $\widetilde{B}$ distributed over a broad range of n.

Though strictly helical states are in general unstable when admitting neighboring helicities, it has been observed in numerical simulations that there may be almost helical, almost stationary configurations (Finn et al., 1992). In fact a bifurcation seems to occur between a single- and a multiple-helicity state. Thus two distinct time asymptotic states exist in three dimensions for the same values of I, ψ_t and S, one helical and time-independent, the other three-dimensional and fluctuating. Both are separated by a finite potential barrier. The larger the viscosity, the higher the probability that the system settles in the helical state.

10
Solar flares

Large solar flares are probably the most spectacular eruptive events in cosmical plasmas. Though rather weak in absolute magnitude compared for instance with the enormous energies set free in a supernova explosion, they outshine all other cosmic events for a terrestrial observer. According to the generally accepted picture, a flare constitutes a sudden release of magnetic energy stored in the corona and is therefore primarily an MHD process, though the various nonthermal channels of energy dissipation and deposition, which give rise to the richness of the observations, require a framework broader than MHD theory.

Since the major part of this book is concerned primarily with phenomena in laboratory plasmas, it seems to be convenient for the generally interested reader to find a somewhat broader introduction to this astrophysical topic. The engine driving the magnetic activity in the solar atmosphere is turbulent convection in the solar interior. Section 10.1 therefore gives an overview of our present understanding of the convection zone, in particular magnetoconvection. In section 10.2 we consider the solar atmosphere, its mean stratification, the process of magnetic flux emergence from the convection zone and the magnetic structures in the corona, in particular in active regions. In section 10.3 we then focus in on the MHD modelling of the flare phenomenon.

The presentation of this chapter is necessarily rather concise. The reader interested in more details of the various phenomena observed on the sun and their theoretical interpretation is referred to several good reviews, in particular the monograph by Priest (1984), the collection of review articles comprising the monograph edited by Sturrock et al. (1986), and, concerning more specifically the physics of solar flares, the monograph by Tandberg-Hanssen & Emslie (1988).

10.1 The solar convection zone

The existence of an extended convection zone is a characteristic property of relatively cool, low-mass stars such as the sun. While in the bulk of the stellar interior the energy generated in the thermonuclear hot core is transported diffusively by radiation (besides the direct energy loss by neutrinos), leading to hydrodynamically stable conditions, the radiative transport coefficient κ is reduced in the cooler outer regions where photons are reabsorbed more rapidly in ionization processes, which results in a steeper temperature gradient, since the heat flux $\kappa d T/dr = $ const. When dT/dr becomes superadiabatic,

$$-\frac{dT}{dr} > -(\gamma - 1)\frac{T}{\rho}\frac{d\rho}{dr},\tag{10.1}$$

convective instability sets in. Here $\rho(r)$ is the mass density and $\gamma = c_p/c_v$ the ratio of specific heats, and the ideal gas law $p = (R/\mu)\rho T$ is assumed (see section 2.1). Relation (10.1) can readily be understood. Consider a fluid element δV originally located at radius r being slightly displaced outward to $r + \delta r$. It is thus decompressed to the ambient pressure at the new radius $p(r + \delta r) = p(r) + p'\delta r$, $p' \equiv dp/dr < 0$. If the fluid element is large enough for the heat exchange to be negligible, which is satisfied in the solar interior for even relatively very small scales, then the change of state $\delta p, \delta\rho$ is adiabatic,

$$\frac{\delta\rho}{\rho} = \frac{1}{\gamma}\frac{\delta p}{p} = \frac{1}{\gamma}\frac{p'}{p}\delta r.\tag{10.2}$$

If the density of the displaced fluid element is smaller than the surrounding unperturbed density,

$$\rho + \delta\rho < \rho(r + \delta r) = \rho + \rho'\delta r,\tag{10.3}$$

there is a buoyancy force on the fluid element pushing it still further out, away from the original position, i.e. the fluid is convectively unstable. Inserting eq. (10.2) into eq. (10.3) gives

$$\frac{1}{\gamma}\left(\frac{p'}{p} + \frac{T'}{T}\right) < \frac{\rho'}{\rho},$$

which is identical with eq. (10.1).

The most conspicuous evidence of the solar convection zone is the granular structure (Fig. 10.1) of the photosphere, the "surface" of the sun. The basic cells, called granules, have an irregular shape with a typical

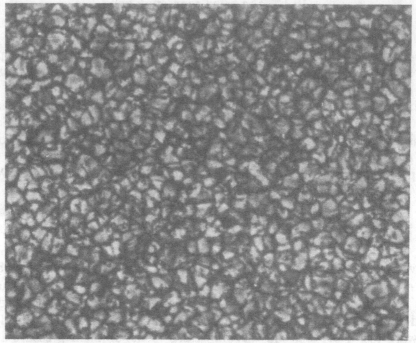

Fig. 10.1.   Solar granulation (courtesy of R. Müller).

diameter of 10^3 km. The central part of a cell, where plasma is rising, is brighter, hence hotter, than the cell boundaries, where the cooler plasma is falling back. The cell pattern changes in a period of about 10 minutes which is of the order of a cell turnover time. The granules are, however, not representative for the internal turbulent motions in the convection zone as we see in section 10.1.2, but are determined by the radiation transport properties in the outermost layer of less than 10^3 km. There is still some uncertainty about the depth of the solar convection zone. Earlier studies predicted a depth $d \simeq 1.5 \times 10^5$ km, i.e. 20 per cent of the solar radius. This estimate is based on a most probable value of $Z = 0.02$, where Z is the abundance of elements heavier than helium (Gough & Weiss, 1976). A larger value of Z would lead to a thicker convection zone because of the increased opacity in deeper layers. Measurements of solar oscillations (essentially global acoustic eigenmodes, the most prominent one being the 5 minute oscillation, see e.g. Brown et al., in Sturrock et al., 1986) now suggest a somewhat greater depth $d \simeq 2 \times 10^5$ km. The low surface abundance of lithium, depleted by a factor of 10^2 compared with its cosmic abundance, indicates that convective mixing reaches down to the level where lithium burning takes place, which would give a still greater depth of the zone.

10.1.1 Phenomenological description of thermal convection

The most important difference between the solar convection zone and terrestrial convectively unstable systems is the strong stratification in the former. Between the top and the bottom of the convection zone the pressure increases by a factor of $\sim 10^9$ and the density by $\sim 10^7$, hence the extent of the zone comprises many pressure and density scale heights $(d \ln p/dr)^{-1}$, $(d \ln \rho/dr)^{-1}$, which cannot be modelled adequately by laboratory experiments. Hence our understanding of convection zone physics rests mainly on numerical modelling.

The simplest phenomenological approach is based on the mixing length concept introduced by Prandtl (1925) and Taylor (1932). The method, which is the only one available for global stellar computations, has been described in detail by Gough & Weiss (1976). One decomposes the equation of motion (2.5) with $\mathbf{B} = 0$ but including the gravitational force, the continuity equation (2.12) and the energy equation, into equilibrium and fluctuating parts, $p = p_0 + \tilde{p}, \rho = \rho_0 + \tilde{\rho}, T = T_0 + \tilde{T}$ and $\mathbf{v} = \tilde{\mathbf{v}}$. With $\mathbf{g} = -g\mathbf{e}_z$ the equilibrium pressure balance is

$$\frac{dp_0}{dz} = -\rho_0 g \, , \tag{10.4}$$

which gives $p_0(z)$ in terms of $\rho_0(z)$. Since $p = (R/\mu)\rho T$, the equilibrium stratification is determined if $T_0(z)$ is known. Including heat diffusion the equation for the internal energy $p/(\gamma-1)$ can be written in the form (using the continuity equation (2.12) and the equation of state (2.14)):

$$(\partial_t + \mathbf{v} \cdot \nabla) \frac{p}{\gamma - 1} - \frac{\gamma(R/\mu)T}{\gamma - 1} (\partial_t + \mathbf{v} \cdot \nabla) \rho$$
$$= \rho(c_p/\mu) (\partial_t + \mathbf{v} \cdot \nabla) T - (\partial_t + \mathbf{v} \cdot \nabla) p = \nabla \cdot \kappa \nabla T \, . \tag{10.5}$$

We now introduce the *Boussinesq approximation*, consisting of two assumptions: (a) The spatial scales of the fluctuations are small compared with the equilibrium pressure or density scale heights; (b) the fluctuating velocities are small compared with the sound speed, i.e. the flow is incompressible, $\nabla \cdot \tilde{\mathbf{v}} = 0$. As a consequence horizontal pressure variations are negligible, $\tilde{p} = 0$, such that local density and temperature fluctuations follow the relation

$$\tilde{\rho}/\rho_0 = -\tilde{T}/T_0 \, . \tag{10.6}$$

Insertion into eq. (10.5), integration and averaging over horizontal fluctuations gives the equilibrium heat flux q:

$$q = F_c + F_r = \text{const.} \, ,$$

where F_c is the convective heat flow

$$F_c = (c_p \rho_0/\mu)\langle w\tilde{T}\rangle \,, \tag{10.7}$$

w denoting $\tilde{v}_z$, and F_r the diffusive heat flow due to radiative transport (heat conduction is negligible in the dense plasma of the convection zone),

$$F_r = -(c_p \rho_0/\mu)\chi d\,T_0/dz \,, \tag{10.8}$$

with $\chi = \kappa/(c_p\rho_0/\mu)$, the heat diffusivity. The problem is to express F_c in terms of the equilibrium quantities, in particular the superradiabatic temperature gradient

$$\beta_s = -\left[\frac{dT_0}{dz} - \left(\frac{\partial T}{\partial p}\right)_{ad}\frac{dp_0}{dz}\right]$$

$$= -\left[\frac{dT_0}{dz} - (\gamma-1)\frac{T_0}{\rho_0}\frac{d\rho_0}{dz}\right] \,, \tag{10.9}$$

$\beta_s > 0$ driving the convection; see eq. (10.1). The equations for the fluctuating quantities are

$$\rho_0\,(\partial_t + \tilde{\mathbf{v}}\cdot\nabla)\,\tilde{\mathbf{v}} = -\nabla\tilde{p} + \mathbf{g}\tilde{\rho} \simeq \mathbf{g}\tilde{\rho} \tag{10.10}$$

$$(\partial_t + \tilde{\mathbf{v}}\cdot\nabla)\,\tilde{\rho} = -w\,d\rho_0/dz \,, \tag{10.11}$$

$$(\partial_t + \tilde{\mathbf{v}}\cdot\nabla)\,\tilde{T} = -w\beta_s + \nabla\cdot\chi_0\nabla\tilde{T} + \nabla\cdot\tilde{\chi}\nabla T_0 \,, \tag{10.12}$$

neglecting the pressure fluctuation. For sufficiently weak fluctuations, the equations can be linearized, corresponding to the quasi-linear approximation. In practice, however, an even simpler, heuristic approach is followed. In a buoyantly turbulent fluid the vertical velocity w can be estimated by the kinetic energy a fluid element gains when being shifted upward by a distance l by the buoyancy force. From eq. (10.10) we find (omitting the subscripts on the equilibrium quantities)

$$\rho w^2 \simeq |\tilde{\rho}|gl \simeq |\tilde{T}|\,(\rho/T)\,gl \,, \tag{10.13}$$

using eq. (10.6). To obtain an estimate of $\tilde{T}$ one introduces the *mixing length* l as the average vertical distance a fluid element moves with a turbulence eddy until the eddy loses its identity by turbulent mixing. In the case of negligible heat diffusion eq. (10.12) leads to a temperature fluctuation

$$\tilde{T} \simeq \beta_s l \,, \tag{10.14}$$

while in the opposite case of strong heat diffusion, where the temperature gain by buoyant expansion is finally lost by diffusion, $w\beta_s \simeq \chi\partial_z^2\tilde{T}$, one has

$$\tilde{T} \simeq \beta_s w l^2/\chi \,. \tag{10.15}$$

We now insert the estimates (10.13)–(10.15) in the turbulent heat flux, eq. (10.7), noting that $\langle w\tilde{T}\rangle > 0$, since $\tilde{T} > 0$ for a rising fluid element, $w > 0$, and $\tilde{T} < 0$ for a falling one, $w < 0$. In the limit of negligible heat diffusion we find, up to factors of order unity,

$$F_c/(\rho\,c_p/\mu) \simeq S^{1/2}\chi\beta_s, \tag{10.16}$$

while in the opposite limit

$$F_c/(\rho\,c_p/\mu) \simeq S^2\chi\beta_s . \tag{10.17}$$

For intermediate conditions some interpolation between expressions (10.16) and (10.17) is used. S is a dimensionless quantity,

$$S = g\beta_s l^4/T\chi^2 . \tag{10.18}$$

Relating the mixing length to the pressure scale height h,

$$l = \alpha h , \tag{10.19}$$

S is proportional to the product of the (local) Rayleigh number Ra and the Prandtl number Pr,

$$S = \alpha^4 Ra \cdot Pr ,$$
$$Ra = g\beta_s h^4/T\chi\nu , \quad Pr = \nu/\chi , \tag{10.20}$$

where ν is the kinematic viscosity. The Rayleigh number is the ratio of the convective heat flux $\sim gh^3\beta_s/\nu$ and the diffusive heat flux $\sim \chi T/h$ and is hence a measure of the nonlinearity of the flow. Convective instability occurs if Ra exceeds a threshold value, $Ra_c \sim 10^3$ typically. The Prandtl number is very small in the solar convection zone, which justifies neglecting the viscosity in eq. (10.10). For a given Rayleigh number the Prandtl number is roughly inversely proportional to the Reynolds number Re of the convection flow. Only if Re is large enough is convection expected to be turbulent.

The essential free parameter is the normalized mixing length α, which has to be determined by comparing observations of surface properties with predictions from computations of the solar interior using the convective heat flux $F_c(S)$, $S = S(\alpha)$. One of the general results of such computations is that, rather independently of the particular choice of the function $F_c(S)$ and the value of α, convection is very efficient over most of the convection zone, making the temperature gradient practically adiabatic, except for a thin outer surface layer of thickness $\lesssim 10^3$ km, where direct radiation losses are important. The value of α is important to fit the properties of the solar interior to the observed surface luminosity for plausible element abundances. The resulting α is slightly above unity. This violates, at least formally, the condition of small fluctuation scales, i.e. $\alpha \ll 1$, inherent in the Boussinesq approximation, on which the mixing length concept is based.

10.1.2 *Compressible convection in a strongly stratified fluid*

The most promising tool for obtaining a better understanding of the physics in convectively unstable stratified systems is numerical simulations using the full compressible fluid equations. In spite of their limitations in Rayleigh number and Prandtl number ($Ra \lesssim 10^6$ and $Pr \gtrsim 10^{-1}$ compared with 10^{20} and 10^{-9}, respectively, in the sun) simulations have revealed interesting features characteristic of compressible convection. Two-dimensional computations (e.g. Graham, 1975; Hurlburt et al., 1984) show significant deviations from the properties assumed in the mixing length approach. Though for sufficiently high Ra (e.g. $Ra/Ra_c \sim 10^3$ used by Hurlburt) convective motions are strongly nonstationary, they are nevertheless found to span coherently the full extent of the unstable layer, covering many scale heights instead of breaking up into a series of eddies with vertical sizes of the order of the scale height. In addition, flows exhibit a pronounced up–down asymmetry, with broad regions of relatively slow upward motion and narrow regions of fast downflow. This asymmetry, which is absent in Boussinesq flows, is caused by the effect of horizontal pressure fluctuations. Since $\tilde{p} \propto \tilde{T}$, upflows with $\tilde{T} > 0$ are slowed down by the gradient of the local pressure enhancement (called buoyancy braking), while downflows are accelerated. As a consequence there is a negative (downward) mean flux of kinetic energy, while the heat (more precisely the enthalpy) flux is of course positive (upward).

The extended coherent convection cells seen in 2-D simulations are, however, not representative of 3-D turbulent convection. The reason is self-organization, i.e. the tendency to form large-scale coherent structures, which dominate the dynamics in 2-D, though this has rigorously been shown only for the incompressible case (inverse energy cascade, see section 7.3). Hence only 3-D simulations can give a reliable picture of turbulent convection. Such simulations have recently been performed by several groups. On the one hand these are "physics simulations" using idealized physical models and geometry (Toomre et al., 1990) characterized by constant transport coefficients χ, ν and simple boundary conditions at bottom and top (horizontally periodic boundary conditions are always chosen) in order to allow the largest possible Rayleigh and Reynolds numbers. Let us first consider the behavior close to the top. Hot material slowly rising in broad cells of irregular, nonstationary shape is turned around and cooled by the fixed boundary. The cool material flows down at a much higher velocity and is concentrated in narrow sheets along the cell boundaries, which form a (horizontally) interconnected downflow network isolating the upflow cells from each other. This apparent difference of connectivity between rising and falling flows is, however, restricted to the uppermost layer. Further down, the downflow sheets break up rapidly into filaments, thus restoring the topological symmetry between up- and downflows. With increasing depth the horizontal cellular scale increases, probably because of an increase of the vertical density scale height, and the downflow filaments tend to coalesce into fewer filaments of larger horizontal separation (Stein & Nordlund, 1989). In contrast to the coherent vertical convection cells observed in 2-D simulations, 3-D convection looks more random or turbulent, which is to be expected because of the greater freedom of the motion. Whether for much higher Rayleigh and Reynolds numbers turbulence reduces vertical coherence length to the scale height or even a fraction thereof is still an open question. It appears that the mixing length theory may after all not be completely wrong.

On the other hand there are "real-world simulations" of the upper layer of the convection zone including the photosphere, using the compressible MHD equations together with a sophisticated equation of state and radiative energy transport (see for instance Stein et al., 1989). One of the questions such simulations should answer is: What determines the size of the granules, the uppermost convection cells, conspicuously visible as a brightness modulation in the photosphere? Since high-resolution observations reveal that there are granules of arbitrarily small size (e.g. Müller, 1989), the question refers to the maximum granule size. In fact there is no directly visible indication of photospheric convection cells of radii r exceeding about 10^3 km. No simple quantitative explanation of this scale exists. We can, however, obtain a consistency condition relating r to the density scale height h_ρ. Assuming that the mass ascending along a circular cell of radius r with vertical velocity v_z is turned over within one scale height, leaving the cell with a horizontal velocity v_r, mass conservation gives

$$\pi r^2 \rho v_z \simeq 2\pi r h_\rho \rho v_r \,,$$

$$r \simeq 2 h_\rho v_r / v_z \,. \tag{10.21}$$

With the observed velocities $v_z \simeq 0.5$ km/s, $v_r \simeq 1.4$ km/s and the computed value of $h_\rho \simeq 200$ km close to the photosphere one obtains $r \simeq 10^3$ km. Simulations clearly show a preferred granule size, though their radius is somewhat larger than observed, probably owing to the presence of an unrealistically large numerical viscosity. In general the radius of the largest cells seems to scale with the scale height, much larger cells would break up into smaller ones. Simulations also indicate that the maximum cell size increases with depth, probably because the scale height increases. Overshoot effects of these larger internal convection cells into the uppermost layer may be responsible for the *supergranular* convection in the photosphere with cell diameters of about 3×10^4 km, which is not visible as a luminosity variation but can only be observed by velocity and magnetic field measurements (see section 10.2.2).

10.1.3   Solar magnetoconvection

It was noted in section 7.8 that in a turbulent, electrically conducting fluid a weak magnetic field will under certain conditions be amplified until the Lorentz force leads to saturation. Theoretical arguments and numerical simulations show that the magnetic field is concentrated in thin, rope-like structures. Though the total magnetic energy is small compared with the kinetic energy, resistive dissipation may be comparable with viscous dissipation. While the field seems to have only a minor effect on the average dynamics of the thermally driven turbulence in the convection zone, it completely dominates the properties of the dilute atmosphere when it emerges through the solar surface. Before considering the magnetohydrodynamics of the solar atmosphere, the proper focus of this chapter, it is useful to obtain a qualitative picture of magnetic field behavior in the convection zone. On the global scale of the sun this is intimately connected with the solar dynamo, responsible for the various magnetic phenomena characterizing the 22-year solar magnetic cycle, which are not addressed. (For an introduction to this field see for instance Priest, 1984, chapter 9.) Here we concentrate on the local magnetic properties, generalizing the results of section 7.8 to the case of stratification.

Magnetoconvection in stratified systems has been investigated by direct numerical simulations using the full compressible MHD equations. Despite their inherent limitations 2-D simulations already reveal interesting features. Consider an unstably stratified layer with a superimposed vertical magnetic field of constant total flux (Hurlburt & Toomre, 1988; Weiss, 1990). It is found that the field is rapidly swept aside and concentrated in sheets along the separatrices of the convection cells, the interiors of the latter carrying only weak fields. The phenomenon is called flux expulsion. An important effect, not encountered in the incompressible case, is a substantial density reduction in the sheets, where the magnetic pressure is usually significantly higher than the kinetic energy density ρv^2 and may become of the order of the thermal pressure.

Dynamo action, i.e. internal sustainment of magnetic field (as opposed to external sustainment by prescribed magnetic fluxes through the boundaries), can occur only in three-dimensional fluid motions. Three-dimensional simulations have been performed by Brandenburg et al. (1990) and Nordlund et al. (1991). The system considered by the latter authors consists of a lower stable layer (enforced by a higher heat diffusion coefficient) and an upper convectively unstable layer, similar to the solar convection zone above the stable radiative interior. The motivation is to include the effect of the overshooting convective motions penetrating into the stably stratified region beneath, which is believed to be important for the dynamo process. Dynamo action is enhanced by the Coriolis force $2\rho v \times \mathbf{\Omega}$ to be added to the r.h.s. of the equation of motion (2.5) in a rotating system.

The local dynamo properties are found to be very similar to those described in section 7.8.1 for homogeneous isotropic turbulence. Starting with a weak (horizontal) seed field immersed into fully developed thermally driven turbulence, the magnetic flux is rapidly concentrated in thin tubes or ropes. For sufficiently large magnetic Prandtl number, $Pr_m = v/\eta > O(1)$, the magnetic energy $E^M = \langle B^2 \rangle$ as well as the mean field $\langle B \rangle$ grow exponentially owing to a continuous random stretching, twisting and folding of the flux ropes, a process resembling the simple geometric dynamo model by Vainshtein & Zeldovich (1972). Saturation occurs when the Lorentz force is strong enough to affect the flow, which leads to a statistically stationary magnetic state. It is interesting to note that of the three components of the Lorentz force,

$$\mathbf{j} \times \mathbf{B} \equiv -\nabla B^2/2 + \mathbf{bb} \cdot (\mathbf{B} \cdot \nabla \mathbf{B}) + (\mathbf{I} - \mathbf{bb}) \cdot (\mathbf{B} \cdot \nabla \mathbf{B}) , \qquad (10.22)$$

i.e. pressure, tension and curvature forces, the last one gives the dominant contribution. The total magnetic energy is small, $E^M/E^V \ll 1$. The simulations show that this ratio is increasing with Pr_m, but it may become independent of Pr_m in the limit $Pr_m \gg 1$ (and sufficiently strong turbulence $Re \gg 1$).

Inhomogeneity of the dynamo process arises because of the stratification. Consider the change of the horizontally averaged magnetic energy

$$\partial_t \langle B^2/2 \rangle = -\partial_z \langle \mathbf{e}_z \cdot \mathbf{E} \times \mathbf{B} \rangle - \langle \mathbf{v} \cdot \mathbf{j} \times \mathbf{B} \rangle - \langle \eta j^2 \rangle , \qquad (10.23)$$

where the brackets denote the average over a plane z = const. and the z-direction is such that the gravitational acceleration is $-g\mathbf{e}_z$. This equation is obtained from the induction law by integrating by parts and using Ohm's law $\mathbf{E} = -\mathbf{v} \times \mathbf{B} + \eta\mathbf{j}$. In the stationary state the terms on the r.h.s. balance each other. In particular the second term, the work done by the fluid against the Lorentz force, $\langle \mathbf{v} \cdot \mathbf{j} \times \mathbf{B} \rangle < 0$, is found to balance approximately the resistive dissipation, which indicates the quasi-local character of the dynamo process. In the unstable layer the balance is slightly positive, $-\langle \mathbf{v} \cdot \mathbf{j} \times \mathbf{B} \rangle > \eta \langle j^2 \rangle$. The excess field is transported down by a net Poynting flux, $\langle \nabla \cdot (\mathbf{E} \times \mathbf{B}) \rangle = \partial_z \langle \mathbf{e}_z \cdot \mathbf{E} \times \mathbf{B} \rangle < 0$, and deposited in the stable layer by overshooting fluid motions, where it is dissipated resistively.

As a result the magnetic field energy $\langle B^2/2 \rangle$ increases downward, reaching a maximum just below the interface between stable and unstable regions.

This effective magnetic stratification is a consequence of the up–down asymmetry of the fluid motions characterized by slowly rising and rapidly descending flows. The strong downdrafts set into rotation by the Coriolis force pull flux ropes down, winding them up along a vortex tube. Though some of these are subsequently caught by updrafts, there remains a net downflow of magnetic flux piling up against the convectively stable region. These results indicate that most of the magnetic field generated in the solar convection zone is located at the bottom of the zone, an important effect for the global solar dynamo phenomenon.

10.2   Magnetic fields in the solar atmosphere

10.2.1   Structure of the solar atmosphere

The solar atmosphere consists of three main layers governed by different physical processes. The lowest one is the thin transparent surface layer of the dense sun of about 500 km, measured from the temperature minimum (see Fig. 10.2) down to the point where the optical depth becomes unity. The white continuum light (with superimposed absorption lines) is emitted from this layer. The overlying chromosphere of about 2000 km is the region of strong line emission. While the transition from the photosphere is gradual, the transition to the outer atmosphere, the corona, is very sharp, the density dropping and the temperature rising by two orders of magnitude. The emission from the corona is mainly SX continuum radiation, the thermal radiation corresponding to the coronal temperature $T > 10^6$ K. The origin of this rather unexpectedly high temperature is still not fully understood, several different mechanisms being discussed, in particular resonant Alfvén wave heating and coronal current sheet dissipation (see e.g. Hollweg, 1990; Goossens, 1991; Berger, 1991). The vertical structure of the solar atmosphere is mainly determined by the hydrostatic equilibrium equation

$$\frac{dp}{dz} = -g\,\rho \,, \tag{10.24}$$

the plasma equation of state for a dilute hydrogen plasma

$$p = p_e + p_i \simeq 2 n_e k_B\, T \,, \quad \rho \simeq n_e m_p \,, \tag{10.25}$$

and the temperature transport equation

$$\frac{d}{dz} \kappa \frac{dT}{dz} = P_r - P_h \,, \tag{10.26}$$

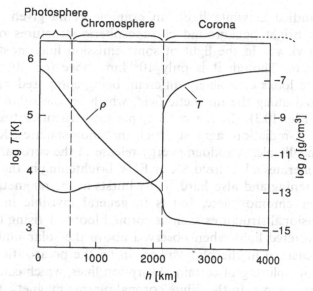

Fig. 10.2.   Mean density ρ and temperature T profiles in the quiet solar atmosphere as functions of height h above the solar surface.

where $\kappa \propto T^{5/2}$ is the electron heat conductivity (parallel to the magnetic field), P_r is the radiative loss and P_h the heating power. Note that in the dilute hot atmosphere heat conduction is very important, in contrast to the dense convection zone plasma. The sharp transition from the corona to the photosphere is a consequence of the strong temperature dependence of κ. Neglecting radiation and heating, $\kappa dT/dz =$ const. implies

$$T \propto (z - z_0)^{2/7} . \tag{10.27}$$

While the photosphere and lower chromosphere are rather uniformly stratified, the corona is highly nonuniform because of the strong influence of the magnetic field. In the dense photosphere the thermal pressure is much higher than the average magnetic pressure, $\beta = 2\langle p \rangle / \langle B^2 \rangle \gg 1$, but the opposite is true in the dilute corona, $\beta \ll 1$, since in spite of the higher temperature the thermal pressure is much lower. Hence in the photosphere the magnetic field is essentially moved around passively by the plasma flows, while in the upper chromosphere and the corona it determines the plasma dynamics.

Observational techniques have been enormously broadened and refined during the past twenty years. In fact it is fair to say that progress in solar physics was due mainly to the improvements in the observations. In order to visualize the processes connected with the flare phenomenon it might be helpful to have at least a rough idea of how physical conditions in the different parts of the solar atmosphere are observed. The photosphere is usually studied in white light, an example being given in Fig. 10.1. Against this bright background chromospheric structures only become visible when viewed in the light of some emission line, mostly Hα, such as in Fig. 10.6. Though it is only 10^3 km above the photosphere the chromosphere looks completely different, being composed of myriads of fibrils oriented along the magnetic field, which on the solar limb appear as spicules (Fig. 10.3). In the corona, plasma structures become visible using their SX-radiation, against which the photospheric SX intensity is completely negligible. A sudden energy release in the corona, for instance by a flare, generates a localized SX or EUV brightening in the corona, and an Hα-brightening and also hard X-ray bursts at the magnetic footpoints in the denser chromosphere, but is in general invisible in white light. Three-dimensional structures such as coronal loops or rising filaments are visible in scattered light when observed above the solar limb. Magnetic fields are measured in the photosphere using the polarization properties of the Zeeman splitting of certain absorption lines, which can provide full vector magnetograms. In the dilute coronal plasma magnetic fields cannot be measured, but may be inferred from bright coronal structures and (in principle) be computed from the photospheric values, assuming force-free equilibrium conditions.

10.2.2 Active regions

The surface of the sun is not homogeneous with constant luminosity. Apart from the small-scale fluctuations of the granulation which covers the entire surface, generating a uniform coarseness of the luminosity, there are the much stronger luminosity variations due to sunspots, which are visible even to the naked eye if viewed through a darkened glass. Sunspots appear in a latitude belt of $\pm 30°$ about the equator. During the solar 11-year optical cycle their number varies from a minimum with almost no spots to a maximum with about one hundred. Though their effect on the total solar luminosity is rather small (~ 0.3 per cent), they essentially determine the magnetic properties of the sun and therefore the plasma dynamics in the solar atmosphere. An *active region* is the bright region around a local group of sunspots with a diameter of about 2×10^5 km. Viewed in white light, this is an area of increased brightness called photospheric *faculae*, while in Hα it is called (chromospheric) *plage*, and further up in the corona it manifests itself by an X-ray enhancement. An active region is formed when magnetic flux emerges through the surface which is rapidly organized into a single flux tube giving rise to a pair of sunspots, the footpoint areas where the flux tube pierces the surface. Sunspots have typical diameters of 2×10^4 km and a central field of 0.3 T. The presence of the strong field suppresses the convective heat transport from the interior, hence the temperature is lower than in the surrounding photosphere, 3700 K as compared to 5700 K, which implies a luminosity ratio of 5 so that sunspots are in fact dark. The field intensity in a sunspot corresponds to the maximum magnetic pressure that can be supported by the photospheric plasma pressure. Relatively simple two-dimensional magnetostatic equilibrium models $\mathbf{B}(r, z)$, $p(r, z)$ can be constructed (e.g. Meyer et al., 1977) by neglecting the internal plasma pressure, which is small compared with the photospheric pressure $p_{in}(z) \ll p_{ex}(z)$. Since the temperatures are not much different, one has $\rho_{in}(z) \ll \rho_{ex}(z)$, i.e. the flux tube is almost evacuated. After their rapid formation sunspots gradually disintegrate (time scales varying between days and months), the so-called following spot vanishing more rapidly than the leading one such that an older active region is usually governed by one large sunspot. The flux fragments are carried along with the photospheric flows and concentrated along the supergranular cell boundaries and corners, forming a highly intermittent *network* of thin flux tubes of diameter ~ 200 km (probably even smaller) and field intensities $\simeq 0.15$ T, the maximum field intensity of a thin tube in the photosphere. The concentration of the field in the cell boundaries can easily be understood as a consequence of passive convection by supergranular flows which can be visualized by simple numerical simulations (e.g. Simon & Weiss, 1989). The final state of an active region is a simple bipolar magnetic region exhibiting no enhanced activity. In its neighborhood a new active region may appear.

Since sunspots form and disintegrate easily, it has been suggested (Parker, 1979b) that the thick flux tube does not reach very far down into the convection zone but divides into many individual loose flux strands. Such behavior is possible since the flux constituents in a sunspot do not seem to be strongly twisted around each other.

10.2.3 Magnetic buoyancy

Numerical simulations of magnetoconvection as discussed in section 10.1.3 show that the magnetic field is concentrated in thin flux ropes being whirled around by the turbulent fluid motions. The simulations do not, however, account for the organization of such ropes into the thick flux tubes associated with the sunspots. This process requires large-scale flows in the convection zone (in addition to the small-scale buoyant turbulence) which also determine the strictly regular field direction of the emerging flux. Sunspot pairs on one hemisphere have the same, mainly toroidal field orientation, while on the other hemisphere the orientation is just the opposite. After a sunspot minimum the polarities of the new sunspot pairs are reversed, such that the full magnetic solar cycle is 22 years, twice the optical cycle. These global magnetic features of the sun are modelled in the framework of mean field dynamo theory (for a recent review see e.g. Roberts & Soward, 1992).

A particular question concerns the process of flux emergence. The way flux tubes pop up through the surface is reminiscent of lighter objects rising in a heavier fluid owing to buoyancy. The concept of magnetic buoyancy has been introduced by Parker (1955). Since in a flux tube the gas pressure p_{in} is smaller than the ambient external pressure p_{ex}, while temperatures are about equal,

$$
\begin{aligned}
B^2/2 &= p_{ex} - p_{in} \\
&= (R/\mu)\,T\,(\rho_{ex} - \rho_{in}) \;,
\end{aligned}
\tag{10.28}
$$

there is a buoyancy force per unit volume,

$$
f_b = g\,(\rho_{ex} - \rho_{in}) \;.
\tag{10.29}
$$

Buoyant motion will be counteracted by magnetic tension and by the friction due to the small-scale turbulent eddies. The former is inversely proportional to the length L of the buoyantly rising flux element, $\mathbf{B}\cdot\nabla\mathbf{B} \sim B^2/L$, while the latter is proportional to the surface of the tube. Hence longer and thicker flux tubes rise faster. A quantitative estimate of the time scale for the rise of flux tubes from the bottom of the convection zone, where they are probably formed, is very difficult.

10.2.4   Magnetic structures in active regions

The magnetic field distribution in the corona differs fundamentally from that observed in the dense photosphere. In the photosphere the high plasma pressure compresses the magnetic flux into thin tubes distributed very intermittently over the surface. In the corona, where the plasma pressure is much smaller than in the photosphere, the magnetic flux expands filling the entire space (Fig. 10.3). Since β is small, $\beta \lesssim 10^{-2}$, the field intensity varies only slightly, even if the plasma is distributed very inhomogeneously, forming bright X-ray loops observed as the typical structures in the corona. Also, the distribution of the current density is expected to be very spotty, often concentrated in sheets, for instance at the boundaries between field lines originating from different photospheric flux tubes. The dissipation of such localized currents gives rise to plasma heating along a narrow bundle of field lines. As a result the equilibrium vertical pressure scale height h_p along the bundle is increased, since

$$
\frac{dp}{dz} = -g\rho
$$

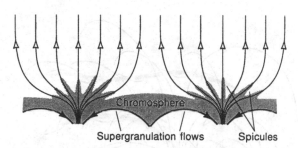

Fig. 10.3. Expansion of the thin photospheric flux tubes in the corona (schematic drawing).

implies

$$h_p = \left(\frac{d \ln p}{dz} \right)^{-1} \propto T \,, \tag{10.30}$$

which leads to an increase of the density flowing up from the photosphere. In fact the enhanced radiation from individual loops is due mainly to a local increase of density, not of temperature.

The topological field structure in an active region is very complex, forming high-rising loops filled with hot coronal plasma or low-lying fibrils and *filaments* filled with cooler chromospheric plasma. The latter are often covered by an *arcade* of loops. Viewed from above, the photospheric line-of-sight component of the magnetic field vanishes along the filament, which is therefore often (misleadingly) called a neutral line. Figure 10.4 illustrates schematically the field distribution in a sunspot region.

In active regions the fields differ substantially from potential fields, the state of lowest magnetic energy for a given photospheric field distribution. Since $\beta \ll 1$ in the corona, fields are usually assumed to be force-free, $\mathbf{j} \times \mathbf{B} = 0$. While most models considered to date are symmetric (plane or rotational), some interesting 3-D analytical models of solar magnetic configurations have recently been developed (see for instance Low & Lou, 1990; Low, 1991). Field lines are continuously moved around, sheared and twisted around each other, which increases the free, i.e. nonpotential field energy. In addition new flux may emerge leading to strong magnetic field gradients. As a result there is a firework display of relaxation events, which are called flares.

10.3 Solar flares

After this brief introduction to the generation of magnetic fields in the solar convection zone and their properties when emerging into the solar atmosphere we now turn our attention to the flare phenomenon. We

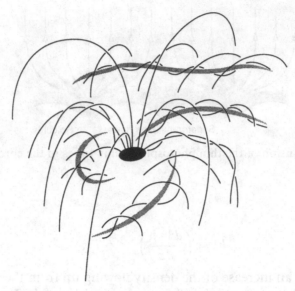

Fig. 10.4.   Artist's drawing of the magnetic field behavior in an active region around a sunspot. The lightly shaded regions are dense low-lying filaments, above which magnetic arcades tend to form.

first discuss the observational morphology of flares and present some popular pictorial models (section 10.3.1). In section 10.3.2 the problem of magnetic energy storage is addressed. Rigorous theoretical modelling has mainly been restricted to symmetric configurations, cylindrical models of coronal loops and two-dimensional arcades. The linear stability of loops, which differs from the toroidal case treated in chapter 4 because of photospheric line-tying, is considered in section 10.3.3. Section 10.3.4 deals with catastrophe theory of magnetic arcades, while finally in section 10.3.5 we discuss dynamical computations of plasmoid generation and eruption, which are essential features of a two-ribbon flare.

10.3.1   Phenomenology of flares

Flares result mainly from a sudden release of magnetic free energy. The maximum energy available is the difference between the energy of the actual configuration and that of the constant-μ, force-free state with the same photospheric flux distribution, i.e. the lowest energy state for a given value of the magnetic helicity as discussed in chapter 9.1. In contrast to the rather well-defined predisruption configuration in a tokamak plasma, solar magnetic fields, evolving due to the random photospheric motions, have very irregular shapes, where we can only distinguish certain characteristic elements as indicated in Fig. 10.4. A broad range of relaxation processes occur with energies between 1 and 5×10^{25} J for the largest flares and between 10^{22} and 10^{23} J for smaller ones. There is no lower energy limit. While very small flares (nanoflares with $\sim 10^{17}$ J) occuring at a continuous rate seem to be a major coronal heating source (Parker, 1988), we are interested here in the major sporadic events with say $> 10^{22}$ J. Different classification schemes of flare magnitudes are in use corresponding to the different observational windows, mainly X-ray and $H\alpha$ classifications; for details see Tandberg-Hanssen & Emslie (1988).

A large flare consists of typically three distinct phases:

(a) The preflare phase (minutes to hours) = flare precursor;

(b) The impulsive phase (seconds to minutes) = rapid coherent energy release by electron and ion acceleration, producing microwave radiation due to collective interactions in the corona and hard X-rays and γ-rays due to collisions in the chromosphere;

(c) The subsequent gradual phase (hours) consisting of a faster rise (flash phase) and a slower decay (main phase) = thermal energy release, producing enhanced optical ($H\alpha$) and SX emissions.

Roughly half the energy released appears in fast particles and half in the bulk plasma motion generated by the blast wave of the flare explosion.

Morphologically, two main types of flares are distinguished, the smaller *compact flares*, confined to a single loop, and the huge *two-ribbon flares*, extended over a larger region. Loosely speaking, compact flares are caused by an instability in the flux tube of a loop or also by interaction between two different loops, while a two-ribbon flare is usually associated with an erupting filament. The name derives from the characteristic two bright $H\alpha$ ribbons appearing in the flash phase, a feature of only minor importance in the flare energetics. Figure 10.5 gives a particularly instructive example, with two almost symmetric ribbons and a system of dark loops, the so-called post-flare loops, interconnecting the bright ribbons.

The standard conception of a two-ribbon flare, illustrated in Fig. 10.6, is due to a number of researchers, e.g. Hirayama (1974), Van Tend & Kuperus (1978), Pneuman (1980), Martens & Kuin (1989). Figure and figure caption are self-explanatory. However, this pictorial model does not explain the two crucial features of a two-ribbon flare:

(a) Why does the filament suddenly start to rise? In fact lifting the filament material against gravity and upward stretching of the arcade magnetic field primarily *consumes* energy which has to be supplied from outside. Furthermore the filament itself is not strictly two-dimensional but also anchored in the photosphere, such that its

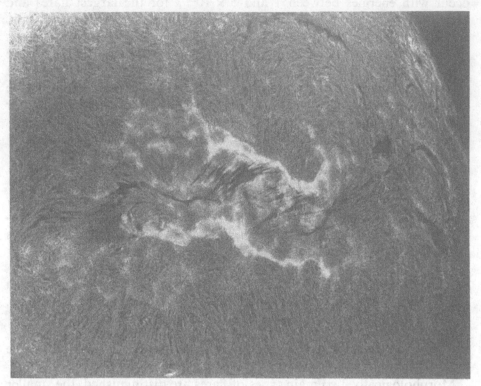

Fig. 10.5.   Two-ribbon flare observed in Hα (courtesy of H. Morishita).

upward motion implies the stretching of field lines along the filament. In fact single arcades do not seem to erupt, as is discussed in
section 10.3.4. The energy driving the eruption and along with it the
reconnection beneath must come from the surrounding field configuration not drawn in Fig. 10.6. As we will find, in a two-dimensional
model the process seems to be similar to the formation and acceleration of a plasmoid (see section 6.6.3) either by neighboring arcades or
along open field lines. In a real three-dimensional situation the magnetic configuration is presumably very complicated, but there may
be some *topological* resemblance with the two-dimensional models.

(b) A substantial fraction of the free magnetic energy is released, very
probably by reconnection, in the impulsive phase within minutes or
even only seconds. This is much shorter than any time scale depending, however weakly, on the collisional resistivity. Hence a truly fast,
i.e. η-independent, reconnection process is required, which has been,
and sometimes still is, regarded as the major problem in flare theory.
However, this is mainly a consequence of the idealized stationary

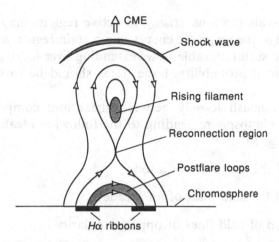

Fig. 10.6. Standard model of a two-ribbon flare. The rising filament produces a coronal mass ejection (CME). Reconnection of the arcade fields leads to the formation of the postflare loops. Energy dissipated in the reconnection region gives rise to Hα brightening (and other effects) in the dense chromospheric plasma at the field line feet.

models, to which reconnection theory was restricted previously, as we have seen in section 6.2. It now appears that the time scale problem is largely overemphasized. Fluids become turbulent if stirred rapidly enough, and a plasma may develop many forms of turbulence, MHD turbulence or current-driven micro-turbulence, which give rise to effective dissipation rates independent of the collisional dissipation coefficients.

10.3.2 Energy storage

The magnetic stress released in a flare may either have been generated in the convection zone and hence is already present when the field emerges from the photosphere, or is built up gradually in the corona due to the effect of photospheric motions at the field line feet. In the following we are mainly concerned with the latter form of magnetic energization. Photospheric velocities are small (~ 1 km/s) compared with typical coronal Alfvén velocities (~ 500 km/s), such that the configuration evolves quasi-statically. In a random velocity field motions increasing the magnetic energy are as probable as those decreasing it, so that on the average no energy build-up seems to be possible. The essential point, however, is the presence of a broad range of different spatial and temporal scales, large-scale motions having a longer correlation time than smaller ones. While the small-scale granular motions seem to have little effect, supergranular or even larger scale motions arising in active regions may give rise to a secular increase of the magnetic energy in a certain region which continues over days. Since such favorable flows extending over large areas and long times have a small probability, large flares should be rare events, as in fact they are.

One can distinguish loosely between shear and compressive surface motions in the photosphere, leading to the following idealized situations, illustrated in Fig. 10.7:

(a) Twisting of the field lines of a loop;

(b) Shearing of the field lines of an arcade;

(c) Compression of field lines of opposite polarity.

Let us derive the equations describing the sequence of magnetic equilibrium states for a given photospheric velocity pattern, restricting ourselves to 2-D force-free fields. First consider case (b), an arcade with plane symmetry sheared in the direction of symmetry. The configuration is determined by the flux function $\psi(x, z)$ obeying eq. (3.23)

$$\nabla^2 \psi = -B_y B_y' , \qquad (10.31)$$

where the axial field $B_y = B_y(\psi)$ is a flux function; for notations see Fig. 10.8. The plane $z = 0$ represents the photosphere. The photospheric flux distribution $\psi(x, z = 0)$ is assumed constant in time, corresponding to constant vertical photospheric field $B_z = -\partial_x \psi.$* The axial component B_y is determined by the axial displacement of the field line foot points. Integrating the field line equation (3.62),

$$\frac{dy}{B_y} = \frac{ds}{|\nabla \psi|} , \qquad (10.32)$$

between foot points, where ds is the line element along the poloidal magnetic field $\mathbf{B}_p = \mathbf{e}_y \times \nabla \psi$, gives the relation

$$\Delta y = B_y \int_{s_1}^{s_2} \frac{ds}{|\nabla \psi|} , \qquad (10.33)$$

where $s_{1,2} = s(x_{1,2})$. The r.h.s. is the differential flux $d\Phi/d\psi$, $\Phi(\psi)$ being the axial flux across the arcade area $\Sigma(\psi)$ bounded by the flux surface ψ. Equations (10.31) and (10.33) supplemented by lateral and upper boundary conditions for ψ determine $\psi(x, z)$ and $B_y(\psi)$ for given photospheric flux distribution $\psi(x, 0)$ and field line shear $\Delta y(x)$. (Note that the result is

---

* A compressive motion, case (c), or more generally a superposition of shearing and compressive motions is simply accounted for by a change of the photospheric flux distribution, but some care is required to distinguish between reconnected flux and flux submerged into the photosphere.

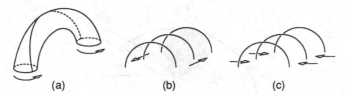

Fig. 10.7.   Activation of coronal magnetic configurations by photospheric flows: (a) twisting of a coronal loop, (b) arcade shearing, (c) arcade compression.

independent of the actual time evolution $v_y(x,t)$ but depends only on $\Delta y = \int v_y(x,t)dt$, the time being only a parameter.)

The set of coupled equations (10.31), (10.33) is called a generalized differential equation, which we have already encountered in section 9.3.2 and which in general requires a numerical treatment. Let us discuss a simple analytical solution, the periodic arcade configuration, eq. (3.52),

$$\psi(x,z) = e^{-z\sqrt{k^2-\mu^2}}\cos kx \,, \tag{10.34}$$

$$B_y(\psi) = -\mu\psi \,. \tag{10.35}$$

This configuration is generated from the potential field $\psi = e^{-kz}\cos kx$ by the foot point displacement

$$\Delta y = 2x\mu/\sqrt{k^2-\mu^2} \,, \tag{10.36}$$

as can be verified by inserting expressions (10.34), (10.35) into eq. (10.33). Asymptotically we have $\mu \to k$ for $\Delta y \to \infty$, i.e. B_y becomes constant, while the arcade scale height $(k^2-\mu^2)^{-1/2}$ increases. The energy of an arcade can easily be calculated

$$\tfrac{1}{2}\int B^2 dxdz = \frac{\pi k}{4\sqrt{k^2-\mu^2}} \propto \Delta y \quad \text{for} \quad \mu \to k \,. \tag{10.37}$$

Case (a) of a twisted loop can be treated like the case of a sheared arcade when ignoring the loop curvature. Instead of plane symmetry we now have an axisymmetric system, extending between the planes $z=0$ and $z=L$, the length of the loop or flux tube, which is twisted by axisymmetric flows in these planes. Equation (10.31) is replaced by eq. (3.22),

$$\Delta^*\psi = -ff' \,, \tag{10.38}$$

where the function $f(\psi) = rB_\theta$ is determined by the net twist of the field line between the foot points in the planes $z=0, L$,

$$\frac{rd\theta}{B_\theta} = \frac{ds}{B_p} = \frac{rds}{|\nabla\psi|} \,,$$

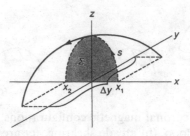

Fig. 10.8.   Magnetic field line of an arcade of cross-section Σ sheared by foot point displacement Δy.

$$\theta = \int_{s_1}^{s_2} \frac{B_\theta\, ds}{|\nabla\psi|} = f \int_{s_1}^{s_2} \frac{ds}{|\nabla\psi|r} \;, \tag{10.39}$$

where B_p and s are the projections in the rz plane along the flux tube, and $s_{1,2} = s(r_{1,2})$. θ can be called the rotational transform of the field, which for a cylindrical tube becomes the familiar expression (eq. (3.9))

$$\theta = \frac{L\,B_\theta}{r\,B_z} \;,$$

a crucial parameter for the stability of the tube.

10.3.3   Stability of twisted flux tubes

It is clear from the stability considerations given in chapter 4 that the twist of a flux tube is limited by the onset of a kink instability. Compared with the stability of a cylindrical pinch (sections 4.4, 4.5), where the perturbation is assumed to be periodic in axial direction corresponding to toroidal topology, a higher instability threshold is expected for coronal loops owing to the effect of photospheric line-tying. Because of its large inertia the dense photospheric plasma does not respond to the dynamics in the dilute corona, hence any perturbation of the loop vanishes at the foot points.   The ideal stability of a force-free straight flux tube has been investigated by Hood & Priest (1981), using the energy principle for line-tied displacements vanishing at $z = 0, L$,

$$\xi \propto e^{i(m\theta+kz)} \sin \pi z/L \;.$$

For the case of uniform twist (Gold & Hoyle, 1960),

$$B_\theta = \frac{rB_0/b}{1 + r^2/b^2} \;, \quad B_z = \frac{B_0}{1 + r^2/b^2} \;, \tag{10.40}$$

$$\theta = \frac{L}{b} \;,$$

one finds the threshold for the ideal $m=1$ mode

$$\theta \simeq 2.5\pi , \tag{10.41}$$

which is somewhat higher than the Kruskal-Shafranov limit $\theta = 2\pi$ or $q = 2\pi/\theta = 1$ in a topological torus. In the central part of the plasma column away from the line-tying boundaries the velocity of the unstable mode has essentially the structure of a helical kink mode, but the line-tying effect leads to an additional axial flow (see Mikic et al., 1990).

It is of course interesting to consider flux tube configurations more closely related to coronal loops twisted by photospheric motions. Mikic et al. (1990) have computed such configurations by applying radially localized twisting flows to a cylindrical column with initially uniform plasma density embedded in a uniform axial field $\mathbf{B} = B_0\mathbf{e}_z$. The flow profile at $z = 0$ is chosen to be (flowing in the opposite sense at $z = L$)

$$v_\theta(r) = \begin{cases} v_0\dfrac{r}{r_0}(1 - r^2/r_0^2)^2 & r \leq r_0 \\ 0 & r > r_0 , \end{cases} \tag{10.42}$$

leading to a twist profile:

$$\theta(r) = \int_0^t dt\, v_\theta(r)/r = \theta_0(1 - r^2/r_0^2)^2 . \tag{10.43}$$

Since the twist and hence B_θ vanish for $r > r_0$, the total axial current is zero. The axial current density $j_z(r)$ is peaked in the center and reverses sign further out, vanishing for $r > r_0$. Away from the axial boundaries $z = 0, L$ the central part of the plasma contracts slightly, leading to an increase of the plasma density, while it expands in the outer region, remaining of course unchanged for $r > r_0$. The effect is due to the repelling force between the regions of antiparallel currents. Since the initial homogeneous axial field is frozen in at $z = 0, L$, boundary layers are generated close to these surfaces, while the main part of the plasma column away from these axial inhomogeneities is nearly cylindrically symmetric. Note that the magnetic configuration generated by such twisting motions differs both from a tokamak configuration, because the total current vanishes, and from a reversed-field pinch configuration, because the current density reverses sign, but not the axial field.

Stability analysis of the sequence of equilibria generated in this way, which are characterized by different values of the central twist $\theta(0) = \theta_0$, gives the threshold for the ideal $m = 1$ mode

$$\theta_0 \simeq 4.8\pi , \tag{10.44}$$

which is significantly higher than the threshold $\theta_0 = 2\pi$, i.e. $q_0 = 1$ for the ideal kink mode in the cylindrical tokamak. It is interesting to note that the average twist corresponding to the central value $\theta_0 = 4.8\pi$ is

$$\langle\theta\rangle = \theta_0 \int_0^{r_0} (1 - r^2/r_0^2)^2 dr/r_0 = 2.56 \,, \tag{10.45}$$

which is close to the critical value for a uniform twist profile eq. (10.41) as pointed out by Hood (1991). This fact may be significant since even near threshold the mode is distributed over the entire twisted column $r \lesssim r_0$, the $q = 1$ radius being $r_s \simeq 0.6r_0$. Inclusion of finite resistivity can, however, lower the threshold, possibly to $\theta_0 \simeq 2.5\pi$.

The nonlinear evolution of the instability has not yet been studied. Two types of behavior are in principle possible: (a) There is a neighboring quasi-helical equilibrium. The slowly twisted system will show an equilibrium bifurcation at $\theta_0 \simeq 4.8\pi$ with a helical amplitude increasing as the twisting continues, similar to the behavior of the reversed-field pinch at a value of the pinch parameter $\simeq 1.5$ (section 9.1). In this case reaching the critical twist $\theta_0 \simeq 4.8\pi$ in a coronal loop would not result in any observable effects.   (b) The nonlinear instability evolution removes the $q = 1$ surface flattening the twist profile, like the Kadomtsev picture of the sawtooth collapse (see section 8.1.2). This process would be associated with magnetic reconnection and hence requires finite resistivity.

The problem of explaining the sudden onset of a compact flare in a coronal loop is certainly as difficult as understanding the sawtooth oscillation in a tokamak. Similar to toroidal curvature in a tokamak plasma the curvature of a coronal loop, not accounted for in the work by Mikic et al., may play an important role in determining the stability and nonlinear properties. In addition to $m=1$ higher-m tearing modes may become unstable, their interaction leading to processes similar to the major disruption in tokamaks (Spicer, 1977). Because of the rather different current profiles and the photospheric line-tying effect the analogy between tokamak plasmas and coronal loops is probably not very close. The main observed phenomena are rather different in both cases. Due to the inherently good thermal confinement in a tokamak a disruption manifests itself primarily as a sudden *loss* of *thermal energy* deposited somewhere outside the plasma column, such that the in situ deposition of the relatively small fraction of free magnetic energy has practically no observable effect. By contrast a coronal loop has much worse inherent plasma energy confinement because of the heat contact with the cool chromosphere. Hence the free energy in a flare is mainly magnetic energy which is both thermalized in the loop plasma leading to a local brightening, i.e. a *gain* of thermal energy, and transformed into fast-particle energy, dissipated in the chromosphere.

10.3.4 *Current sheet formation and catastrophe theory in a sheared arcade*

It is conceptually more convincing to associate an eruptive event with a loss of equilibrium or a catastrophe in the equilibrium evolution rather than the onset of a linear instability, which in many cases corresponds only to an equilibrium bifurcation. For the two-ribbon flare this concept has been followed by numerous investigators. The model of a two-dimensional arcade energized by photospheric shear flows in the direction of symmetry is particularly convenient. Since shear motions appear to increase the axial field $B_y(\psi)$, the simplest approach is a one-parameter model $B_y = \alpha f(\psi)$, where $f(\psi)$ is some "reasonable", fixed profile function and $\alpha = \alpha(t)$ describes the slow equilibrium evolution. Equation (10.31) assumes the form

$$\nabla^2 \psi = \alpha^2 F(\psi) \,, \tag{10.46}$$

which can readily be solved using the techniques described in section 3.4. It is found that for rather general classes of functions $F(\psi)$ there is some value $\alpha = \alpha_c$, beyond which no solutions satisfying the prescribed boundary conditions exist. The free energy W of the configuration has two branches, a lower and a higher branch, as indicated in Fig. 10.9a. Increasing α from the unsheared ($\alpha = 0$) potential field ($W = 0$) on the lower branch, it is tempting to associate the critical value α_c with the onset of a rapid dynamical process. Of course equilibrium theory cannot predict the extent of such a process, which may be slow or insignificant. But more importantly monotonic shearing does *not* lead to a monotonic increase of $B_y(\psi)$, i.e. of α. Jockers (1978) has shown for a simple function $F(\psi)$ that the high energy branch of $W(\alpha)$ corresponds to larger field line shear than the lower branch, both branches joining smoothly at $\alpha = \alpha_c$, which suggests that on reaching α_c continued shearing will drive the solution on the upper branch with W increasing but B_y decreasing. This has been verified by solving the full equilibrium equations (10.31), (10.33) (e.g. Zwingmann, 1987) and by dynamical simulation (e.g. Biskamp & Welter, 1989b). The arcade behavior can in fact be discussed analytically for rather general conditions (Aly, 1990). For notational convenience assume stationary photospheric flows, $\Delta y \propto t$. Two different phases can be distinguished:

(a) In the initial phase the axial field increases, while $\psi(x, y)$ and hence the arcade cross-section $\Sigma(\psi)$ do not change,

$$B_y(\psi) \propto t \,, \ \Sigma(\psi) \simeq \text{const.} \,, \\ W \propto t^2 \,; \tag{10.47}$$

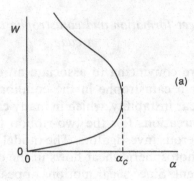

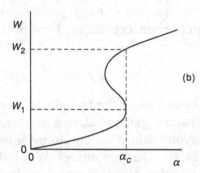

Fig. 10.9. Critical behavior in the equilibrium evolution characterized by a monotonically increasing (or decreasing) parameter α. W is some relevant equilibrium quantity not necessarily the energy. (a) loss of equilibrium for $\alpha > \alpha_c$, (b) catastrophe = jump of W at $\alpha = \alpha_c$.

(b) In the subsequent asymptotic phase the arcade expands, while the field B_y decreases such that the increase of the energy is very slow,

$$B_y(\psi) \propto t^{-1} , \quad \Sigma(\psi) \propto t^2 ,$$
$$W \propto \log t . \tag{10.48}$$

The process continues smoothly to arbitrarily large foot point displacements. Hence shearing an arcade does not lead to a critical point which can be identified with the onset of an eruptive event.

The theory has been generalized by Zwingmann (1987) to include finite plasma pressure p, replacing eq. (10.31) by

$$\nabla^2 \psi = -\partial_\psi [(B_y^2(\psi)/2 + p(\psi, \phi_g))] . \tag{10.49}$$

As in the force-free case, $B_y(\psi)$ is determined by the field line shear, eq. (10.33). The pressure is a function of both the flux function ψ and the gravitational potential ϕ_g, which for an isothermal loop can be written in the form

$$p = \alpha_p \, p_0(\psi) \, e^{-z/h} \,, \tag{10.50}$$

where h is the pressure scale height and α_p a parameter measuring the coronal plasma β, which depends on the local heating and the thermal coupling to the photosphere. For an unsheared arcade, $B_y = 0$, equation (10.49) has essentially the form of eq. (10.46). Fixing $p_0(\psi)$ and h in eq. (10.50) and changing α_p, Zwingmann finds a catastrophe behavior, Fig. 10.9b. The critical value α_{pc}, however, corresponds to unrealistically high coronal β. In addition the system evolution must proceed in the direction of decreasing α_p to allow a drop in energy which would be released. It is interesting to note that for a sufficiently strongly sheared arcade field the catastrophe behavior (the S in the equilibrium curve $W(\alpha_p)$) vanishes, which underlines the robust smoothness of the evolution of a sheared arcade.

It appears that an important feature in the equilibrium evolution toward a catastrophe is the appearance of a current sheet, indicating the transition to a different topology of the magnetic configuration. Parker (1983) has argued that in a general three-dimensional coronal configuration smooth photospheric flows will give rise to current sheets between flux tubes in the corona; this appears rather plausible, though there are certain reservations regarding the generality of the process (see e.g. Van Ballegooijen, 1988; Zweibel & Li, 1987). Let us again restrict consideration to two-dimensional arcade configurations sheared in the axial direction, where the situation is much clearer than in the general three-dimensional case. Following Zwingmann et al. (1985) it can easily be shown that in a force-free configuration subject to shearing a regular neutral point as in Fig. 10.10a does not exist, which is a consequence of eq. (10.33). In the vicinity of a neutral point (x_0, z_0), where

$$\psi - \psi_s = \tfrac{1}{2} [a(x - x_0)^2 - b(z - z_0)^2] \,, \tag{10.51}$$

the line integral taken for instance along the surface B, B' diverges for $\psi \to \psi_s$ (see eq. (3.33)),

$$\int_B^{B'} \frac{ds}{|\nabla \psi|} \propto \ln(\psi - \psi_s) \,,$$

while $\Delta y_{BB'}$ has a finite limit $\Delta_s = \Delta y(\psi_s)$. From eq. (10.33) it follows that

$$B_y(\psi) \propto \Delta_s / \ln(\psi - \psi_s) \,. \tag{10.52}$$

Hence the r.h.s. of the equilibrium equation (10.31) diverges for $\psi \to \psi_s$,

$$B_y \, B_y' \propto \frac{\Delta_s}{(\psi - \psi_s) \, [\ln(\psi - \psi_s)]^3} \,, \tag{10.53}$$

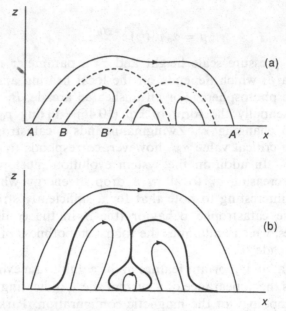

Fig. 10.10.   Impossibility of a normal *X*-point in a sheared arcade, (a). Instead a cusp-like behavior is expected together with a singular sheet current along the separatrix, (b) (from Vekstein & Priest, 1992).

while the l.h.s. is finite,

$$\nabla^2 \psi = a - b \tag{10.54}$$

using eq. (10.51), which implies that an *X*-point configuration cannot be in equilibrium. Hence shearing produces a qualitative change of the *X*-point geometry, much as does a non-vanishing electric field in Syrovatskii's theory (section 6.2.3) or finite reconnecting flows in Cowley's theory (section 6.4.1). It has been shown that in general even in the absence of a neutral point current sheets are generated along the entire separatrix and that the configuration at the end points of the individual separatrix branches has a cusp-like behavior (Fig. 10.10b) (Low & Wolfson, 1988; Vekstein et al., 1990).

An interesting analytical model giving rise to a catastrophe behavior has been proposed by Forbes & Isenberg (1991). The configuration consists of a filament of radius *r* carrying a current *I* which is suspended in an arcade at a height $z = h$ above the photosphere. Instead of foot point shearing the model implies compressive horizontal photospheric motions (Fig. 10.7c) and reconnection of the photospheric flux at the point below the filament. As a result the filament current and height increases. At a certain point in the evolution a vertical current sheet appears below the filament. Shortly afterwards a catastrophe occurs, which gives rise to a sudden rise of the filament to some greater height, a crucial feature of a two-ribbon flare. However, the model is highly idealized and hence of more aesthetic than practical value, since the photospheric plasma flows are singular, all coronal currents are assumed to flow in the filament, whose radius r is assumed to be constant, and the catastrophe occurs only for an extremely thin filament, $r/h < 10^{-3}$. For a more diffuse filament the S in the equilibrium curve vanishes, which indicates that the catastrophe mechanism is a somewhat artificial effect. It should be emphasized, however, that compressive motions, which bring fields of opposite polarity closer together, are more likely to generate an explosive situation than pure shearing motions, which have the opposite tendency because of the arcade expansion.

10.3.5 *Dynamical models of plasmoid generation and eruption*

Instead of solving the equilibrium problem given by the generalized differential equations (10.31) and (10.33) or (10.38) and (10.39) it is often more convenient to consider the evolution of *quasi*-equilibrium states by solving the full (compressible) dynamic equations subject to motions in the photospheric boundary which are slow compared with the internal relaxation time $\sim \tau_A$. The advantages of the direct simulation approach are:

(a) Flexibility. Dissipative effects, above all finite resistivity, can easily be incorporated;

(b) Physical realizability. The states considered are stable (within the spatial dimensions considered). If instability arises it can be followed dynamically;

(c) Numerical simplicity. The semi-implicit scheme, in particular in the simple form proposed by Harned & Schnack (1987), greatly reduces the computer time requirements.

First the evolution of single arcades has been considered, either laterally restricted by vertical walls or freely expanding, with various photospheric flux distributions and shear flow profiles (Mikic et al., 1988; Biskamp & Welter, 1989b). In all cases arcades are found to grow quasi-statically without exhibiting any fast dynamics, in agreement with the results of equilibrium theory discussed in the previous section.[†]

[†] Single arcade eruption, however, may occur in spherical geometry. Numerical simulations show that a large-scale magnetic arcade sheared by differential solar rotation erupts, modelling a coronal mass ejection (CME) (Steinolfson, 1991). A CME is a sudden inflation and bursting of

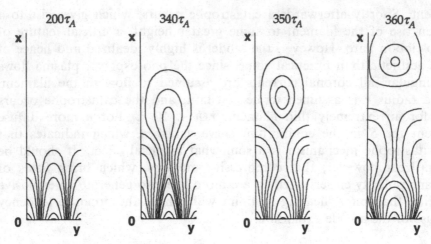

Fig. 10.11. Rapid plasmoid formation in a (laterally restricted) pair of sheared arcades (Mikic et al., 1988).

However, if several interacting arcades are energized simultaneously by shearing, a rapid dynamic reconnection process may set in at a certain arcade height leading to the formation of a plasmoid (Fig. 10.11), which is ejected upward modelling the filament eruption, an essential feature of a two-ribbon flare. Let us therefore look at these simulations more closely. Mikic et al. consider a somewhat artificial configuration consisting of two arcades out of the periodic initial arcade system $\psi(x, y, t = 0) =$ $e^{-\pi z} \cos \pi x$, by restricting the computational system to the interval $-1 \leq$ $x \leq 1$ with periodic boundary conditions. By applying a shear flow $v_y(x) =$ $v_0 \sin 2\pi x$, both arcades are energized in a symmetric way. Symmetry is broken only by numerical discreteness effects, which gives rise to a sudden growth and expansion of one arcade over the other one. This rapid *ideal* dynamical process vanishes if the arcades are sheared in a more realistic nonsymmetric way, where the larger arcade gradually leans over the smaller one. However, the fast *resistive* process of plasmoid formation and ejection in the larger arcade, driven by the lateral expansion of the smaller one, seems to occur under rather general shearing conditions, where only the threshold height of the arcades depends on the shear flow profile (Biskamp and Welter, 1989b). For large heights, the arcades may even break down into a sequence of several plasmoids.

---

a magnetic helmet streamer configuration covering a quiescent filament or prominence, which is also ejected but does not seem to drive the CME eruption. The energy involved in a CME is huge, often exceeding that of the largest flares, but it usually occurs outside active regions in higher solar latitudes and has little relation to a typical active region flare.

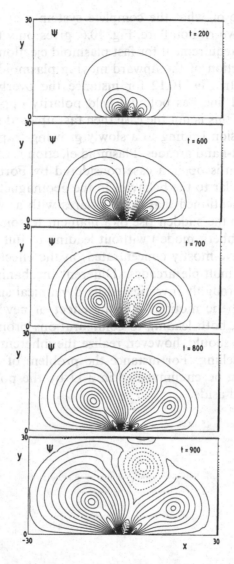

Fig. 10.12. Plasmoid formation and ejection in a triple of sheared arcades (from Biskamp & Welter, 1989b).

Plasmoid ejection only occurs if the interaction between the arcades is strong enough. In an isolated pair of arcades the partners may avoid such interaction by leaning away from each other as they grow in size, even if the foot points are close together. Only a triple of arcades, i.e. a sufficiently packed and complex configuration, leads to fast plasmoid ejection (Fig. 10.12), which is in fact very similar to the behavior of a laterally restricted pair of arcades shown in Fig. 10.11. The configuration gives an example of what the complete system may look like, of which the sketch of a two-ribbon flare, Fig. 10.6, gives only the central part.

An important requirement for fast plasmoid ejection is the possibility of efficient reconnection of the upward-moving plasmoid with the overlying magnetic field. In Fig. 10.12 for instance the overlying field indicated only by one field line has poloidal field polarity opposite to that of the plasmoid field. If this is not the case then fast upward motion is prevented by field compression leading to a slowly growing trapped plasmoid as in the right- and left-hand arcades. Plasmoid ejection is always possible if the field configuration is open, a case considered by Forbes & Priest (1983), which is very similar to the behavior in the geomagnetic tail (Fig. 6.24). It should also be mentioned that weaker flares with a two-ribbon signature may occur due to the sudden rise of a filament to some greater height (as in the Forbes-Isenberg model) without leading to full filament eruption.

Simulations have mostly concentrated on the effect of field line shearing in simple or multiple arcades. Different mechanisms for two-ribbon flares not yet thoroughly investigated by dynamical simulations are compressive photospheric motions and the effect of newly emerging flux of opposite polarity, both leading to configurations prone to onset of rapid reconnection. We should, however, realize the inherent limitations of two-dimensional modelling. For instance the problem of the compression of overlying flux can be circumvented in 3-D by the possibility of pushing the overlying field aside.

Outlook

In conclusion it seems appropriate to attempt a general assessment of where we are in understanding magnetohydrodynamic processes. The coherent nonlinear evolution of unstable configurations seems to be quite well understood, particularly in the case of spatial symmetry. We have a rather precise picture of the nature of resistive magnetic reconnection, the basic dissipative mechanism in nonlinear MHD. It has become clear that the problem of too-slow reconnection rates has been due mainly to the restriction to stationary models, which in reality become unstable at sufficiently large Reynolds number and ultimately turbulent, allowing reconnection rates essentially independent of the collisional resistivity. It should also be noted that in many hot plasmas reconnection rates are enhanced due to noncollisional processes, which tend to be more effective than resistivity.

We have a good comprehension of MHD turbulence, in particular in two dimensions. Knowledge is still restricted for fully three-dimensional nonperiodic dynamical systems, such as the turbulent phase in disruptive processes. Further progress will come mainly from numerical modelling. Since 'realistic' simulations including both the global geometry, which accounts for the free energy of the system, and small-scale dynamics responsible for dissipation, will not be feasible in the near future, the problem has to be split into two.

The focus will be on global simulations, since the strongest interest is in understanding particular phenomena such as major disruptions or two-ribbon flares. Since the possibility of rapid turbulent dissipation seems to be guaranteed, one should be less worried about the missing fine-scale resolution, but use a physically reasonable and numerically convenient subgrid model. These models can in principle be provided by studying appropriate small-scale systems such as homogeneous turbulence. Expectations should not be too high, however. It has turned out that the interface between such idealized small-scale systems and global systems is

349

a complex subject, and that turbulence studies tend rather to concentrate on the intrinsic properties of turbulence, primarily the different facets of intermittency, which is a fascinating and gratifying topic in its own right.

A further subject of major interest concerns the onset conditions, for instance, for the sawtooth disruption. How can energy be stored efficiently without being released prematurely, and if it is finally released, what is the trigger mechanism? Depending on the phenomenon, it may be necessary to include non-MHD effects such as diamagnetic effects in the sawtooth process or the onset of anomalous resistivity in the flare. It is also interesting to obtain a better understanding of the transition from the coherent precursor to the turbulent dissipative part of a disruptive event. Any modelling should of course be guided by experimental and observational results. However, there are also disadvantages caused by the steadily growing quantity of data of steadily improved resolution. With all these details each event differs from the others, so that it becomes increasingly difficult to define the "typical" event which the theory is intended to describe.

References

References added in the paperback edition are at the end of the reference list.

Abramowitz, M. & Stegun, I.A. (1965). *Handbook of Mathematical Functions* (Dover Publications, New York).

An, Z.G., Diamond, P.H., Hahm, T.S., Lee, G.S., Scott, B.D., Carreras, B.A., Garcia, L., Holmes, J.A. & Lynch, V.E. (1987). Energy confinement and nonlinear tearing mode dynamics in a high current reversed field pinch, *Proceedings of the Eleventh Conference on Plasma Physics and Controlled Nuclear Fusion Research* (IAEA, Vienna), vol. II, pp. 663–72.

Anselmet, F., Gagne, Y., Hopfinger, E.J. & Antonia, R.A. (1984). High-order velocity structure functions in turbulent shear flows, *J. Fluid Mech.* **140**, 63–89.

Appert, K., Gruber, R. & Vaclavik, J. (1974). Continuous spectra of a cylindrical magnetohydrodynamic equilibrium, *Phys. Fluids* **17**, 1471–2.

Ara, G., Basu, B., Coppi, B., Laval, G., Rosenbluth, M.N. & Waddell, B.V. (1978). Magnetic reconnection and $m = 1$ oscillation in current carrying plasmas, *Ann. Phys.* **112**, 443–76.

Aydemir, A.Y. (1990). MHD modes driven by anomalous electron viscosity and their role in fast sawtooth crashes, *Phys. Fluids* **B2**, 2135–42.

Aydemir, A.Y. & Barnes, D.C. (1984). Sustained self-reversal in the reversed-field pinch, *Phys. Rev. Lett.* **52**, 930–3.

Aydemir, A.Y., Wiley, J.C. & Ross, D.W. (1989). Toroidal studies of sawtooth oscillations in tokamaks, *Phys. Fluids* **B1**, 774–87.

Bacry, E., Arneodo, A., Frisch, U., Gagne, Y. & Hopfinger, E. (1991). Wavelet analysis of fully developed turbulence data and measurement of scaling exponents, in *Turbulence and Coherent Structures*, ed. O. Métais and M. Lesieur (Kluwer, Dordrecht), pp. 203–15.

Barsukov, A.G., Kovrov, P.E., Kulygin, V.M., et al. (1983). Investigation of plasma confinement and injection heating in the T-11 tokamak, in *Proceedings of the Ninth Conference on Plasma Physics and Controlled Nuclear Fusion Research* (IAEA, Vienna), vol. I, pp. 83–94.

351

Barston, E.M. (1969). Stability of the resistive sheet pinch, *Phys. Fluids* **12**, 2162–74.

Batchelor, G.K. (1950). On the spontaneous magnetic field in a conducting liquid in turbulent motion, *Proc. Roy. Soc. London* **A 201**, 405–16.

Batchelor, G.K. (1969). Computation of the energy spectrum in homogeneous two-dimensional turbulence, *Phys. Fluids* **12**, II 233–9.

Bateman, G. (1980). *MHD Instablities* (MIT Press, Cambridge, Mass.).

Baty, H., Luciani, J.F. & Bussac, M.N. (1992). Asymmetric reconnection and stochasticity in tokamaks, *Nucl. Fusion* **32**, 1217–23.

Bauer, F., Betancourt, O. & Garabedian, P. (1978). *A Computational Method in Plasma Physics* (Springer, New York).

Bennett, W.H. (1934). Magnetically self-focussing streams, *Phys. Rev.* **45**, 890–7.

Benzi, R., Biferale, L., Paladin, G., Vulpiani, A. & Vergassola, M. (1991). Multifractality in the statistics of the velocity gradients in turbulence, *Phys. Rev. Lett.* **67**, 2299–3002.

Benzi, R., Paladin, G., Parisi, G. & Vulpiani, A. (1984). On the multifractal nature of fully developed turbulence and chaotic systems, *J. Phys.* **A 17**, 3521–31.

Benzi, R., Patarnello, S. & Santangelo, P. (1988). Self-similar coherent structures in two-dimensional decaying turbulence, *J. Phys.* **A 21**, 1221–37.

Berger, M.A. (1991). Magnetic heating of the solar corona, *Advances in Solar System Magnetohydrodynamics*, ed. E. Priest and A. Hood (Cambridge University Press, Cambridge), pp. 241–56.

Berger, M.A. & Field, G.B. (1984). The topological properties of magnetic helicity, *J. Fluid Mech.* **147**, 133–48.

Bernstein, I.B., Frieman, E.A., Kruskal, M.D. & Kulsrud, R.M. (1958). An energy principle for hydromagnetic stability theory, *Proc. Roy. Soc. London* **A 244**, 17–40.

Bertin, G. (1982). Effects of local current gradients on magnetic reconnection, *Phys. Rev.* **25A**, 1786–9.

Bhattacharjee, A. & Hameiri, E. (1986). Self-consistent dynamo-like activity in turbulent plasmas, *Phys. Rev. Lett.* **57**, 206–9.

Bickerton, R.J. (1958). The amplification of a magnetic field by a high current discharge, *Proc. Phys. Soc.* **72**, 618–24.

Birn, J. & Hones, Jr., E.W. (1981). Three-dimensional computer modeling of dynamic reconnection in the geomagnetic tail, *J. Geophys. Res.* **86**, 6802–8.

Biskamp, D. (1979). Nonlinear quenching of diamagnetic and gyroviscous effects in tearing modes, *Nucl. Fusion* **19**, 777–83.

Biskamp, D. (1982). Dynamics of a resistive sheet pinch, *Z. Naturforsch.* **37a**, 840–7.

Biskamp, D. (1984). Anomalous resistivity and viscosity due to small scale magnetic turbulence, *Plasma Phys. Contr. Fusion* **26**, 311–19.

Biskamp, D. (1986a). Magnetic reconnection via current sheets, *Phys. Fluids* **29**, 1520–31.

Biskamp, D. (1986b). Natural current profiles in tokamaks, *Comments Plasma Phys. Controlled Fusion* **10**, 165–72.

Biskamp, D. (1991). Algebraic nonlinear growth of the resistive kink instability, *Phys. Fluids* **B3**, 3353–6.

Biskamp, D. & Welter, H. (1977). Numerical studies of resistive instabilities, in *Proceedings of the Sixth Conference on Plasma Physics and Controlled Nuclear Fusion Research* (IAEA, Vienna), vol. I, pp. 579–89.

Biskamp, D. & Welter, H. (1980). Coalescence of magnetic islands, *Phys. Rev. Lett.* **44**, 1069–72.

Biskamp, D. & Welter, H. (1983). Studies of MHD turbulence in low-β current-driven plasmas, *Proceedings of the Ninth Conference on Plasma Physics and Controlled Nuclear Research* (IAEA, Vienna), vol. III, pp. 373–82.

Biskamp, D. & Welter, H. (1989a). Dynamics of decaying two-dimensional magnetohydrodynamic turbulence, *Phys. Fluids* **B1**, 1964–79.

Biskamp, D. & Welter, H. (1989b). Magnetic arcade evolution and instability, *Solar Phys.* **120**, 49–77.

Biskamp, D. & Welter, H. (1990). Magnetic field amplification and saturation in two-dimensional magnetohydrodynamic turbulence, *Phys. Fluids* **B2**, 1787–93.

Biskamp, D., Welter, H. & Walter, M. (1990). Statistical properties of two-dimensional magnetohydrodynamic turbulence, *Phys. Fluids* **B2**, 3024–31.

Bodin, H.A.B. & Newton, A.A. (1980). Reversed-field pinch research, review paper, *Nucl. Fusion* **20**, 1255–324.

Bondeson, A., Parker, R.D., Hugon, M. & Smeulders, P. (1991). MHD modelling of density limit disruptions in tokamaks, *Nucl. Fusion* **31**, 1695–716.

Boozer, A.H. (1986). Ohm's law for mean magnetic fields, *J. Plasma Phys.* **35**, 133–9.

Brachet, M.E., Meneguzzi, M., Politano, H. & Sulem, P.L. (1988). The dynamics of freely decaying two-dimensional turbulence, *J. Fluid Mech.* **194**, 333–49.

Braginskii, S.I. (1965). Transport processes in a plasma, *Reviews of Plasma Physics*, ed. M.A. Leontovich (Consultants Bureau, New York), vol. 1, pp. 205–311.

Brandenburg, A., Nordlund, A., Pulkkinen, P., Stein, R.F. & Tuominen, I. (1990). 3-D simulation of turbulent cyclonic magneto-convection, *Astron. Astrophys.* **232**, 277–91.

Bulanov, S.V., Sakai, J. & Syrovatskii, S.I. (1979). Tearing mode instability in approximately steady MHD configurations, *Sov. J. Plasma Phys.* **5**, 157–63.

Buneman, O. (1969). A compact non-iterative Poisson-solver, SUIPR report No. 294, Stanford University.

Burlage, L.F. & Turner, J.M. (1976). Microscale Alfvén waves in the solar wind at 1 AU, *J. Geophys. Res.* **81**, 73–7.

Burrell, K.H. et al. (1983). Attainment of reactor level volume-averaged toroidal beta in Doublet III, *Nucl. Fusion* **23**, 536–40.

Burrell, K.H. et al. (1989). Confinement physics of *H*-mode discharges in DIII–D, *Plasma Phys. Contr. Fusion* **31**, 1649–64.

Bussac, M.N., Pellat, R., Edery, D. & Soulé, J.L. (1975). Internal kink modes in a toroidal plasma with circular cross section, *Phys. Rev. Lett.* **35**, 1638–41.

Camargo, S.J. & Tasso, H. (1992). Renormalization group in magnetohydrodynamic turbulence, *Phys. Fluids* **B4**, 1199–212.

Campbell, D.J. et al. (1986). Sawtooth activity in ohmically heated JET plasmas, *Nucl. Fusion* **26**, 1085–92.

Campbell, D.J. et al. (1987). Sawteeth and disruptions in JET, *Proceedings of the Eleventh Conference on Plasma Physics and Controlled Nuclear Fusion Research* (IAEA, Vienna), vol. I, pp. 433–45.

Campbell, D.J. et al. (1989). Sawtooth activity and current density profiles in JET, *Proceedings of the Twelfth Conference on Plasma Physics and Controlled Nuclear Fusion Research* (IAEA, Vienna), vol. I, pp. 377–85.

Caramana, E.J., Nebel, R.A. & Schnack, D.D. (1983). Nonlinear, single-helicity magnetic reconnection in the reversed-field pinch, *Phys. Fluids* **26**, 1305–19.

Carreras, B., Waddell, B.V. & Hicks, H.R. (1979a). Poloidal magnetic field fluctuation, *Nucl. Fusion* **19**, 1423–30.

Carreras, B., Hicks, H.R. & Waddell, B.V. (1979b). Tearing-mode activity for hollow current profiles, *Nucl. Fusion* **19**, 583–96.

Carreras, B., Hicks, H.R., Holmes, J.A. & Waddell, B.V. (1980). Nonlinear coupling of tearing modes with self-consistent resistivity evolution in tokamaks, *Phys. Fluids* **23**, 1811–26.

Carreras, B., Hicks, H.R. & Lee, D.K. (1981a). Effect of toroidal coupling on the stability of tearing modes, *Phys. Fluids* **24**, 66–77.

Carreras, B., Rosenbluth, M.N. & Hicks, H.R. (1981b). Nonlinear destabilization of tearing modes, *Phys. Rev. Lett.* **46**, 1131–4.

Charlton, L.A., Holmes, J.A., Hicks, H.R., Lynch, V.E. & Carreras, B. (1986). Numerical calculations using the full MHD equations in toroidal geometry, *J. Comput. Phys.* **63**, 107–29.

Chen, S., Doolen, G.D., Kraichnan, R.H. & She, Z.S. (1993). On statistical correlations between velocity increments and locally averaged dissipation in homogeneous turbulence, *Phys. Fluids* **A5**, 458–63.

Chodura, R. & Schlüter, A. (1981). A 3-D Code for MHD equilibrium and stability, *J. Comput. Phys.* **41**, 68–88.

Coppi, B., Greene, J.M. & Johnson, J.L. (1966). Resistive instabilities in a diffusive pinch, *Nucl. Fusion* **6**, 101–17.

Cowley, S.W.H. (1975). Magnetic field line reconnection in a highly-conducting incompressible fluid: properties of the diffusion region, *J. Plasma Phys.* **14**, 475–90.

Cowling, T.G. (1934). The magnetic field of sunspots. *Monthly Notices Roy. Astron. Soc.* **94**, 39–48.

Degtyarev, L.M. & Drozdov, V.V. (1985). An inverse variable technique in the MHD-equilibrium problem, *Comput. Phys. Rep.* **2**, 341–87.

Denton, R.E., Drake, J.F., Kleva, R.G. & Boyd, D.A. (1986). Skin currents and compound sawteeth in tokamaks, *Phys. Rev. Lett.* **56**, 2477–80.

Denton, R.E., Drake, J.F. & Kleva, R.G. (1987). The $m = 1$ convection cell and sawteeth in tokamaks, *Phys. Fluids* **30**, 1448–51.

Diamond, P.H. & Craddock, G.G. (1990). On the Alfvén effect in MHD turbulence, *Comments Plasma Phys. Controlled Fusion* **13**, 287–97.

Diamond, P.H., Hazeltine, R.D., An, Z.G., Carreras, B.A. & Hicks, H.R. (1984). Theory of anomalous tearing mode growth and the major disruption, *Phys. Fluids* **27**, 1449–62.

DiMarco, J.N. (1983). The ZT-40M experiment, in *Mirror-Based and Field-Reversed Approaches to Magnetic Fusion*, ed. R.F. Post, D.E. Baldwin, D.D. Ryutov & H.A.B. Bodin (International School of Plasma Physics, Varenna, Italy) vol. II, pp. 681–703.

Dnestrovskij, Yu.N., Kostomarov, D.P., Pedorenko, A.V. & Popov, A.M. (1987). Bifurcation of the equilibrium of a plasma with a free boundary, *Sov. J. Plasma Phys.* **13**, 683–8.

Dnestrovskij, Yu.N., Kostomarov, D.P., Popov, A.M. & Shagirov, E.A. (1983). Nonlinear evolution of helical modes in a tokamak with low q, *Proceedings of the XI European Conference on Controlled Fusion and Plasma Physics*, vol. II, pp. 173–6.

Dnestrovskij, Yu.N., Kostomarov, D.P., Popov, A.M. & Shagirov, E.A. (1985). Nonlinear growth of external modes at small q in a tokamak, *Sov. J. Plasma Phys.* **11**, 616–21.

Dobrott, D., Barnes, D.C., Mikić, Z. & Schnack, D.D. (1985). Reversed-field pinch model with self-consistent pressures, fields and flows, *Bull. Am. Phys. Soc.* **30**, 1399.

Dobrowolny, M., Mangeney, A. & Veltri, P. (1980). Fully developed anisotropic hydromagnetic turbulence in interplanetary space, *Phys. Rev. Lett.* **45**, 144–7.

Doyle, E.J., Groebner, R.J., Burrell, K.U., Gohil, P., Lehecka, T., Luhmann, N.C., Matsumoto, H., Osborne, T.H., Peebles, W.A. & Philipona, R. (1991). Modifications in turbulence and edge electric fields at the *L-H* transition in the DIII-D tokamak, *Phys. Fluids* **B3**, 2300–7.

Drake, J.F. & Kleva, R.G. (1984). Are vacuum bubbles a cause of major disruptions in tokamaks?, *Phys. Rev. Lett.* **53**, 1465–8.

Dreicer, H. (1959). Electron and ion runaway in a fully ionized gas, *Phys. Rev.* **115**, 242–9.

Dubois, M.A., Pecquet, A.L. & Reverdin, C. (1983). Internal disruptions in the TFR tokamak: A phenomenological analysis, *Nucl. Fusion* **23**, 147–62.

Edenstrasser, J.W. (1980a). Unified treatment of symmetric MHD equilibria, *J. Plasma Phys.* **24**, 299–313.

Edenstrasser, J.W. (1980b). The only three classes of symmetric MHD equilibria, *J. Plasma Phys.* **24**, 515–18.

Edwards, A.W. et al. (1986). Rapid collapse of a plasma sawtooth oscillation in the JET tokamak, *Phys. Rev. Lett.* **57**, 210–13.

Elsässer, W.M. (1950). The hydromagnetic equations, *Phys. Rev.* **79**, 183.

Fadeev, V.M., Kvartskhava, I.F. & Komarov, N.N. (1965). Self-focusing of local plasma currents, *Nucl. Fusion* **5**, 202–9.

Feller, W. (1966). *An Introduction to Probability Theory and its Applications* (J. Wiley, New York), vol. II.

Fielding, S.J., Hugill, J., McCracken, G.M., Paul, J.W.M., Prentice, R. & Stott, P.E. (1977). High-density discharges with gettered torus walls in DITE, *Nucl. Fusion* **17**, 1382–5.

Finn, J.M. (1975). The destruction of magnetic surfaces in tokamaks by current perturbations, *Nucl. Fusion* **15**, 845–54.

Finn, J.M. & Antonsen, T.M. (1985). Magnetic helicity: what is it and what is it good for?, *Comments Plasma Phys. Controlled Fusion* **26**, 111–26.

Finn, J.M. & Kaw, P.K. (1977). Coalescence instability of magnetic islands, *Phys. Fluids* **20**, 72–8.

Finn, J.M., Nebel, R.A. & Bathke, C.G. (1992). Single and multiple helicity ohmic states in reversed-field pinches, *Phys. Fluids* **B4**, 1262–79.

Forbes, T.G. & Isenberg, P.A. (1991). A catastrophe mechanism for coronal mass ejections, *Astrophys. J.* **373**, 294–307.

Forbes, T.G. & Priest, E.R. (1983). A numerical experiment relevant to line-tied reconnection in two-ribbon flares, *Solar Phys.* **84**, 169–88.

Forbes, T.G. & Priest, E.R. (1987). A comparison of analytical and numerical models for steadily driven magnetic reconnection, *Rev. Geophys.* **25**, 1587–607.

Freidberg, J.P. (1987). *Ideal Magnetohydrodynamics* (Plenum Press, New York).

Frisch, U. & Morf, R. (1981). Intermittency in nonlinear dynamics and singularities at complex times, *Phys. Rev.* **A 23**, 2673–705.

Frisch, U., Pouquet, A., Léorat, J. & Mazure, A. (1975). Possibility of an inverse cascade of magnetic helicity in magnetohydrodynamic turbulence, *J. Fluid Mech.* **68**, 769–78.

Frisch, U., Pouquet, A., Sulem, P.L. & Meneguzzi, M. (1983). Dynamics of two-dimensional ideal MHD, *J. Méc. Théor. Appl.* Special issue on two-dimensional turbulence, pp. 191–216.

Frisch, U., Sulem, P.L. & Nelkin, M. (1978). A simple dynamical model of intermittent fully developed turbulence, *J. Fluid Mech.* **87**, 719–36.

Fukao, S., Masayuki, U. & Takao, T. (1975). Topological study of magnetic field near a neutral point, *Rep. Ionosph. Res. Jpn* **29**, 133–9.

Furth, H.P., Killeen J. & Rosenbluth, M.N. (1963). Finite resistivity instabilities of a sheet pinch, *Phys. Fluids* **6**, 459–84.

Furth, H.P., Killeen, J., Rosenbluth, M.N. & Coppi, B. (1966). Stabilization by shear and negative V'', in *Proceedings of a Conference on Plasma Physics and Controlled Nuclear Fusion Research* (IAEA, Vienna), vol. I, 103–26.

Furth, H.P., Rutherford, P.H. & Selberg, H. (1973). Tearing mode in the cylindrical tokamak, *Phys. Fluids* **16**, 1054–63.

Fyfe, D. & Montgomery, D. (1976). High-beta turbulence in two-dimensional magnetohydrodynamics, *J. Plasma Phys.* **16**, 181–91.

Gajewski, R. (1972). Magnetohydrodynamic equilibrium of an elliptical plasma cylinder, *Phys. Fluids* **15**, 70–4.

Gibson, R.D. & Whiteman, K.J. (1968). Tearing mode instability in the Bessel function model, *Plasma Phys.* **10**, 1101–4.

Gill, R.D., et al. (1986). Sawtooth activity during additional heating in JET, *Proceedings of the Thirteenth European Conference on Controlled Fusion and Plasma Heating* (EPS, Geneva), vol. I, pp. 21–4.

Gill, R.D., Edwards, A.W., Pasini, D. & Weller, A. (1992). Snake-like density perturbations in JET, *Nucl. Fusion* **32**, 723–35.

Glasser, A.H., Greene, J.M. & Johnson, J.L. (1975). Resistive instabilities in general toroidal plasma configurations, *Phys. Fluids* **18**, 875–88.

Glasser, A.H., Furth, H.P. & Rutherford, P.H. (1977). Stabilization of resistive kink modes in the tokamak, *Phys. Rev. Lett.* **38**, 234–7.

Goedbloed, J.P. (1975). Spectrum of ideal magnetohydrodynamics of axisymmetric toroidal systems, *Phys. Fluids* **18**, 1258–68.

Goedbloed, J.P. & Hagebeuk, H.J.L. (1972). Growth rates of instabilities of a diffuse linear pinch, *Phys. Fluids* **15**, 1090–101.

Goedbloed, J.P. & Sakanaka, P.H. (1974). New approach to magneto-hydrodynamic stability I. A practical stability concept, *Phys. Fluids* **17**, 908–18.

Goeler, von, S., Stodiek, W. & Sauthoff, N. (1974). Studies of internal disruptions and $m = 1$ oscillations in tokamak discharges with soft X-ray techniques, *Phys. Rev. Lett.* **33**, 1201–3.

Gold, T. & Hoyle, F. (1960). The origin of solar flares, *Monthly Notices Roy. Astron. Soc.* **120**, 89–105.

Goossens, M. (1991). Magnetohydrodynamic waves and wave heating in non-uniform plasmas, *Advances in Solar System Magnetohydrodynamics,* ed. E. Priest and A. Hood (Cambridge University Press, Cambridge), pp. 137–72.

Gough, D.O. & Weiss, N.O. (1976). The calibration of stellar convection theories, *Monthly Notices Roy. Astron. Soc.* **176**, 589–607.

Grad, H. & Rubin, H. (1958) Hydromagnetic equilibria and force-free fields, *Proceedings of the Second United Nations International Conference on the Peaceful Uses of Atomic Energy,* vol. 31, pp. 190–7 (United Nations, Geneva).

Grad, H., Hu, P.N. & Stevens, D.C. (1975). Adiabatic evolution of plasma equilibrium, *Proc. Natl. Acad. Sci. (USA)* **72**, 3789–93.

Graham, E. (1975). Numerical simulations of two-dimensional compressible convection, *J. Fluid Mech.* **70**, 689–703.

Grappin, R., Frisch, U., Léorat, J. & Pouquet, A. (1982). Alfvénic fluctuations as asymptotic states of MHD turbulence, *Astron. Astrophys.* **105**, 6–14.

Grappin, R., Pouquet, A. & Léorat, J. (1983). Dependence of MHD turbulence spectra on the velocity magnetic field correlation, *Astron. Astrophys.* **126**, 51–8.

Greene, J.M. (1988). Geometrical properties of three-dimensional reconnecting magnetic fields with nulls, *J. Geophys. Res.* **93**, 8583–90.

Greene, J.M. & Johnson, J.L. (1961). Determination of hydromagnetic equilibria, *Phys. Fluids* **4**, 875–90.

Greene, J.M. & Johnson, J.L. (1968). Interchange instabilities in ideal hydromagnetic theory, *Plasma Phys.* **10**, 729–45.

Gruber, R. & Rappaz, J. (1985). *Finite Element Methods in Linear Ideal Magnetohydrodynamics* (Springer, Berlin).

Hain, K., Lüst, R. & Schlüter, A. (1957). Zur Stabilität eines Plasmas, *Z. Naturforschung* **12a**, 833–41.

Hain, K. & Lüst, R. (1958). Zur Stabilität zylindersymmetrischer Plasmakonfigurationen mit Volumenströmen, *Z. Naturforschung* **13a**, 936–40.

Hamada, S. (1962). Hydromagnetic equilibria and their proper coordinates, *Nucl. Fusion* **2**, 23–7.

Harned, D.S. & Kerner, W. (1985). Semi-implicit method for three-dimensional compressible magnetohydrodynamic simulations, *J. Comput. Phys.* **60**, 62–75.

Harned, D.S. & Schnack, D.D. (1986). Semi-implicit method for long-time-scale magnetohydrodynamic computations in three dimensions, *J. Comput. Phys.* **65**, 57–70.

Harned, D.S. & Schnack, D.D. (1987). Semi-implicit magnetohydrodynamic calculations, *J. Comput. Phys.* **70**, 330–54.

Hasegawa, A. (1985). Self-organization in continuous media, *Advances in Physics* **34**, 1–42.

Hastie, R.J., Hender, T.C., Carreras, B.A., Charlton, L.A. & Holmes, J.A. (1987). Stability of ideal and resistive internal kink modes in toroidal geometry, *Phys. Fluids* **30**, 1756–66.

Hautz, R. & Scholer, M. (1987). Numerical simulation on the structure of plasmoids in the deep tail, *Geophys. Res. Lett.* **14**, 969–72.

Hatori, T. (1984). Kolmogorov-style argument for decaying homogeneous MHD turbulence, *J. Phys. Soc. Jpn* **53**, 2539–45.

Hayashi, T., Sato, T. & Takei, A. (1990). Three-dimensional studies of helical equilibria and magnetic surface breaking due to the finite beta effect, *Phys. Fluids* **B2**, 329–37.

Hender, T.C., Hastie, R.J. & Robinson, D.C. (1987). Finite-β effects on tearing modes in tokamaks, *Nucl. Fusion* **27**, 1389–400.

Hicks, H.R., Carreras, B., Holmes, J.A., Lee, D.K. & Waddell, B.V. (1981). 3-D nonlinear calculations of resistive tearing modes, *J. Comput. Phys.* **44**, 46-69.

Hicks, H.R., Holmes, J.A., Carreras, B.A., Tetreault, D.J., Berge, G., Freidberg, J.P., Politzer, P.A. & Sherwell, D. (1981). Resistive MHD calculations on the disruptive instability, *Proceedings of the Eighth Conference on Plasma Physics and Controlled Nuclear Fusion Research* (IAEA, Vienna), vol. I, pp. 259–68.

Hirayama, T. (1974). Theoretical model of flares and prominences, *Solar Phys.* **34**, 323–38.

Hirshman, S.P. & Whitson, J.C. (1983). Steepest descent moment method for three-dimensional magnetohydrodynamic equilibria, *Phys. Fluids* **26**, 3553–68.

Hockney, R.W. & Eastwood, J.W. (1988). *Computer Simulation Using Particles* (Adam Hilger, Bristol).

Hollweg, J.V. (1990). Heating of the solar corona, *Comput. Phys. Rep.* **12**, 205–32.

Holmes, J.A. (1991). Nonlinear evolution of resistive modes for $q < 1$ in tokamaks, *Phys. Fluids* **B3**, 594–600.

Holmes, J.A., Carreras, B.A. & Charlton, L.A. (1989). Magnetohydrodynamic stability and nonlinear evolution of the $m = 1$ mode in toroidal geometry for safety factor profiles with an inflection point, *Phys. Fluids* **B1**, 788–97.

Holmes, J.A., Carreras, B.A., Hender, T.C., Hicks, H.R., Lynch, V.E., An, Z.G. & Diamond, P.H. (1985). Nonlinear interaction of tearing modes: A comparison between the tokamak and the reversed-field pinch configurations, *Phys. Fluids* **28**, 261–70.

Hood, A.W. (1991). MHD of solar flares, *Advances in Solar System Magnetohydrodynamics,* ed. E. Priest and A. Hood (Cambridge University Press, Cambridge), pp. 307–26.

Hood, A.W. & Priest, E.R. (1979). Kink instability of solar coronal loops as the cause of solar flares, *Solar Phys.* **64**, 303–21.

Hood, A.W. & Priest, E.R. (1981). Critical conditions for magnetic instabilities in force-free coronal loops, *Geophys. Astrophys. Fluid Dynamics* **17**, 297–318.

Hosokawa, I. (1989). An advanced model of dissipation cascade in locally isotropic turbulence, *Phys. Fluids* **A1**, 186–9.

Hosokawa, I. & Yamamoto, K. (1990). Intermittency of dissipation in directly simulated fully developed turbulence, *J. Phys. Soc. Jpn* **59**, 401–4.

Hurlburt, N.E. & Toomre, J. (1988). Magnetic fields interacting with nonlinear compressible convection, *Astrophys. J.* **327**, 920–32.

Hurlburt, N.E., Toomre, J. & Massaguer, J.M. (1984). Two-dimensional compressible convection extending over multiple scale heights, *Astrophys. J.* **282**, 557–73.

Iroshnikov, P.S. (1964). Turbulence of a conducting fluid in a strong magnetic field, *Sov. Astron.* **7**, 566–71.

Izzo, R., Monticello, D.A., Strauss, H.R., Park, W., Manickam, J., Grimm, R.C. & Delucia, J. (1983). Reduced equations for internal kinks in tokamaks, *Phys. Fluids* **26**, 3066–9.

JET team (1989). Latest JET results and future prospects, *Proceedings of the Twelfth Conference on Plasma Physics and Controlled Nuclear Fusion Research* (IAEA, Vienna), vol. I, pp. 41–65.

JET team (1991). High density regimes and beta limits in JET, *Proceedings of the Thirteenth Conference on Plasma Physics and Controlled Nuclear Fusion Research* (IAEA, Vienna), vol. I, pp. 219–27.

Jockers, K. (1978). Bifurcation of the force-free solar magnetic fields : a numerical approach, *Solar Phys.* **56**, 37–53.

Kadomtsev, B.B. (1966). Hydromagnetic stability of a plasma, *Reviews of Plasma Physics*, ed. M.A. Leontovich (Consultants Bureau, New York), vol. 2, pp. 153–99.

Kadomtsev, B.B. (1975). Disruptive instability in tokamaks, *Sov. J. Plasma Phys.* **1**, 389–91.

Kadomtsev, B.B. (1987). Tokamak plasma self-organization, *Comments Plasma Phys. Controlled Fusion* **11**, 153–63.

Kadomtsev, B.B. & Pogutse, O.P. (1974). Nonlinear helical perturbations of a plasma in the tokamak, *Sov. Phys.-JETP* **38**, 283–90.

Kageyama, A., Watanabe, K. & Sato, T. (1990). Global simulation of the magnetosphere with a long tail: The formation and ejection of plasmoids, National Institute of Fusion Studies, Res. Report NIFS-49, Nagoya, Japan.

Karger, F., Wobig, H., Corti, S., Gernhardt, J., Klüber, O., Lisitano, G., McCormick, K., Meisel, D. & Sesnic, S. (1975). Influence of resonant helical fields on tokamak discharges, *Proceedings of the Fifth Conference on Plasma Physics and Controlled Nuclear Fusion Research* (IAEA, Vienna), vol. I, pp. 207–15.

Kaw, P.K., Valeo, E.J. & Rutherford, P.H. (1979). Tearing modes in a plasma with magnetic braiding, *Phys. Rev. Lett.* **43**, 1398–401.

Kerner, W., Gruber, R. & Troyon, F. (1980). Numerical study of the internal kink mode in tokamaks, *Phys. Rev. Lett.* **44**, 536–40.

Kerner, W. & Tasso, H. (1982). Tearing mode stability for arbitrary current distribution, *Plasma Phys.* **24**, 97–107.

Kerner, W. & Tasso, H. (1983). Stability computation of multihelical resistive tearing modes, *Proceedings of the Ninth Conference on Plasma Physics and Controlled Nuclear Fusion Research* (IAEA, Vienna), vol. III, pp. 49–55.

Kerr, R.M. (1985). Higher-order derivative correlations and the alignment of small-scale structures in isotropic numerical turbulence, *J. Fluid Mech.* **153**, 31–58.

Kida, S. & Murakami, Y. (1989). Statistics of velocity gradients in turbulence at moderate Reynolds numbers, *Fluid Dyn. Res.* **4**, 347–70.

Kida, S., Yamada, M. & Ohkitani, K. (1988). The energy spectrum in the universal range of two-dimensional turbulence, *Fluid Dyn. Res.* **4**, 271–301.

Kolesnichenko, Y.I., Yakovenko, Y.V., Anderson, D., Lisak, M. & Wising, F. (1992). Sawtooth oscillations with central safety factor below unity, *Phys. Rev. Lett.* **68**, 3881–4.

Kolmogorov, A.N. (1941a). Local structure of turbulence in an incompressible fluid at very large Reynolds numbers, *Dokl. Akad. Nauk SSSR* **30**, 299–303.

Kolmogorov, A.N. (1941b). Energy dissipation in locally isotropic turbulence, *Dokl. Akad. Nauk SSSR* **32**, 19–21.

Kolmogorov, A.N. (1962). A refinement of previous hypotheses concerning the local structure of turbulence in a viscous incompressible fluid at high Reynolds number, *J. Fluid Mech.* **13**, 82–5.

Kotschenreuther, M., Hazeltine, R.D. & Morrison, P.J. (1985). Nonlinear dynamics of magnetic islands with curvature and pressure, *Phys. Fluids* **28**, 294–302.

Kraichnan, R.H. (1959). The structure of isotropic turbulence at very high Reynolds numbers, *J. Fluid Mech.* **5**, 497–543.

Kraichnan, R.H. (1965a). Lagrangian-history closure approximation for turbulence, *Phys. Fluids* **8**, 575–98.

Kraichnan, R.H. (1965b). Inertial range spectrum in hydromagnetic turbulence, *Phys. Fluids* **8**, 1385–7.

Kraichnan, R.H. (1967). Inertial ranges in two-dimensional turbulence, *Phys. Fluids* **10**, 1417–23.

Kraichnan, R.H. (1973). Helical turbulence and absolute equilibrium, *J. Fluid Mech.* **59**, 745–52.

Kraichnan, R.H. (1990). Models of intermittency in hydrodynamic turbulence, *Phys. Rev. Lett.* **65**, 575 8.

Kraichnan, R.H. & Nagarajan, S. (1967). Growth of turbulent magnetic fields, *Phys. Fluids* **10**, 859–70.

Kraichnan, R.H. & Montgomery, D. (1980). Two-dimensional turbulence, *Rep. Prog. Phys.* **43**, 547–619.

Krause, F.K. & Rädler, K.H. (1980). *Mean-field Magnetohydrodynamics and Dynamo Theory* (Pergamon Press, Oxford).

Kruskal, M.D. & Kulsrud, R.M. (1958). Equilibrium of a magnetically confined plasma in a toroid, *Phys. Fluids* **1**, 265–74.

Kusano, K. & Sato, T. (1990). Simulation study of the self-sustainment mechanism in the reversed-field pinch configuration, *Nucl. Fusion* **30**, 2075–96.

Kuvshinov, B.N. & Savrukhin, P.V. (1990). Internal disruptions in tokamaks, *Sov. J. Plasma Phys.* **16**, 353–63.

Lackner, K. (1976). Computation of ideal MHD equilibria, *Comput. Phys. Commun.* **12**, 33–44.

LaHaye, R.J., Carlstrom, T.N., Goforth, R.R., Jackson, G.L., Schaffer, M.J., Tamano, T. & Taylor, P.L. (1984). Measurements of magnetic field fluctuations in the OHTE toroidal pinch, *Phys. Fluids* **27**, 2576–82.

Lao, L.L. (1984). Variational moment method for computing magnetohydrodynamic equilibria, *Comput. Phys. Commun.* **31**, 201–12.

Lao, L.L., Hirshman, S.P. & Wieland, R.M. (1981). Variational moment solutions to the Grad-Shafranov equation, *Phys. Fluids* **24**, 1431–41.

Lau, Y.T. & Finn, J.M. (1990). Three-dimensional kinematic reconnection in the presence of field line nulls and closed field lines, *Astrophys. J.* **350**, 672–91.

Laval, G., Mercier, C. & Pellat, R. (1965). Necessity of the energy principle for magnetostatic stability, *Nucl. Fusion* **5**, 156–8.

Laval, G., Pellat, R. & Soulé, J.S. (1974). Hydromagnetic stability of a current-carrying pinch with noncircular cross-section, *Phys. Fluids* **17**, 835–45.

Lee, L.C., Fu, Z.F. & Akasofu, S.I. (1985). A simulation study of forced reconnection processes and magnetospheric storms and substorms, *J. Geophys. Res.* **90**, 10896–910.

Lerbinger, K. & Luciani, J.F. (1991). A new semi-implicit method for MHD computations, *J. Comput. Phys.* **97**, 444–59.

Leslie, D.C. (1973). *Developments in the Theory of Turbulence* (Clarendon Press, Oxford).

Lichtenberg, A.J. & Lieberman, M.A. (1983). *Regular and Stochastic Motion* (Springer, New York).

Liewer, P.C. (1985). Measurements of micro-turbulence in tokamaks and comparison with theories of turbulence and anomalous transport, *Nucl. Fusion* **25**, 543–621.

Lifshitz, A.E. (1989). *Magnetohydrodynamics and Spectral Theory* (Kluwer Academic Publishers, Dordrecht).

Longcope, D.W. & Sudan, R.N. (1991). Renormalization group analysis of reduced magnetohydrodynamics with application to subgrid modeling, *Phys. Fluids* **B3**, 1945–62.

Lortz, D. (1968). Exact solutions of the hydromagnetic dynamo problem, *Plasma Phys.* **10**, 967–72.

Lortz, D. (1989). Axisymmetric dynamo solutions, *Z. Naturforsch.* **44a**, 1041–5.

Lortz, D. & Spies, G.O. (1984). Spectrum of a resistive plasma slab, *Phys. Lett.* **101A**, 333–7.

Low, B.C. (1991). Three-dimensional structures of magnetostatic atmospheres III. A general formulation, *Astrophys. J.* **370**, 427–34.

Low, B.C. & Lou, Y.Q. (1990). Modelling solar force-free magnetic fields, *Astrophys. J.* **352**, 343–52.

Low, B.C. & Wolfson, R. (1988). Spontaneous formation of electric current sheets and the origin of solar flares, *Astrophys. J.* **324**, 574–81.

Lüst, R. & Schlüter, A. (1957). Axialsymmetrische magnetohydrodynamische Gleichgewichtskonfigurationen, *Z. Naturforschung* **12a**, 850–4.

Lumley, J.L. (1970). *Stochastic Tools in Turbulence* (Academic Press, New York).

Maeda, H., Sengoku, S., Kimura, H. et al. (1979). Experimental study of magnetic divertor in DIVA, in *Proceedings of the Seventh Conference on Plasma Physics and Controlled Nuclear Fusion Research* (IAEA, Vienna), vol. I, pp. 377–85.

Marchuk, G.I. (1975). *Methods of Numerical Mathematics* (Springer, New York).

Martens, P.C.H. & Kuin, N.P.M. (1989). A circuit model for filament eruptions and two-ribbon flares, *Solar Phys.* **122**, 263–302.

Martin, T.J. & Taylor, J.B. (1974). Helically deformed states in toroidal pinches, Culham Laboratory report (unpublished).

Matthaeus, W.H., Goldstein, M.L. & Smith, C. (1982). Evaluation of magnetic helicity in homogeneous turbulence, *Phys. Rev. Lett.* **48**, 1256–9.

McWilliams, J.C. (1984). The emergence of isolated, coherent vortices in turbulent flow, *J. Fluid Mech.* **146**, 21–43.

Meneguzzi, M., Frisch, U. & Pouquet, A. (1981). Helical and nonhelical turbulent dynamos, *Phys. Rev. Lett.* **47**, 1060–4.

Mercier, C. (1960). Un critère nécessaire de stabilité hydromagnétique pour un plasma en symétrie de révolution, *Nucl. Fusion* **1**, 47–53.

Merlin, D. & Biskamp, D. (1988). Numerical studies of MHD turbulence in the reversed-field pinch, Max-Planck-Institut für Plasmaphysik report No. IPP6/276.

Meyer, F., Schmidt, H.U. & Weiss, N.O. (1977). Stability of sunspots, *Monthly Notices Roy. Astron. Soc.* **179**, 741–61.

Mikic, Z., Barnes, D.C. & Schnack, D.D. (1988). Dynamical evolution of a solar coronal magnetic field arcade, *Astrophys. J.* **328**, 830–47.

Mikic, Z., Schnack, D.D. & Van Hoven, G. (1990). Dynamical evolution of twisted magnetic flux tubes. I. Equilibrium and linear stability, *Astrophys. J.* **361**, 690–700.

Moffat, H.K. (1961). The amplification of a weak applied magnetic field by turbulence in fluids of moderate conductivity, *J. Fluid Mech.* **11**, 625–35.

Moffat, H.K. (1969). The degree of knottedness of tangled vortex lines, *J. Fluid Mech.* **35**, 117–29.

Moffat, H.K. (1978). *Magnetic Field Generation in Electrically Conducting Fluids* (Cambridge University Press, Cambridge).

Monin, A.S. & Yaglom, A.M. (1975). *Statistical Fluid Mechanics: Mechanics of Turbulence* (MIT Press, Cambridge, Mass.), vol. II.

Montgomery, D. & Hatori, T. (1984). Analytical estimates of turbulent MHD transport coefficients, *Plasma Phys. Control. Fusion* **26**, 717–30.

Montgomery, D., Turner, L. & Vahala, G. (1979). Most probable states in magnetohydrodynamics, *J. Plasma Phys.* **21**, 239–51.

Morrison, P.J. & Hazeltine, R.P. (1984). Hamiltonian formulation of reduced magnetohydrodynamics, *Phys. Fluids* **27**, 886–97.

Müller, R. (1989). Solar granulation: overview, *Proc. Third Internat. Workshop of the OAC and NATO Advanced Workshop on Solar and Stellar Granulation*, NATO ASI Series C, vol. 263 (Kluwer, Dordrecht), pp. 101–23.

Murakami, M., Callen, J.D. & Berry, L.A. (1976). Some observations on maximum densities in tokamak experiments, *Nucl. Fusion* **16**, 347–8.

Nagayama, Y. et al. (1991). Analysis of sawtooth oscillations using simultaneous measurements of electron cyclotron emission imaging and X-ray tomography on TFTR, *Phys. Rev. Lett.* **67**, 3527–30.

Nave, M.F.F. & Wesson, J. (1988). Stability of the ideal $m = 1$ mode in a tokamak, *Nucl. Fusion* **28**, 297–301.

Nebel, R.A., Caramana, E.J. & Schnack, D.D. (1989). The role of the $m = 0$ modal components in the reversed-field pinch dynamo effect in the single-fluid magnetohydrodynamic model, *Phys. Fluids* **B1**, 1671–4.

Nordlund, A., Brandenburg, A., Jennings, R.L., Rieutord, M., Ruokolainen, J., Stein, R.F. & Tuominen, I. (1992). Dynamo action in stratified convection with overshoot, *Astrophys. J.* **392**, 647–52.

Novikov, E.A. (1970). Intermittency and scale similarity of the structure of turbulent flow, *Prikl. Mat. Mekh.* **35**, 266–77; (1971), *J. Appl. Math. Mech.* **35**, 231–41.

Novikov, E.A. & Stewart, R.W. (1964). Intermittency of turbulence and spectrum of fluctuations in energy dissipation, *Izv. Akad. Nauk SSSR, Ser. Geofiz.*, No. 3, 408–13.

Obukhov, A.M. (1941). Energy distribution in the spectrum of a turbulent flow, *Izv. Akad. Nauk SSSR, Ser. Geofiz.*, No. 4–5, 453–66.

Obukhov, A.M. (1962). Some features of atmospheric turbulence, *J. Fluid Mech.* **13**, 77–81.

O'Rourke, J. (1991). The change in the safety factor profile at a sawtooth collapse, *Plasma Phys. Controlled Fusion* **33**, 289–96.

Orszag, S.A. (1977). In *Fluid Dynamics*, ed. R. Balian and J.J. Peube (Gordon and Breach, New York), pp. 235–374.

Otto, A., Schindler, K. & Birn, J. (1990). Quantitative study of the nonlinear formation and acceleration of plasmoids in the earth's magnetotail, *J. Geophys. Res.* **95**, 15023–37.

Paladin, G. & Vulpiani, A. (1987). Anomalous scaling laws in multifractal objects, *Phys. Reports* **156**, 147–225.

Park, W., Monticello, D.A., White, R.B. & Jardin, S.C. (1980). Nonlinear saturation of the internal kink mode, *Nucl. Fusion* **20**, 1181–5.

Park, W., Monticello, D.A. & White, R.B. (1984). Reconnection rates of magnetic fields including the effects of viscosity, *Phys. Fluids* **27**, 137–49.

Parker, E.N. (1955). The formation of sunspots from the solar toroidal field, *Astrophys. J.* **121**, 491–507.

Parker, E.N. (1963). The solar flare phenomenon and the theory of reconnection and annihilation of magnetic fields, *Astrophys. J. Suppl. Ser.* **8**, 177–211.

Parker, E.N. (1979a). *Cosmical Magnetic Fields* (Clarendon Press, Oxford).

Parker, E.N. (1979b). Sunspots and the physics of magnetic flux tubes. I. The general nature of the sunspot, *Astrophys. J.* **230**, 905–13.

Parker, E.N. (1983). Magnetic neutral sheets in evolving fields. I. General theory, *Astrophys. J.* **264**, 635–41.

Parker, E.N. (1988). Nanoflares and the solar X-ray corona, *Astrophys. J.* **330**, 474–9.

Pegoraro, F., Porcelli, F., Coppi, B., Detragiache, P. & Migliuolo, S. (1989). Theory of sawtooth stabilization in the presence of energetic ions, *Proceedings of the Twelfth Conference on Plasma Physics and Controlled Nuclear Fusion Research* (IAEA, Vienna), vol. II, pp. 243–9.

Pegoraro, F. & Schep, T.J. (1986). Theory of resistive modes in the ballooning representation, *Plasma Phys. Controlled Fusion* **28**, 647–67.

Petschek, H.E. (1964). Magnetic field annihilation, *AAS/NASA Symposium on the Physics of Solar Flares*, ed. W.N. Hess (NASA, Washington, DC), pp. 425–37.

Phan, T.D. & Sonnerup, B.U.O. (1991). Resistive tearing-mode instability in a current sheet with equilibrium viscous stagnation-point flow, *J. Plasma Phys.* **46**, 407–21.

Pneuman, G.W. (1980). Reconnection-driven coronal transients, *Solar and Interplanetary Dynamics, IAU Symp. 91*, ed. M. Dryer and E. Tandberg-Hanssen, pp. 317–21.

Politano, H., Pouquet, A. & Sulem, P.L. (1989). Inertial ranges and resistive instabilities in two-dimensional magnetohydrodynamic turbulence, *Phys. Fluids* **B1**, 2330–9.

Pouquet, A. (1978). On two-dimensional magnetohydrodynamic turbulence, *J. Fluid Mech.* **88**, 1–16.

Pouquet, A., Frisch, U. & Léorat, J. (1976). Strong MHD helical turbulence and the nonlinear dynamo effect, *J. Fluid Mech.* **77**, 321–54.

Pouquet, A., Sulem, P.L. & Meneguzzi, M. (1988). Influence of velocity magnetic field correlations on decaying magnetohydrodynamic turbulence with neutral X-points, *Phys. Fluids* **31**, 2635–43.

Prager, S.C. (1990). Transport and fluctuations in reversed-field pinches, *Plasma Phys. Control. Fusion* **32**, 903–16.

Prager, S.C., Almagri, A.F., Assadi, S., et al. (1990). First results from the Madison Symmetric Torus reversed-field pinch, *Phys. Fluids* **B2**, 1367–71.

Prandtl, L. (1925). Bericht über Untersuchungen zur ausgebildeten Turbulenz, *Z. angew. Math. Mech.* **5**, 136–9.

Praskovsky, A.A. (1992). Experimental verification of the Kolmogorov refined similarity hypothesis, *Phys. Fluids* **A4**, 2589–91.

Priest, E.R. (1984). *Solar Magnetohydrodynamics* (Reidel, Dordrecht).

Priest, E.R. & Forbes, T.G. (1986). New models for fast steady state magnetic reconnection, *J. Geophys. Res.* **91**, 5579–88.

Pritchett, P.L. & Wu, C.C. (1979). Coalescence of magnetic islands, *Phys. Fluids* **22**, 2140–6.

Proudman, I. & Reid, W.H. (1954). On the decay of a normally distributed and homogeneous turbulent field, *Phil. Trans. Roy. Soc.* **A 247**, 163–89.

Reiman, A. (1980). Minimum energy state of a toroidal discharge, *Phys. Fluids* **23**, 230–1.

Roberts, P.H. & Soward, A.M. (1992). Dynamo theory, *Annual Rev. Fluid Mech.* **24**, 459–512.

Robinson, D.C. (1971). High-β diffuse pinch configurations, *Plasma Phys.* **13**, 439–62.

Robinson, D.C. (1978). Tearing-mode-stable diffuse-pinch configurations, *Nucl. Fusion* **18**, 939–53.

Rosenbluth, M.N., Dagazian, R.Y. & Rutherford, P.H. (1973). Nonlinear properties of the internal $m = 1$ kink instability in the cylindrical tokamak, *Phys. Fluids* **16**, 1894–902.

Rosenbluth, M.N., Monticello, D.A., Strauss, H.R. & White, R.B. (1976). Numerical studies of nonlinear evolution of kink modes in tokamaks, *Phys. Fluids* **19**, 1987–96.

Rutherford, P.H. (1973). Nonlinear growth of the tearing mode, *Phys. Fluids* **16**, 1903–8.

Santangelo, P., Benzi, R. & Legras, B. (1989). The generation of vortices in high-resolution, two-dimensional decaying turbulence and the influence of initial conditions on the breaking of selfsimilarity, *Phys. Fluids* **A1**, 1027–34.

Sato, T. & Hayashi, T. (1979). Externally driven magnetic reconnection and a powerful magnetic energy converter, *Phys. Fluids* **22**, 1189–202.

Savrukhin, P.V., Vasin, N.L., Bagdasarov, A.A. & Tarasyan, K.N. (1991). Trigger mechanism of the sawtooth crash in T-10, *Plasma Phys. Controlled Fusion* **33**, 1347–61.

Schindler, K., Hesse, M. & Birn, J. (1988). General magnetic reconnection, parallel electric fields, and helicity, *J. Geophys. Res.* **93**, 5547–57.

Schmalz, R. (1981). Reduced three-dimensional nonlinear equations for high-β plasmas including toroidal effects, *Phys. Lett.* **82A**, 14–17.

Schnack, D.D., Barnes, D.C., Mikić, Z., Harned, D.S., Caramana, E.J. & Nebel, R.A. (1986). Numerical simulation of reversed-field pinch dynamics, *Comput. Phys. Commun.* **43**, 17–28.

Schnack, D.D., Caramana, E.J. & Nebel, R.A. (1985). Three-dimensional magnetohydrodynamic studies of the reversed-field pinch, *Phys. Fluids* **28**, 321–33.

Shafranov, V.D. (1958). Magnetohydrodynamical equilibrium configurations, *Sov. Phys. JETP* **6**, 545–54.

Shafranov, V.D. & Yurchenko, E.I. (1968). Condition for flute instability of a toroidal-geometry plasma, *Sov. Phys. JETP* **26**, 682–6.

Shafranov, V.D. (1970). Hydromagnetic stability of a current-carrying pinch in a longitudinal magnetic field, *Sov. Phys. Tech. Phys.* **15**, 175–83.

Shaing, K.C., Lee, G.S., Carreras, B.A., Houlberg, W.A. & Crume, E.C. (1989). Model for the *L-H* transition in tokamaks, *Proceedings of the Twelfth Conference on Plasma Physics and Controlled Nuclear Fusion Research* (IAEA, Vienna), vol. II, pp. 13–22.

She, Z.S. (1991). Physical model of intermittency in turbulence: Near-dissipation-range non-Gaussian statistics, *Phys. Rev. Lett.* **66**, 600–3.

She, Z.S. & Orszag, S.A. (1991). Physical model of intermittency in turbulence: Inertial range non-Gaussian statistics, *Phys. Rev. Lett.* **66**, 1701–4.

Shivamoggi, B.K. (1985). Magnetohydrodynamic properties near an *X*-type magnetic neutral line, *J. Plasma Phys.* **31**, 333–5.

Siggia, E.D. (1981). Numerical study of small-scale intermittency in three-dimensional turbulence, *J. Fluid Mech.* **107**, 375–406.

Simon, G.W. & Weiss, N.O. (1989). A simple model of mesogranular and supergranular flows, *Proc. Third Internat. Workshop of the OAC and NATO Advanced Workshop on Solar and Stellar Granulation*, NATO ASI Series C, vol. 263 (Kluwer, Dordrecht), pp. 595–9.

Snider, R.T. (1990). Scaling of the sawtooth inversion radius and the mixing radius on DIII–D, *Nucl. Fusion* **30**, 2400–5.

Solov'ev, L.S. (1968). The theory of hydromagnetic stability of toroidal plasma configurations, *Sov. Phys. JETP* **26**, 400–7.

Soltwisch, H. (1986). Current distribution measurement in a tokamak by FIR polarimetry, *Rev. Sci. Instrum.* **57**, 1939–44.

Soltwisch, H. (1988). Measurement of current-density changes during sawtooth activity in a tokamak by far-infrared polarimetry, *Rev. Sci. Instrum.* **59**, 1599–604.

Soltwisch, H., Stodiek, W., Manickam, J. & Schlüter, J. (1987). Current density profiles in the TEXTOR tokamak, *Proceedings of the Eleventh Conference on Plasma Physics and Controlled Nuclear Fusion Research* (IAEA, Vienna), vol. I, pp. 263–73.

Somon, J.P. (1984). Curvature effects in the nonlinear growth of the cylindrical tearing mode, Energy Research Center, Frascati, report No. RF/FUS/84/13.

Sonnerup, B.U. (1970). Magnetic field reconnection in a highly conducting incompressible fluid, *J. Plasma Phys.* **4**, 161–74.

Spicer, D.S. (1977). An unstable arch model of a solar flare, *Solar Phys.* **53**, 305–45.

Steenbeck, M., Krause, F. & Rädler, K.H. (1966). Berechnung der mittleren Lorentz-Feldstärke $\langle v \times B \rangle$ für ein elektrisch leitendes Medium in turbulenter, durch Coriolis Kräfte beeinflußter Bewegung. *Z. Naturforsch.* **21a**, 369–76.

Stein, R.F. & Nordlund, A. (1989). Topology of convection beneath the solar surface, *Astrophys. J.* **342**, L95–8.

Stein, R.F., Nordlund, A. & Kuhn, J.R. (1989). Convection and Waves, *Proc. Third Internat. Workshop of the OAC and NATO Advanced Workshop on Solar and Stellar Granulation*, NATO ASI Series C, vol. 263 (Kluwer, Dordrecht), pp. 453–70.

Steinolfson, R.S. (1991). Coronal evolution due to shear motion, *Astrophys. J.* **382**, 677–87.

Strauss, H.R. (1976). Nonlinear three-dimensional magnetohydrodynamics of noncircular tokamaks, *Phys. Fluids* **19**, 134–40.

Strauss, H.R. (1977). Dynamics of high-β tokamaks , *Phys. Fluids* **20**, 1354–60.

Strauss, H.R. (1986). Hyperresistivity produced by tearing mode turbulence, *Phys. Fluids* **29**, 3668–71.

Strauss, H.R., Monticello, D.A. & Manickam, J. (1989). Stabilization of the resistive internal kink mode in tokamaks with non-circular cross-section, *Nucl. Fusion* **29**, 320–4.

Sturrock, P.A., Holzer, T.E., Mihalas, D.M. & Ulrich, R.K, editors (1986). *Physics of the Sun* (D. Reidel, Dordrecht).

Sulem, P.L., Frisch, U., Pouquet, A. & Meneguzzi, M. (1985). On the exponential flattening of current sheets near neutral X-points in two-dimensional ideal MHD flow, *J. Plasma Phys.* **33**, 191–8.

Suydam, B.R. (1958). Stability of a linear pinch, *Proceedings of the Second United Nations International Conference on the Peaceful Uses of Atomic Energy* (United Nations, Geneva), vol. 31, pp. 157–9.

Sweet, P.A. (1958). The production of high energy particles in solar flares, *Nuovo Cimento Suppl.* **8**, *Ser. X*, 188–96.

Sykes, A., Turner, M.F. & Patel, S. (1983). Beta limits in tokamaks due to high-n ballooning modes, *Proceedings of the 11th European Conference on Controlled Fusion and Plasma Physics* (Aachen), vol. II, pp. 363–6.

Sykes, A. & Wesson, J.A. (1976). Relaxation instability in tokamaks, *Phys. Rev. Lett.* **37**, 140–3.

Sykes, A. & Wesson, J.A. (1977). Field reversal in pinches, *Proceedings of the 8th European Conference on Controlled Fusion and Plasma Physics* (Prague), vol. I, p. 80.

Sykes, A. & Wesson, J.A. (1980). Major disruptions in tokamaks, *Phys. Rev. Lett.* **44**, 1215–18.

Syrovatskii, S.I. (1971). Formation of current sheets in a plasma with a frozen-in strong magnetic field, *Sov. Phys. JETP* **33**, 933–40.

Syrovatskii, S.I. (1981). Pinch sheets and reconnection in astrophysics, *Annual Rev. Astron. Astrophys.* **19**, 163–229.

Takeda, T. & Tokuda, S. (1991). Computation of MHD equilibrium of tokamak plasma, *J. Comput. Phys.* **93**, 1–107.

Tandberg-Hanssen, E. & Emslie, A.G. (1988). *The Physics of Solar Flares* (Cambridge University Press, Cambridge).

Tasso, H. (1975). Energy principle for two-dimensional resistive instabilities, *Plasma Phys.* **17**, 1131–4.

Tasso, H. & Virtamo, J.T. (1980). Energy principle for 3-D resistive instabilities in shaped cross-section tokamaks, *Plasma Phys.* **22**, 1003–13.

Taylor, G.I. (1932). The transport of vorticity and heat through fluids in turbulent motion, *Proc. Roy. Soc. London* **A135**, 685–705.

Taylor, J.B. (1974). Relaxation of toroidal plasma and generation of reversed magnetic fields, *Phys. Rev. Lett.* **33**, 1139–41.

Taylor, J.B. (1986). Relaxation and magnetic reconnection in plasmas, *Rev. Mod. Phys.* **53**, 741–63.

Taylor, J.B. (1990). Natural current profiles in a tokamak, Institute for Fusion Studies report No. 447, University of Texas.

Toomre, J., Brummell, N. & Cattaneo, F. (1990). Three-dimensional compressible convection at low Prandtl numbers, *Comput. Phys. Commun.* **59**, 105–17.

Troyon, F., Gruber, R., Saurenmann, H., Semenzato, S. & Succi, S. (1984). MHD limits to plasma confinement, *Plasma Phys. Controlled Fusion* **26**, 209–15.

Turner, M.F. & Wesson, J.A. (1982). Transport, instability and disruptions in tokamaks, *Nucl. Fusion* **22**, 1069–78.

Ugai, M. (1989). Computer studies of a large-scale plasmoid driven by spontaneous fast reconnection, *Phys. Fluids* **B1**, 942–8.

Vainshtein, S.I. & Zeldovich, Y.B. (1972). Origin of magnetic fields in astrophysics, *Sov. Phys. Usp.* **15**, 159–72.

Van Ballegooijen, A.A. (1988). Force-free fields and coronal heating, *Geophys. Astrophys. Fluid Dyn.* **41**, 181–211.

Van Tend, W. & Kuperus, M. (1978). The development of coronal electric current systems in active regions and their relation to filaments and flares, *Solar Phys.* **59**, 115–27.

Vasyliunas, V.M. (1975). Theoretical models of magnetic field line merging, *Rev. Geophys. Space Phys.* **13**, 303–36.

Vekstein, G.E., Priest, E.R. & Amari, T. (1990). Formation of current sheets in force-free fields, *Astron. Astrophys.* **243**, 492–500.

Vekstein, G.E. & Priest, E.R. (1992). Magnetohydrodynamic equilibria and cusp formation at an X-type neutral line by foot point shearing, *Astrophys. J.* **384**, 333–40.

Vincent, A. & Meneguzzi, M. (1991). The spatial structure and statistical properties of homogeneous turbulence, *J. Fluid Mech.* **225**, 1–20.

Vlad, G. & Bondeson, A. (1989). Numerical simulations of sawteeth in tokamaks, *Nucl. Fusion* **29**, 1139–52.

Voslamber, D. & Callebaut, D.K. (1962). Stability of force-free magnetic fields, *Phys. Rev.* **128**, 2016–21.

Waddell, B.V. et al. (1979). Nonlinear interaction of tearing modes in highly resistive tokamaks, *Phys. Fluids* **22**, 896–910.

Waddell, B.V., Carreras, B., Hicks, H.R., Holmes, J.A. & Lee, D.K. (1978) Mechanism for major disruptions in tokamaks, *Phys. Rev. Lett.* **41**, 1386–9.

Waddell, B.V., Rosenbluth, M.N., Monticello, D.A. & White, R.B. (1976). Nonlinear growth of the $m = 1$ tearing mode, *Nucl. Fusion* **16**, 528–32.

Waelbroeck, F.L. (1989). Current sheets and nonlinear growth of the $m = 1$ kink-tearing mode, *Phys. Fluids* **B1**, 2372–80.

Wagner, F. et al. (1982). Regime of improved confinement and high beta in neutral-beam-heated divertor discharges in the ASDEX tokamak, *Phys. Rev. Lett.* **49**, 1408–12.

Wagner, F. et al. (1984). Development of an edge transport barrier at the *H*-mode transition of ASDEX, *Phys. Rev. Lett.* **53**, 1453–6.

Wagner, F. et al. (1991). Recent results of *H*-mode studies on ASDEX, *Proceedings of the Thirteenth Conference on Plasma Physics and Controlled Nuclear Fusion Research* (IAEA, Vienna), vol. I, pp. 277–90.

Ward, D.J., Gill, R.D., Morgan, P.D. & Wesson, J.A. (1988). The final phase of JET disruptions, *Proceedings of the Fifteenth European Conference on Controlled Fusion and Plasma Heating* (EPS, Geneva), vol. I, pp. 330–3.

Weiss, N.O. (1990). *Comput. Phys. Rep.* **12**, 233–45.

Weller, A. et al. (1987). Persistent density perturbations at rational-*q* surfaces following pellet injection in the Joint European Tokamak, *Phys. Rev. Lett.* **59**, 2303–6.

Wesson, J.A. (1978). *Nucl. Fusion* **18**, 87–132.

Wesson, J.A. (1986). *Plasma Phys. Controlled Fusion* **28**, 243–8.

Wesson, J.A. (1987). *Tokamaks* (Clarendon Press, Oxford).

Wesson, J.A. (1990). Sawtooth reconnection, *Nucl. Fusion* **30**, 2545–9.

Wesson, J.A. et al. (1989). Disruptions in JET, *Nucl. Fusion* **29**, 641–66.

Wesson, J.A., Ward, D.J. & Rosenbluth, M.N. (1990). Negative voltage spike in tokamak disruptions, *Nucl. Fusion* **30**, 1011–14.

White, R.B. (1989). *Theory of Tokamak Plasmas* (North Holland, Amsterdam).

White, R.B., Bussac, M.N. & Romanelli, F. (1989). High-β sawtooth-free tokamak operation using energetic trapped particles, *Phys. Rev. Lett.* **62**, 539–42.

White, R.B., Monticelli, D.A., Rosenbluth, M.N. & Waddell, B.V. (1977). Saturation of the tearing mode, *Phys. Fluids* **20**, 800–5.

White, R.B., Rutherford, P.H., Colestock, P. & Bussac, M.N. (1988). Sawtooth stabilization by energetic trapped particles, *Phys. Rev. Lett.* **60**, 2038–41.

Woltjer, L. (1958). A theorem on force-free magnetic fields, *Proc. Natl. Acad. Sci.* (Washington) **44**, 489–92.

Yaglom, A.M. (1966). Effect of fluctuations in energy dissipation rate on the form of turbulence characteristics in the inertial subrange, *Dokl. Akad. Nauk SSSR* **166**, No. 1, 49–52.

Yakhot, V. & Orszag, S.A. (1986). Renormalization group analysis of turbulence I. Basic theory, *J. Scient. Comput.* **1**, 3–51.

Yamamoto, K. & Hosokawa, I. (1988). Decaying isotropic turbulence pursued by the spectral method, *J. Phys. Soc. Jpn* **57**, 1532–5.

Yanase, S., Mizushima, J. & Kida, S. (1991). Coherent structures in MHD turbulence and turbulent dynamo, *Turbulence and Coherent Structures*, ed. O. Métais and M. Lesieur (Kluwer, Dordrecht), pp. 569–83.

Yeh, T. & Axford, W.I. (1970). On the reconnection of magnetic field lines in conducting fluids, *J. Plasma Phys.* **4**, 207–29.

Zakharov, L.E. & Shafranov, V.D. (1986). Equilibrium of current-carrying plasmas in toroidal configurations, *Reviews of Plasma Physics*, ed. M.A. Leontovich (Consultants Bureau, New York), vol. 11, pp. 153–302.

Zohm, H. et al. (1992). Studies of edge-localized modes on ASDEX, *Nucl. Fusion* **32**, 489–94.

Zweibel, E.G. & Li, He-Sheng (1987). The formation of current sheets in the solar atmosphere, *Astrophys. J.* **312**, 423–30.

Zwingmann, W. (1987). Theoretical study of the onset conditions for solar eruptive processes, *Solar Phys.* **111**, 309–31.

Zwingmann, W., Schindler, K. & Birn, J. (1985). On sheared magnetic field structures containing neutral points, *Solar Phys.* **99**, 133–43.

References added in the paperback edition

Aydemir, A.Y. (1992). *Phys. Fluids* **B4**, 3469–72.

Biskamp, D. (1981). *Phys. Rev. Lett.* **46**, 1522–5.

Biskamp, D. & Bremer, U. (1994). Dynamics and statistics of inverse cascade processes in 2D magnetohydrodynamic turbulence, *Phys. Rev. Lett.* **72**, 3819–22.

Borba, D. et al. (1994). The pseudospectrum of the resistive magnetohydro-dynamics operator: Resolving the resistive Alfvén paradox, *Phys. Plasmas* **1**, 3151–60.

Kleva, R.G. et al. (1995). *Phys. Plasmas* **2**, 23–34.

Moffatt, H.K. & Ricca, R.L. (1992). Helicity and the Calugareanu invariant, *Proc. R. Soc. Lond. A* **439**, 411–29.

Nagayama, Y. et al. (1996). Tomography of full sawtooth crashes on the Tokamak Fusion Test Reactor, *Phys. Plasmas* **3**, 1647–55.

Rogers, B. & Zakharov, L. (1995). Nonlinear ω_*–stabilization of the $m = 1$ mode in tokamaks, *Phys. Plasmas* **2**, 3420–8.

Zakharov, L. et al. (1993). The theory of the early nonlinear stage of $m = 1$ reconnection in tokamaks, *Phys. Fluids* **B5**, 2498–505.

Index

Printed in the United States
By Bookmasters